Real-Time/Parallel Computing
IMAGE ANALYSIS

Real-Time/Parallel Computing
Computing
IMAGE ANALYSIS

EDITED BY
MORIO ONOE
University of Tokyo
Tokyo, Japan

KENDALL PRESTON, JR.
Carnegie-Mellon University
Pittsburgh, Pennsylvania

AND AZRIEL ROSENFELD
University of Maryland
College Park, Maryland

PLENUM PRESS • NEW YORK AND LONDON

Library of Congress Cataloging in Publication Data

Main entry under title:

Real-time/parallel computing.

"Based on proceedings of part of the Japan-United States Seminar on Research Towards Real-Time Parallel Image Analysis and Recognition, held in Tokyo, Japan, October 31, 1978-November 4, 1978."
    Includes bibliographies and indexes.
    1. Image processing—Digital techniques. 2. Real-time data processing. 3. Parallel processing (Electronic computers) I. Onoe, Morio. II. Preston, Kendall, 1927-
III. Rosenfeld, Azriel, 1931-   . IV. Japan-United States Seminar on Research Towards Real-Time Parallel Image Analysis and Recognition, Tokyo, 1978.
    TA1632.R4   001.64   80-28025

ISBN-13: 978-1-4684-3895-6   e-ISBN-13: 978-1-4684-3893-2
DOI: 10.1007/978-1-4684-3893-2

Based on proceedings of part of the Japan—United States Seminar
on Research Towards Real-Time Parallel Image Analysis and Recognition,
held in Tokyo, Japan, October 31, 1978—November 4, 1978

© 1981 Plenum Press, New York
Softcover reprint of the hardcover 1st edition 1981
A Division of Plenum Publishing Corporation
227 West 17th Street, New York, N.Y. 10011

PREFACE

This book is concerned with the aspects of real-time, parallel computing which are specific to the analysis of digitized images including both the symbolic and semantic data derived from such images. The subjects covered encompass processing, storing, and transmitting images and image data. A variety of techniques and algorithms for the analysis and manipulation of images are explored both theoretically and in terms of implementation in hardware and software. The book is organized into four topic areas: (1) theoretical development, (2) languages for image processing, (3) new computer techniques, and (4) implementation in special purpose real-time digital systems.

Computer utilization, methodology, and design for image analysis presents special and unusual problems. One author (Nagao)* points out that, "Human perception of a scene is very complex. It has not been made clear how perception functions, what one sees in a picture, and how one understands the whole picture. It is almost certain that one carries out a very quick trial-and-error process, starting from the detection of gross prominent features and then analyzing details, using one's knowledge of the world." Another author (Duff) makes the observation that, "It is therefore more difficult to write computer programs which deal with images than those which deal with numbers, human thinking about arithmetic being a largely conscious activity. In short, low-level computer languages for arithmetic operations can be similar to the natural language for arithmetic; in conventional computers, a suitable low-

----

*Full names and addresses are given for all authors in the Appendix.

level language for image processing will be far removed from a
natural language."

The above authors spotlight the aspects of image data analysis
that provide uniqueness to work in this field. Another unique
aspect of the computer requirements for image analysis is in the
enormous volume of data which must be stored and processed. The
LANDSAT Earth Resources Satellite, in passing over the entire surface of our planet 20 times each year, scans the equivalent of a
million million bytes of data. Many chapters in this book treat
LANDSAT image processing using both general purpose and special purpose computers. Another chapter (Kodaira, Kato, and Hamada) covers
the elaborate, real-time computer center developed to handle meteorological satellite data. Beyond these problems in satellite image
data handling there is the growing amount of image data now being
generated in medicine. Blood cell and x-ray images generated world
wide each year are equivalent to the output of 1,000 LANDSAT's.
The development of computerized x-ray tomography, blood cell analyzers, and electronic radiographic equipment in the 1980's will produce an outpouring of medical image data in digital form. The
dozen or so magnetic tapes generated per day by satellite image
systems will be eclipsed by the hundreds generated daily in all
types of computerized medical imaging. Some of these problems are
described and analyzed in this volume and others are discussed in
the editors' companion volume, *Real-Time Medical Image Processing*
(Plenum Press, 1980).

Four chapters (Aiso, Sakamura, and Ichikawa; Ichikawa and
Aiso; Iisaka, Ito, Fujisaki, and Takao; Soma, Ida, Inada, and Idesawa) treat the image data base problem by discussing computerized
image data storage and retrieval as regards both the filing of raw
image data as well as searching the semantic data base which results from processing the raw input. In these chapters associative
relevancy estimation, virtual plane methods, relational techniques
and query languages are treated. In a related set of chapters
(Enomoto, Katayama, Yonezaki, and Miyamura; Chang and Fu; Levialdi,
Maggiolo-Schettini, Napoli, and Uccella; Preston) the basic aspects
of image manipulative languages are explored with attention directed
upon image processing and feature extraction rather than solely on
storage and retrieval. Many existing command languages are sur-
veyed, languages developed for specific hardware systems are discussed, syntactic techniques are reviewed, and a new general-purpose, high-level language (PIXAL) is proposed. Two other chapters
(Fukada; Ichioka and Kawata) present certain special techniques for
multispectral image segmantation and for high-speed, hybrid (i.e.,
part optical, part electronic) image enhancement, respectively.

To allay the burden placed on the image analyzing computer by
the enormous volume of image data, the hierarchical organization

of processing algorithms is emphasized using cone or pyramid methods. Three chapters (Ichikawa and Aiso; Nagin, Hanson, and Riseman; Rosenfeld) exploit this approach. General formulae for estimating execution times, examples of applications to practical image analysis problems, and uses in both image segmentation and semantic image data analysis are given. It is recognized that the data burden can be attacked using operators and transforms which are based on binary image primitives and logical neighborhood operators. This leads to high-speed, table lookup implementation thus permitting real-time processing. Five chapters emphasize this approach (Chang and Fu; Haralick; Ichikawa and Aiso; Sternberg; Yokoi, Toriwaki, and Fukumura). These chapters survey and categorize neighborhood operators, present new theoretical analysis, and show how particular image primitives may be constructed and utilized.

Real-time processing implies dedicated, special-purpose computers often running many processing elements in parallel. Some chapters are devoted to four such systems. There is the I^2PAS (Interactive Image Processing and Analysis System) of the Nippon Electric Company (Hanaki) dedicated to multi-spectral data analysis, the PPP (Parallel Pattern Processor) of Toshiba (Kidode, Asada, Shinoda, and Watanabe) characterized by many special processors each dedicated to unique tasks (convolution, fast Fourier transform, affine transform, etc.), the IP (Image Processor) of Hitachi (Matsushima, Uno, and Ejiri) which, in contrast, consists of an array of identical processing elements whose tasks are assigned by a centralized control system, and the POPS (Poly-Processor System) of the Electrotechnical Laboratory (Tojo and Uchida) which has a multi-microprocessor structure. The reader can thus compare the significantly different architectures with each other and with the Environmental Research Institute of Michigan Cytocomputer (Sternberg).

Finally, four chapters (Hatori and Taki; Kamae, Hoshino, Okada, and Nagura; Onoe, Ishizuka, and Tsuboi; Takagi and Onoe) present new ideas in the important area of image transmission, encoding, and display with emphasis on color imagery. Pulse code modulation for color television transmission, run-length encoding, boundary encoding, shading correction, and display memory handling are treated here. Important advances in efficient bandwidth utilization are described along with new methods for optimally partitioning the large refresh memories now used for display.

This book and the above mentioned companion text *Real-Time Medical Image Processing* evolved from presentations made by the authors at the Japan-United States seminar on Research Towards Real-Time, Parallel Image Analysis and Recognition held in Tokyo in 1978. The editors wish to acknowledge the support provided for this meeting by the Japan Society for the Promotion of Science and the United States National Science Foundation. Production of the two resulting books was supported by the authors themselves and/or the organizations with which they are affiliated. The editors gratefully acknowledge this funding. The books were typed by Report Production Associates of Cambridge, Massachusetts, under the direction of Caroline Wadhams. The editors wish to acknowledge the excellent work of Susan Dunham, Johanne Khan, Eve Valentine, and Anne Simpson in preparing the manuscript. Also acknowledged is the work of Rose Kabler of Executive Suite, Tucson, Arizona (headed by Judy Tyree) who organized and drafted the subject index. Finally, the editorial efforts of John Matzka (Managing Editor) and James Busis (Editor), Plenum Press, New York City, as well as the photographic skills of Julius Weber, Mamaronek, New York, in the preparation of this work for publication are greatly appreciated.

M. Onoe
K. Preston, Jr.
A. Rosenfeld

CONTENTS

GENERAL THEORY

The Elements of Digital Picture Processing
    M. J. B. Duff

Some Neighborhood Operators
    R. M. Haralick

Region Relaxation in a Parallel Hierarchical Architecture
   P. A. Nagin, A. R. Hanson, and M. Riseman

Generalized Cellular Automata
   A. Rosenfeld

LANGUAGES

TECHNIQUES

Real Time Region Analysis for Image Data
    Y. Fukada

Hybrid Image Processing Using a Simple Optical Technique
   Y. Ichioka and S. Kawata

Focus of Attention in the Analysis of Complex Pictures
Such as Aerial Photographs
   M. Nagao

The Virtual Plane Concept in Image Processing
   T. Soma, T. Ida, N. Inada, and M. Idesawa

HARDWARE AND SYSTEMS

A Computing System Organization for Image Data Retrieval
    T. Ichikawa and H. Aiso

A Compound Computer System for Image Data Processing
    J. Iisaka, S. Ito, T. Fujisaki, and Y. Takao

Interactive Techniques for Producing and Encoding Color
Graphics
    T. Kamae, T. Hoshino, M. Okada, and M. Nagura

Real-Time Shading Corrector for a Television Camera Using
a Microprocessor
    M. Onoe, M. Ishizuka, and K. Tsuboi

Parallel Architectures for Image Processing
    S. R. Sternberg

THE ELEMENTS OF DIGITAL PICTURE PROCESSING

M. J. B. Duff

Department of Physics and Astronomy

University College London, London, ENGLAND

1. INTRODUCTION

Conventional digital computers have developed along lines
which have been dictated by the essential requirements of arithmetic and logical computation. These computers must, at the very
least, implement one of certain minimal sets of operations in order
that they can execute the tasks presented to them. A typical set
of operations is tabulated in Figure 1 where the term "shifting"
(which is not strictly necessary) implies the multiplication or
division of stored numbers by r, the root of the number system
used to represent numerical information in the computer. Given
these basic operations and facilities for program control, then
any complex calculation can be built up and the computer is describable as "general purpose."

Although a "general purpose" computer will be able to carry
out any computation, no matter how complex, it is worth noting that
a computer is limited to those tasks which can be described and
performed by completely arithmetic or logical methods. All relevant data must be reduced to an ordered set of numbers (or logical
variables) and similarly, all required operations must be represented by equivalent numerical or logical operations. This implies
the need for extreme rigour in the description of problems, and
nowhere is this more noteworthy than in image processing and pattern recognition. Unfortunately (or, perhaps, fortunately), the
rigour which is presumably employed in normal human thinking about
images is subconscious. It is therefore more difficult to write
computer programs which deal with images than those which deal with
numbers, human thinking about arithmetic being a largely conscious

1

ARITHMETIC OPERATIONS

IDENTITY	$Y = X$		COUNTING	$Y = \sum_i X_i$
INVERSION	$Y = -X$		SHIFTING	$Y = r X$
ADDITION	$Y = X + Z$		COMPARISON	$Y = X$ IF $Z > 0$

BOOLEAN OPERATIONS

IDENTITY	$C = A$
INVERSION	$C = \overline{A}$
AND	$C = A.B$

OTHER OPERATIONS

READ	INPUT DATA
WRITE	OUTPUT DATA

Fig. 1.   Basic set of operations.

activity. In short, low level computer languages for arithmetic
operations can be similar to the natural language for arithmetic;
in conventional computers, a suitable low level language for image
processing will be far removed from a natural language.

This chapter attempts to define the essential characteristics
of computers used in image processing and examines distinctions
between serial and parallel architectures. Zero- and multi-
dimensional processing is discussed, with particular reference to
local neighbourhood operators and the process of propagation.
Some brief comments are made concerning the calculation of processing speed in parallel systems.

2.   UNARY AND BINARY OPERATIONS

Zero-dimensional arithmetic and logical operations are illustrated in Figure 2. The unary operators f_1 are + and − (equivalent
to identity and inversion) and can operate either on single bits
or on multiple bit variables. In Figure 3, the equivalent operations on two-dimensional data arrays are shown. Bit planes equate
to single bits in the zero-dimensional case, and bit plane stacks

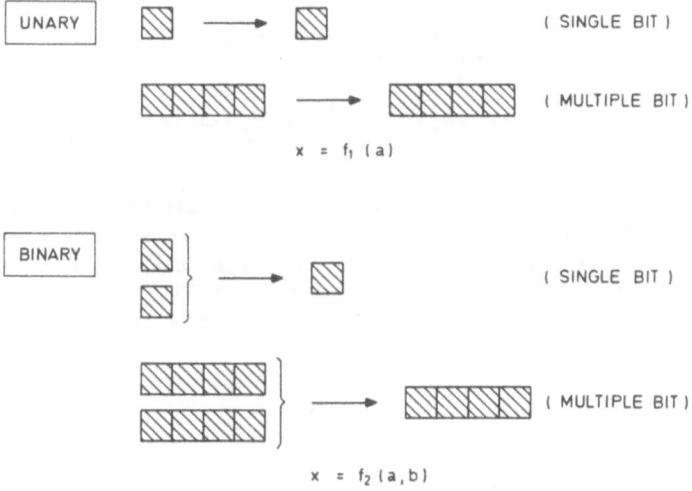

Fig. 2.   Zero-dimensional operations.

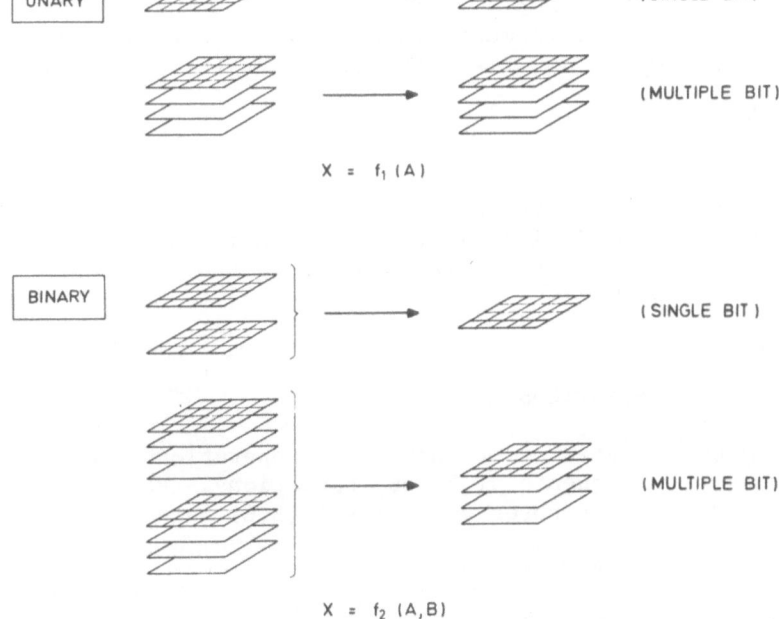

Fig. 3.   Two-dimensional operations.

to multiple bits, or words. Note that no mention has been made here of one-dimensional processing. Confusion concerning the dimensional status of multiple bits often occurs; in this discussion the extension into multiple bits is not regarded as an additional dimension. This approach limits dimensionality to extensions in the spatial dimensions.

It is implicit in the description of two-dimensional binary processing that the state of an image element (pixel) in the output bit-plane or bit-plane stack is a function only of the corresponding pixels in the two input bit-plane stacks. Thus, if the bit-plane stacks A and B each have P planes and elements A_{xy}^p and B_{xy}^p, respectively, where

$$1 \leq p \leq P$$

$$1 \leq x,y \leq n \tag{1}$$

for bit planes containing n^2 elements, then, if C is the output bit plane stack,

$$C_{xy}^q = f \ (A_{xy}^p, \ B_{xy}^p) \tag{2}$$

for all q in the range $1 \leq q \leq P$ (note the particular case for singular bit planes in which P=1).

Often the arrays of elements in bit-plane stacks will be regarded as arrays of pixels, i.e., the binary number represented by

$$A_{xy}^p, \ A_{xy}^{p-1}, \ \ldots\ldots A_{xy}^1 \tag{3}$$

will represent the grey level in the image A at the point (x,y). Equally, however, the arrays of numbers may represent other numerical data such as an optical density correction map or a table of values of variables required whilst processing the image.

3.   HIGHER ORDER OPERATIONS

Operations of higher order than binary operations can always be implemented as a sequence of binary operations. For example, the operation f_{++} on the three variables (A,B,C) such that

$$f_{++} \ (A,B,C) = A+B+C \tag{4}$$

can obviously be performed in two binary operations f_+ where

$$f_+ \ (X,Y) = X+Y \tag{5}$$

putting

$$f_{++}\{(A,B,C)\} = f_+\{A, f_+ (B,C)\} .$$ (6)

In Figure 4, a typical two-dimensional third-order operation is demonstrated for a single bit array. The value of X, a particular element in the output, is a function of three elements in the input : A (the corresponding element to X), and B and C (elements displaced diagonally or vertically from X). In principle, the required computation could be executed by "fetching" the B and C data from addresses calculated with respect to A's address. An alternative approach is also illustrated in which copies of the input data arrays are shifted so as to bring B and C into correspondence with A. This can be seen as another form of fetching.

The elements B and C are members of a set of elements which form a "neighbourhood" of the element A. In high order operations, the value of the output element X will be a function of a subset of a defined neighbourhood of the corresponding input element A. The most commonly used neighbourhoods are shown in Figure 5. For a square array, the local neighbourhood is the 3x3 element block with A in its center; a data defined neighbourhood may be, for example, all the elements in a connected set which comprise an

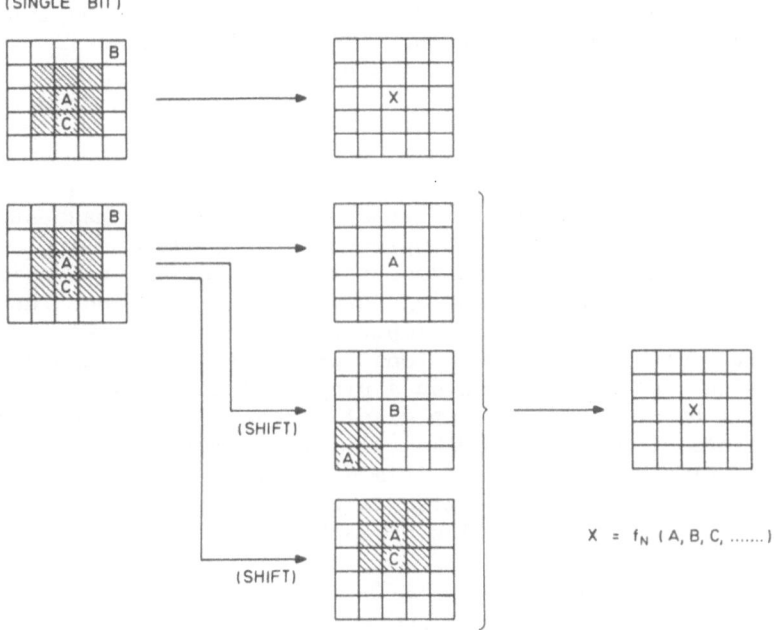

Fig. 4. Higher order operations.

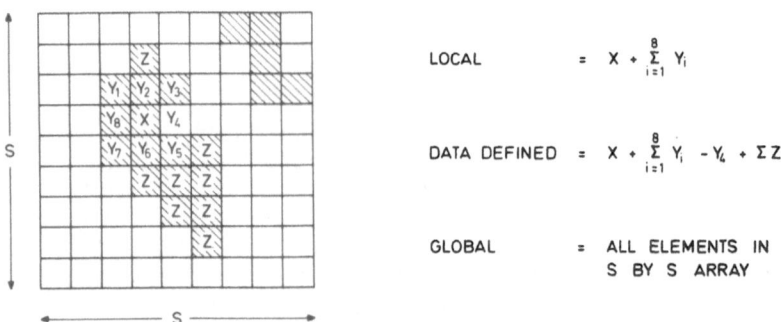

Fig. 5. Defined neighbourhoods.

object; the "global" neighbourhood is the complete set of elements
forming the array.

Other neighbourhoods are employed for particular algorithms;
for example, a 5x5 neighbourhood is often used in digital filtering.
However, it is instructive to consider what is required in the way
of information transfer in order that neighbourhoods of various
extents can be involved in a computation. For a neighbouring pixel
to be involved in a computation, either its value or some quantity
derived from its value must be transferred into the central processor (at pixel A). This transfer can be direct or can be indirect via intermediate pixels. The shifting illustrated in
Figure 4 is an example of such a transfer.

4. SERIAL OR PARALLEL PROCESSING

In the discussions above, no assumptions have been made as to
whether the operations are to be implemented in series or in
parallel. Clearly, the question does not arise in zero-dimensional
operations where parallel processing has no meaning (except insofar
as multiple bit arithmetic can be implemented with parallel action
at each bit). But with two-dimensional processing, a common requirement is that each operation on a pixel and its defined neighbourhood shall be repeated at each pixel in the array, in each case
using the values of pixels at time t to compute the new value of
each pixel at time $t + \Delta t$. Thus, if a defined neighbourhood for a
high order two-dimensional operation is the MxM region surrounding
each pixel A_{xy}, then the new value of each pixel is given by C_{xy}
where

$$C_{xy} = f(A_{jk}) \left| \begin{array}{l} x-\tfrac{1}{2}M \leq j \leq x+\tfrac{1}{2}M \\ y-\tfrac{1}{2}M \leq k \leq y+\tfrac{1}{2}M \end{array} \right. \qquad (7)$$

for all x,y in the array. In contrast, a serial processor will scan through the data, repeating the required operation as each pixel is visited during the scanning process. Consequently, a certain subset of pixels in each neighbourhood may have new values as a result of the operations earlier in the scanning sequence.

It is important to make the distinction between a serial processor executing a serial algorithm, as has just been described, and a serial processor executing a parallel algorithm, as was earlier described. In this latter case, the result is identical to that which would be obtained by using a "dedicated" processor at each pixel. Similarly, it can easily be shown that a parallel processor, whilst obviously being most efficient when executing parallel algorithms, can equally be used to execute serial algorithms. The basis of the method is as follows:

1. Create a mask coincident with the defined neighbourhood surrounding the selected pixel,

2. Abstract the masked region from the input array by ANDing with the mask, so that the region is embedded in an otherwise empty array,

3. Perform the required operation on the new array,

4. Abstract the new pixel from the newly processed array, using a second mask,

5. Replace the new pixel into the original array, using the second mask,

6. Shift both masks in the scanning direction and repeat from Step (2).

In summary, therefore, all operations, whether involving serial or parallel algorithms, can be implemented on both serial and parallel processors.

5.   PROPAGATION

Propagation is a particularly important array operation deserving special attention, not least because it is apparently neither strictly parallel nor strictly serial in quality. Propagation involves the transfer of information between pixels, but the information transferred is not the value of the pixel itself. Instead, the pixel value moderates the flow of the information.

A simple example of propagation can be described for a binary picture, i.e., a single bit-plane. A propagation input is injected into elements at the array edges. The propagation signal is then transmitted to all neighbouring elements by pixels of value zero, but blocked by pixels of value one. If objects in a picture are composed of connected sets of pixels of value one in a background of zero valued pixels, then the propagation will just reach the elements forming the outer edges of the objects; hence a very simple method for abstracting outer edges.

This example demonstrates the ambiguous status of propagation. The output state of each element in the array has two components. One component is the pixel value which is unchanged during propagation, whilst the other is the propagation output which changes at each element as the propagation signal proceeds along its path. The pixel values change in parallel at a subsequent operation in which the propagation component is combined with the original pixel value. Propagation starts as a serial process but may finish with a parallel process. However, if the propagation is constrained to proceed one step at a time, element by element, then the process can be effected in a sequence of parallel operations as discussed in the previous section, but to do this may well be to throw away a considerable speed advantage without any compensation other than obtaining a process which is describable in parallel terms.

Directional Propagation may be produced through elements without references to pixel values (as when a row of the array is to be cleared, for example). Again, the conditions for propagation may be more elaborate than merely the local pixel value, so that propagation will only proceed through elements with a particular state or set of states of a defined neighbourhood. Lastly, multi-valued propagation, although difficult to realise in hardware, offers attractive features. In this, the propagation signal is a continuous variable whose value is changed at each transmitting element in accordance with a predetermined rule. In all these cases, the same considerations concerning the serial or parallel nature of the process can be seen to apply.

6.   CONCLUSIONS

Despite the enormous variety of operations which can be applied to two-dimensional arrays of data, which here are usually taken as representing pictures, it can be concluded that all of them can be implemented as sequences of local neighbourhood operations, executed either in series or in parallel. This has been demonstrated rather than proved formally in this paper.

If special purpose processors are assigned to each pixel in order to permit parallel processing, then a minimum requirement of each processor will be that it must be able to effect a fundamental set of arithmetic and Boolean operations (see Figure 1) and that it must permit data transfer from each of the elements in its local neighbourhood. With these conditions fulfilled, all computational processes can be achieved in a finite sequence of steps. It remains to determine the details of the structure of the processors and the nature of the interconnection structure. Both these factors will, in a practical situation, be a compromise between the opposing considerations of efficiency through complexity on the one hand, and cost on the other. The issue is made more complex by the fact that a cheaper processor will, generally speaking, allow larger arrays to be built, i.e., will lead to a higher degree of parallelism. This will result in increased processing speeds which may more than compensate for the losses due to reduced complexity, especially if the larger array size eliminates the need for repeated scanning of the input picture.

Finally, a note of caution should be sounded concerning over-simplistic approaches to assessment of speed gains obtainable through the introduction of parallelism. In reality, improvements in execution speeds by a factor of as little as two, say, may well make possible real-time processing when otherwise it would not be practicable. Under such circumstances, calculations of the *order* of speed gain achievable in an array are all but valueless. Many factors must be taken into consideration and it is found that even the choice of algorithm may depend critically on apparently insignificant details of the processor hardware. A study by Cordella, Duff and Levialdi (1976) showed that, for a typical set of image processing tasks, an array with 100^2 elements of the CLIP type described by Duff (1976) would achieve speed gains over a conventional serial computer ranging from 10 to more than 10^6. Even so, in simple image processing applications, it will often be the image input/output times which dominate the overall speed which can be achieved. In summary, therefore, it is important to examine each proposed architecture in relationship to the tasks it is expected to perform, and to avoid the use of over-simplified formulae involving only the array dimensions.

7.   REFERENCES

Cordella, L., Duff, M. J. B., and Levialdi, S., "Comparing Sequential and Parallel processing of Pictures," Proc. 3rd Internat'l Joint Conf. Pattern Recog., Coronado (1976).

Duff, M. J. B., "CLIP4: A Large Scale Integrated Circuit Array Parallel Processor," Proc. 3rd Internat'l Joint Conf. Pattern Recog., Coronado (1976).

SOME NEIGHBORHOOD OPERATORS

R. M. Haralick

Virginia Polytechnic Institute and State University

Blacksburg, VA 24061 USA

1. INTRODUCTION

In this paper we review a variety of neighborhood operators in a way which emphasizes their common form. These operators are useful for image segmentation tasks as well as for the construction of primitives involved in structural image analysis. The common form of the operators suggests the possibility of a large scale integration hardware implementation in the VLSI device technology.

Neighborhood operators can be classified according to type of domain, type of neighborhood, and whether or not they are recursive. The two types of domains consist of numeric or symbolic data. Operators having a numeric domain are usually defined in terms of arithmetic operations such as addition, subtraction, computing minimums or maximums, etc. Operators having a symbolic domain are defined in terms of Boolean operations such as AND, OR, NOT, or table look-up operations.

There are two basic neighborhoods a simple operator may use: a 4-connected neighborhood and an 8-connected neighborhood. As illustrated in Figure 1, the 4-connected neighborhood about a pixel consists of the pixel and its north, south, east, and west neighbors. The 8-connected neighborhood about a pixel consists of all the pixels in a 3x3 window whose center is the given pixel.

Recursive neighborhood operators are those which use the same image memory for their input and output. In this way an output from a previously used nearby neighborhood influences the output from its current neighborhood. Non-recursive neighborhood operators are those which use independent image memory for input and

```
+-----+-----+-----+                +-----+-----+-----+
|     | X2  |     |     |          | X7  | X2  | X6  |
+-----+-----+-----+                +-----+-----+-----+
| X3  | X0  | X1  |     |          | X3  | X0  | X1  |
+-----+-----+-----+                +-----+-----+-----+
|     | X4  |     |     |          | X8  | X4  | X5  |
+-----+-----+-----+                +-----+-----+-----+
```

4-connected neighborhood   **8-connected neighborhood**

Fig. 1.   Illustration of the indexing of the pixels in 4-connected and 8-connected neighborhood of X_0.

output.   Previous outputs cannot influence current outputs.   Rosenfeld and Pfaltz (1966) call recursive operators sequential and nonrecursive operators parallel.

2.   REGION GROWING OPERATOR

The region growing operator just described is non-recursive and has a symbolic domain.   It changes all pixels whose label is the background label to the non-background label of its neighboring pixels.   It is based on a two-argument primitive function h which is a projection operator whose output is either its first argument or its second argument depending on their values.   If the first argument is the special symbol "g" for background, then the output of the function is its second argument.   Otherwise, the output is the first argument.   Hence:

$$h(c,d) = \begin{cases} d & \text{if } c = g \\ c & \text{if } c \neq g \end{cases} \tag{1}$$

The region growing operator uses the primitive function h in the following way.   For the operator in the 4-connected mode, let $a_0 = x_0$.   Define $a_n = h(a_{n-1}, x_n)$, $n = 1, \ldots, 4$.   Then the output symbol y is defined by $y = a_4$.   For the operator in the 8-connected mode, let $a_0 = x_0$.   Define $a_n = h(a_{n-1}, x_n)$, $n = 1, \ldots, 8$.   Then the output symbol y is defined by $y = a_8$.

A more sophisticated region growing operator grows background border pixels to the region label a majority of its neighbors have. In the 8-connected mode such an operator sets $a_n = h(x_0, x_n)$, $n = 1, \ldots, 8$ and defines the output symbol y by $y = c$ where $\#\{n | a_n = c\} > \#\{n | a_n = c'\}$ for all c'.

3. NEAREST NEIGHBOR SETS

Given a symbolic image with background pixels labeled "g" and each connected set of non-background pixels labeled with a unique label, it is possible to label each background pixel with the label of its closest non-background neighboring pixel. Just iteratively grow the non-background labels into the background labels using the 8-neighborhood if the max distance is desired, using the 4-neighborhood if city block is desired, using the 4-neighborhood and 8-neighborhood alternately in the ratio of $\sqrt{2}$ for Euclidean distances.

4. REGION SHRINKING OPERATORS

The region shrinking operator is non-recursive and has a symbolic data domain. It changes the label on all border pixels to the background label. The region shrinking operator defined here can change the connectivity of a region and can even entirely delete a region upon repeated application. It is based on a two-argument primitive function h which can recognize whether or not its arguments are identical. If the arguments are the same, h outputs the value of the argument. If the arguments differ, h outputs the special symbol "g" for background. Hence:

$$h(c,d) = \begin{cases} c & \text{if} \quad c = d \\ g & \text{if} \quad c \neq d \end{cases} \tag{2}$$

The region shrinking operator uses the primitive function h in the following way. For the operator in the 4-connected mode, let $a_0 = x_0$. Define $a_n = h(a_{n-1}, x_n)$, $n = 1, \ldots, 4$. Then the output symbol y is defined by $y = a_4$. For the operator in the 8-connected mode, let $a_0 = x_0$. Define $a_n = h(a_{n-1}, x_n)$, $n = 1, \ldots, 8$. Then the output symbol y is defined by $y = a_8$.

A more sophisticated region shrinking operator shrinks border pixels only if they are connected to enough pixels of unlike regions. In the 8-connected mode it sets $a_n = h(x_0, x_n)$, $n = 1, \ldots, 8$ and defines the output symbol y by:

$$y = \begin{cases} g & \text{if} \quad \#\{n \mid a_n = g\} > k \\ x_0 & \text{otherwise} \end{cases} \tag{3}$$

As mentioned in the section on nearest neighbor sets, to obtain a region shrinking (region growing) which is close to a Euclidean distance region shrinking (growing), the 4-neighborhood and the 8-neighborhood must be used alternately approximating as closely as possible the ratio $\sqrt{2}/1$ (Rosenfeld and Pfaltz, 1968). A ratio of

4/3 can be obtained by the sequence 4:3 = <4,8,4,8,4,8,4> and a ratio of 3/2 can be obtained by the sequence 3:2 = <4,8,4,8,4>. Alternating these two sequences will give a ratio of 7/5 just smaller than 2. Using one 4:3 sequence followed by two 3:2 sequences gives a ratio of 10/7, just over $\sqrt{2}$. Alternating between <4:3,3:2,3:2> and <4:3,3:2> gives a ratio of 17/12 which differs from $\sqrt{2}$ by less than 2.5×10^{-3}, an approximation which should be good enough for most purposes.

The choice of 4-neighborhood or 8-neighborhood for the current iteration which best approximates the Euclidean distance can be determined dynamically. Let N4 be the number of uses of the 4-neighborhood so far and N8 be the number of the 8-neighborhood so far. If $|N4-2(N8+1)| < |N4+1-2N8|$, then use the 8-neighborhood for the current iteration; else use the 4-neighborhood.

5.   MARK INTERIOR BORDER PIXELS

The mark interior/border pixels operator is non-recursive and has a symbolic data domain. It marks all interior pixels with the label "i," standing for interior, and all border pixels with the label "b," standing for border. It is based on two primitive functions. One is a two-argument primitive function h very similar to that used in the region shrinking operator. The other one is a one-argument primitive function f. The two argument primitive function h can recognize whether or not its arguments are identical. For identical arguments it outputs the argument. For non-identical arguments it outputs the special symbol "b" standing for border. The one-argument primitive function f can recognize whether or not its argument is the special symbol "b." If it is it outputs b. If not it outputs the special symbol "i" standing for interior. Hence:

$$h(c,d) = \begin{cases} c & \text{if} \quad c = d \\ b & \text{if} \quad c \neq d \end{cases} \tag{4}$$

$$f(c) = \begin{cases} b & \text{if} \quad c = b \\ i & \text{if} \quad c \neq d \end{cases} \tag{5}$$

The mark interior/border pixel operator uses the primitive function h in the following way. For the operator in the 4-connected mode, let $a_0 = x_0$. Define $a_n = h(a_{n-1}, x_n)$, $n = 1, \ldots, 4$. Then the output symbol y is defined by $y = f(a_4)$. For the operator in the 8-connected mode, let $a_0 = x_0$. Define $a_n = h(a_{n-1}, x_n)$, $n = 1, \ldots, 8$. Then the output symbol y is defined by $y = f(a_8)$.

6.   CONNECTIVITY NUMBER OPERATOR

The connectivity number operator is a non-recursive operator
which has a symbolic data domain. Its purpose is to classify the
way a pixel is connected to its like neighbors. As shown in Figure
2, there are six values of connectivity, 5 values for border pixels
and 1 value for interior pixels. The border pixels consist of isolated pixels, edge pixels, connected pixels, branching pixels,
crossing pixels, and interior pixels. The connectivity number operator associates with each pixel a symbol called the connectivity
number of the pixel. The symbol, although a number, has no arithmetic number properties. The number designates which of the six
values of connectivity a pixel has with its like neighbors.

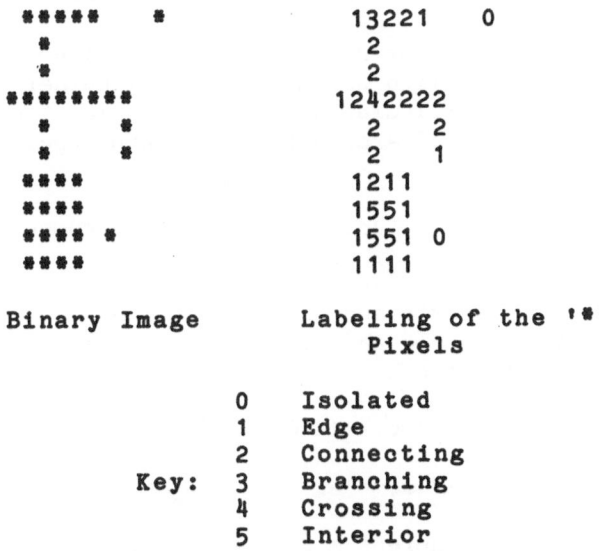

Fig. 2. Illustration of a connectivity number labeling of a binary
image.

6.1 Yokoi Connectivity Number

The definition we give here of connectivity number is based on
a slight generalization of the definitions suggested by Yokoi,
Toriwaki, and Fukumura (1975). This is not the only definition of
connectivity number. Another definition given by Rutovitz (1966)
is based on the number of transitions from one symbol to another as
one travels around the 8-neighborhood of a pixel. The operator, as

defined here, uses an 8-connected neighborhood and can be defined
for either 4-connectivity or 8-connectivity.

For 4-connectivity, a pixel is an interior pixel if its value
and that of each of its 4-connected neighbors is the same. In this
case its 4-connectivity takes the index value 5. Otherwise, the
4-connectivity of a pixel is given by the number of times a 4-con-
nected neighbor has the same value but the corresponding 3 pixel
corner neighborhood does not. These corner neighbors are illustrated in Figure 3.

For 8-connectivity, a pixel is an interior pixel if its value
and that of each of its 8-connected neighbors is the same. Otherwise the 8-connectivity of a pixel is given by the number of times
a 4-connected neighbor has a different value and at least one pixel
in the corresponding 3 pixel neighborhood corner has the same value.

The connectivity operator requires two primitive functions: a
function h which can determine whether a 3-pixel corner neighborhood is connected in a particular way and a function f which basically counts the number of arguments which have a particular
value.

For 4-connectivity, the function h of four arguments is defined by:

$$h(b,c,d,e) = \begin{cases} q \text{ if } b = c \text{ and } (d \neq b \text{ or } e \neq b) \\ r \text{ if } b = c \text{ and } (d = b \text{ and } e = b) \\ s \text{ f } b \neq c \end{cases} \qquad (6)$$

the function f of four arguments is defined by:

$$f(a_1,a_2,a_3,a_4) = \begin{cases} 5 \text{ if } a_1 = a_2 = a_3 = a_4 = r \\ n \text{ where } n = \#\{a_k | a_k = q\}, \text{ otherwise} \end{cases} \qquad (7)$$

The connectivity operator using 4-connectivity is then defined in
the following way. Let

$$a_1 = h(X_0,X_1,X_6,X_2)$$
$$a_2 = h(X_0,X_2,X_7,X_3)$$
$$a_3 = h(X_0,X_3,X_8,X_4)$$
$$a_4 = h(X_0,X_4,X_5,X_1)$$

Define the connectivity number y by $y = f(a_1,a_2,a_3,a_4)$.

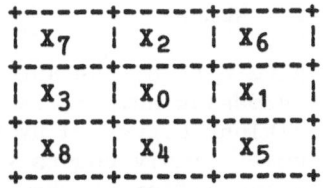

(a)   Indexing pixels in a 3 x 3 neighborhood

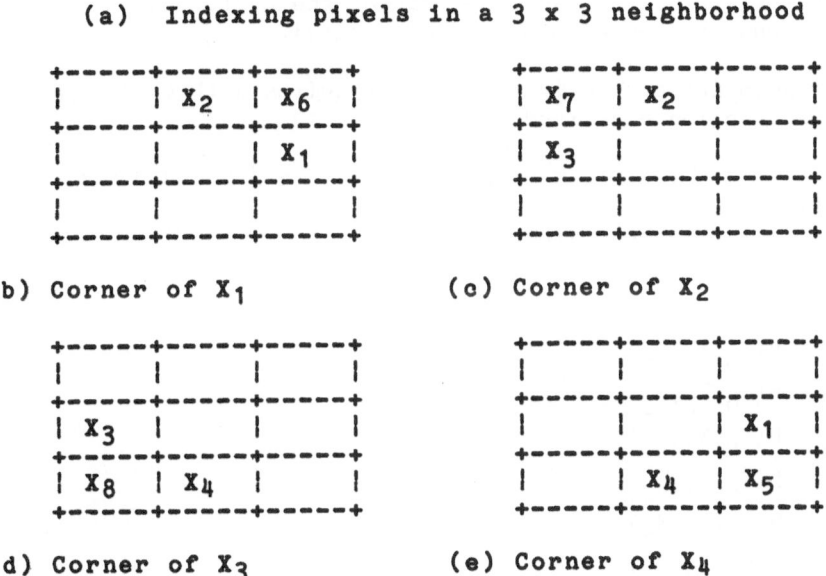

(b) Corner of X_1 (c) Corner of X_2

(d) Corner of X_3 (e) Corner of X_4

Fig. 3.   Illustration of the corner neighborhood corresponding to each of the East, North, West, and South neighbors of the center pixel.

For 8-connectivity, the function h is slightly different.   It is defined by

$$h(b,c,d,e) = \begin{cases} q \text{ if } b \neq c \text{ and } (d = b \text{ or } e = b) \\ r \text{ if } b = c \text{ and } (d = b \text{ and } e = b) \\ s \text{ if } b \neq c \text{ and } (d \neq b \text{ and } e \neq b) \end{cases} \qquad (8)$$

Then, as before, the connectivity number y is defined by $y = f(a_1, a_2, a_3, a_4)$.

6.2 Rutovitz Connectivity Number

The definition we give here of the Rutovitz connectivity number, sometimes called a crossing number, is based on a slight generalization of the definitions suggested by Rutovitz (1966). The
Rutovitz connectivity number simply counts the number of transitions from symbols which are different than that of the center pixel
to symbols which are the same as that of the center pixel as one
travels around the 8-neighborhood of a pixel.

The Rutovitz connectivity number requires a three argument
primitive function h defined by

$$h(a,b,c) = \begin{cases} 1 \text{ if } (a = b \text{ and } a \neq c) \text{ or } (a \neq b \text{ and } a = c) \\ 0 \text{ otherwise} \end{cases} \tag{9}$$

Then set

$$a_1 = h(X_0, X_1, X_6)$$
$$a_2 = h(X_0, X_6, X_2)$$
$$a_3 = h(X_0, X_2, X_7)$$
$$a_4 = h(X_0, X_7, X_3)$$
$$a_5 = h(X_0, X_3, X_8)$$
$$a_6 = h(X_0, X_8, X_4)$$
$$a_7 = h(X_0, X_4, X_5)$$
$$a_8 = h(X_0, X_5, X_1)$$

The output value y is then given by

$$y = \sum_{n=1}^{8} a_n \tag{10}$$

7. CONNECTED SHRINK OPERATOR

The connected shrink operator is a recursive operator having a
symbolic data domain. It is similar in certain respects to the connectivity number operator and the region shrinking operator. Instead of labeling all border pixels with background symbol "g," the
connected shrink operator only labels those border pixels which can
be deleted from a connected region without disconnecting the region.
Since it is applied recursively, pixels which are interior during
one position of the scan may appear as border pixels at another

position of the scan and eventually may be deleted by this operator. After one complete scan of the image, the set of pixels which get labeled as background is a strong function of the way in which the image is scanned with the operator. For example, as illustrated in Figure 4, a top-down left-right scan will delete all edge pixels which are not right boundary edge pixels.

The theoretical basis of the connected shrink operator was explored by Rosenfeld and Pfaltz (1966), Rosenfeld (1970), and Stefanelli and Rosenfeld (1971). Basically, a pixel's label is changed to "g," for background, if upon deleting it from the region it belongs to, the region remains connected. The operator definition given here is due to Yokoi, Toriwaki, and Fukumura (1975). The operator uses an 8-connected neighborhood and can be defined for deleting either 4-deletable or 8-deletable pixels. It requires two primitive functions: a function h which can determine whether the 3-pixel corner of a neighborhood is connected and a function g which basically counts the number of arguments having certain values.

In the 4-connectivity mode, the 4-argument primitive function h is defined:

$$h(b,c,d,e) = \begin{cases} 1 \text{ if } b = c \text{ and } (d \neq b \text{ or } e \neq b) \\ 0 \text{ otherwise} \end{cases} \qquad (11)$$

In the 8-connectivity mode, the 4- argument primitive function h is defined by:

$$h(b,c,d,e) = \begin{cases} 1 \text{ if } c \neq b \text{ and } (d = b \text{ or } e = b) \\ 0 \text{ otherwise} \end{cases} \qquad (12)$$

The 5-argument primitive function f is defined by:

$$f(a_1,a_2,a_3,a_4,x) = \begin{cases} g \text{ if exactly one of } a_1,a_2,a_3,a_4 = 1 \\ x \text{ otherwise} \end{cases} \qquad (13)$$

Using the indexing convention of Figure 3, the connected shrink operator is defined by letting:

$$a_1 = h(X_0,X_1,X_6,X_2)$$
$$a_2 = h(X_0,X_2,X_7,X_3)$$
$$a_3 = h(X_0,X_3,X_8,X_4)$$
$$a_4 = h(X_0,X_4,X_5,X_1)$$

The output symbol y is defined by $y = f(a_1,a_2,a_3,a_4,x_0)$.

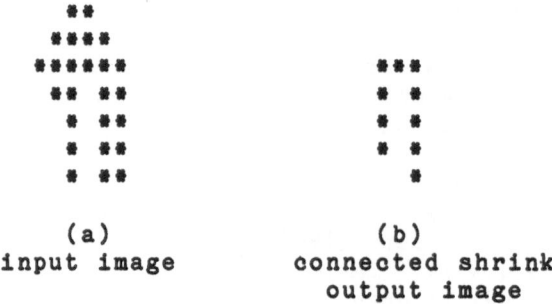

<pre>
        (a)   (b)
   input image   connected shrink
                         output image
</pre>

Fig. 4. Illustration of the connected shrink operator applied in a
top-down left-right scan using 4-connectivity.

        The earliest discussion of connectivity in digital pictures
can be found in Rosenfeld (1971). Rutovitz (1966) preceded Rosenfeld in the use of crossing numbers but did not use connectivity in
his development. Related algorithms and discussion of connectivity
can be found in Levialdi (1972) who introduced a parallel or non-
recursive shrinking algorithm for the purpose of counting the number
of components in a binary image. This iterative algorithm does not
employ the 1-deletability of the Yokoi et al. method; it uses a
2x2 window, rather than a 3x3 window in the shrinking process, but
requires the detection of an isolated element during the iterative
process so that it may be counted before it is made to disappear by
the process. A three-dimensional extension to this non-recursive
algorithm can be found in Arcelli and Levialdi (1972). Lobregt,
Verbeek, and Groen (1980) discuss a recursive operator for three-
dimensional shrinking.

        The first discussions of thinning appeared in Hilditch (1969),
and Deutsch (1969). These initial insights were later expanded by
Fraser (1970), Stefanelli and Rosenfeld (1971), Deutsch (1972),
Rosenfeld (1975), and Rosenfeld and Davis (1976). A brief comparison of thinning techniques can be found in Tamura (1978) who suggests that a smoother 8-connected thinning results if 8-deletable
pixels are removed from thinning 4-connected curves. Tamura also
notes that the thinning of Rosenfeld and Davis (1976) is very sensitive to contour noise when used in the 4-connected mode.

8.   PAIR RELATIONSHIP OPERATOR

The pair relationship operator is a non-recursive and has a symbolic data domain. It is a general operator which labels a pixel on the basis of whether that pixel stands in the specified relationship with some neighborhood pixel. An example of a pair relationship operator is one which relabels with a specified label all border pixels which are next to an interior pixel and either can relabel all other pixels with another specified label or leave their labels alone. Formally, a pair relationship operator marks a pixel with the specified label "p" if the pixel has a specified label "1" and neighbors enough pixels having specified label "m." All other pixels it either marks with another specified label or leaves their original labels unmodified.

The pair relationship operator employs two primitive functions. The two-argument function h is able to recognize if its first argument has the value of its second argument. It is defined by:

$$h(a,m) = \begin{cases} 1 \text{ if } a = m \\ 0 \text{ otherwise} \end{cases} \tag{14}$$

For the 4-connected mode, the output value y is defined by:

$$y = \begin{cases} q \text{ if } \sum_{n=1}^{4} h(X_n,m) < \theta \text{ or } X_0 \neq 1 \\ p \text{ if } \sum_{n=1}^{4} h(X_n,m) \geq \theta \text{ and } X_0 = 1, \end{cases} \tag{15}$$

where q can either be a specified output label or the label X_8.

For the 8-connected mode, the output y is defined by:

$$y = \begin{cases} q \text{ if } \sum_{n=1}^{8} h(X_n,m) < \theta \text{ or } X_0 \neq 1 \\ p \text{ if } \sum_{n=1}^{8} h(X_n,m) \geq \theta \text{ and } X_0 = 1 \end{cases} \tag{16}$$

where q can either be a specified output or the label X_8.

9. THINNING OPERATOR

The thinning operator suggested here is defined as a composition of three operators: the mark interior/border operator, the pair relationship operator, and the marked pixel connected shrink operator. It works by marking all border pixels which are next to interior pixels and then deleting (or shrinking) any marked pixel which is deletable. The result of successively applying the thinning operator on a symbolic image is that all regions are symmetrically shrunk down until there are no interior pixels left. What remains is their centerlines as shown in Figure 5. This operator has the nice property that the centerline is connected in exactly the same geometric and topologic way the original figure is connected. For other similar operators which thin without changing geometry or topology, see Davis and Rosenfeld (1975) or Stefanelli and Rosenfeld (1971) or Arcelli and Sanniti di Baja (1978). To implement the operator as the composition of three operators, the mark interior/border operator examines the original symbolic image to produce an interior/border image. The interior border image is examined by the pair relationship operator which produces an image whose pixels are marked if on the original image they were border pixels and were next to interior pixels. The marked pixel image and the original symbolic image constitute the input to the marked pixel connected shrink operator which is exactly like the connected shrink operator except it only shrinks pixels which are deletable and which are marked.

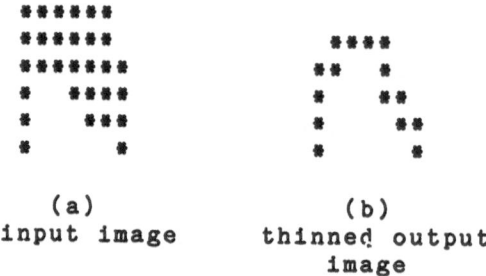

(a)
input image

(b)
thinned output
image

Fig. 5.   Illustration of the result of one application of the thinning operator using the 4-connectivity deletability condition.

10. DISTANCE TRANSFORMATION OPERATOR

The distance transformation operator has an implementation as a recursive or as a non-recursive operator. It requires a binary image whose border pixels are labeled with 0 and whose interior pixels are labeled with "i." The purpose of the distance transformation operator is to produce a numeric image whose pixels are labeled with the distance each of them is to its closest border pixel. Distance between two pixels can be defined by the length of the shortest 4-connected path or 8-connected path between them.

As a non-recursive operator, the distance transform can be achieved by successive application of the pair relationship operator. In the first application the pair relationship labels all pixels whose label is "i" and which are next to a pixel whose label is "0" with the label "1." All other pixels keep their labels. In the nth application, the pair relationship operator labels all pixels whose label is "i" and which are next to a pixel whose label is "n-1" with the label "n." When no pixel has the label "i," an application of the pair relationship operator changes no pixel values and the resulting image is the distance transform image. This implementation is related to the one given by Rosenfeld (1968).

Another way of implementing this operator non-recursively is by the use of the primitive function defined by:

$$
h(a_0,\ldots,a_N) =
\begin{cases}
i & \text{if } a_n = i, \; n = 0,\ldots,N \\
\min \{b \,|\, \text{for some } a_n \leq N, \; a_n \neq i, \; b=a_n+1\} \\
\quad \text{if } a_N = i \text{ and there exists } n \text{ such that } a_n \neq i \\
a_N & \text{if } a_N \neq i
\end{cases}
\tag{17}
$$

In the 8-connected mode the output y is defined by $y = h(x_0,x_1,\ldots,x_8)$. In the 4-connected mode, the output symbol y is defined by $y = h(x_0,x_1,x_2,x_3,x_4)$. See Rosenfeld and Pfaltz (1966).

Another way (Rosenfeld and Pfaltz, 1966) of implementing the distance transform involves the application of two recursive operators, the first operator being applied in a left-right top-bottom scan and the second operator being applied in a right-left bottom-top scan. Both operators are based on similar primitive functions. For the first operator the primitive function h is defined by:

$$h(a_1,\ldots,a_N) = \begin{cases} 0 \text{ if } a_N = 0 \\ \\ \min\{a_1,\ldots,a_N\} + 1 \text{ otherwise} \end{cases} \tag{18}$$

In the 8-connected mode, the output symbol y of the first operator is defined by:

$$y = h(x_2, x_7, x_3, x_0).$$

In the 4-connected mode, the output symbol y is defined by:

$$y = h(x_2, x_3, x_0)$$

For the second operator, the primitive function is simply the minimum function. In the 8-connected mode, the output symbol y of the second operator is defined by:

$$y = \min\{x_0, x_1, x_4\}$$

In the second 4-connected mode, the output symbol y is defined by:

$$y = \min\{x_0, x_1, x_4\}$$

11. CONTACT DISTANCES

Contact distances are related to the radius of fusion defined by Rosenfeld and Pfaltz (1968). The radius of fusion for a pixel is the smallest integer n such that after region growing n iterations and region shrinking n + 1 iterations, the pixels retain a non-background label. The radius of fusion concept has the difficulty that using 4-neighborhoods; for example, it is possible for a pair of pixels or a triangle of pixels never to fuse. Its radius of fusion is, therefore, not defined. Defining it by some large number in these cases is artificial.

Contact distance gives a measure of the distance to a pixel's nearest labeled neighbor and it is always defined. To label every pixel with the distance its associated nearest labeled pixel has with its own closest nearest labeled pixel, begin with the original image having some isolated pixels with unique labels and the remainder pixels having the background label. Exactly as done to determine the nearest neighbor sets, perform a region growing to label every pixel with the label of its nearest labeled neighbor. One iteration of a shrink operation on this image can label all border pixels (pixels which have a neighbor with a different label) with

the label 0. Use this image as the input to the distance transformation operator which labels every pixel with its distance to the
nearest border. Then mask this distance transformation image with
the original image, labeling all pixels with the background label
except those pixels having a non-background label on the original
image. Pixels having a non-background label on the original image
get labeled with the distances associated with their spatial position on the distance transformation image. Finally, region grow the
masked image until there are no more pixels with a background label.
The resulting image has each pixel labeled with the distance its
associated nearest labeled neighbor pixel has with its own nearest
labeled pixel.

12. NON-MINIMA-MAXIMA OPERATOR

The non-minima-maxima operator is a non-recursive operator
that takes a numeric input image and produces a symbolic output
image in which each pixel is labeled with an index 0, 1, 2, or 3
indicating whether the pixel is a non-maximum, non-minimum, interior
to a connected set of equal-valued pixels, or part of a transition
region (a region having some neighboring pixels greater than and
others less than its own value). A pixel whose value is the minimum of its neighborhood and having one neighboring pixel with a
value greater than itself may be a minimum or transition pixel but
it is certainly a non-maximum pixel. Figure 6 illustrates how a
pixel can be its neighborhood maximum, yet not be part of any relative maximum.

The non-minima-maxima operator is based on the primitive function min and max. For the 4-connected case, let $a_0 = b_0 = X_0$ and
define

$$a_n = \min \{a_{n-1}, X_0\} \quad n = 1,2,3,4 \tag{19}$$
$$b_n = \max \{b_{n-1}, X_0\} \quad n = 1,2,3,4$$

The output index 1 is defined by

$$1 = \begin{cases} 0 \text{ (flat)} & \text{if } a_4 = X_0 = b_4 \\ 1 \text{ (non-maximum)} & \text{if } a_4 = X_0 < b_4 \\ 2 \text{ (non-minimum)} & \text{if } a_4 < X_0 = b_4 \\ 3 \text{ (transition)} & \text{if } a_4 < X_0 < b_4 \end{cases} \tag{20}$$

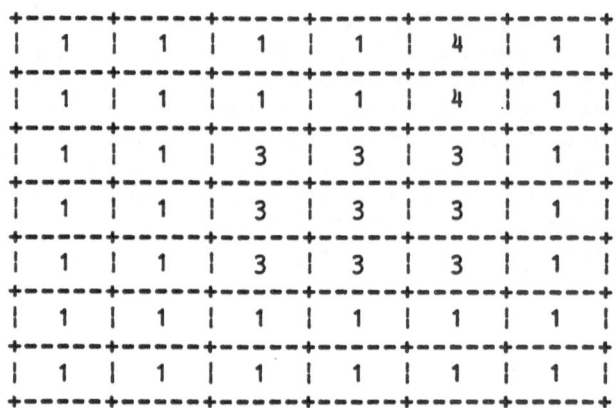

Fig. 6. Illustration of how a pixel can be its neighborhood maximum yet not be a part of any relative maximum. In its 8-neighborhood, the central 3 is a maximum, yet the flat of 3's it belongs to is a transition region.

For the 8-connected case, let $a_0 = b_0 = X_0$ and define

$$a_n = \min \{a_{n-1}, X_0\} \quad n = 1,2,\ldots8$$
$$b_n = \max \{b_{n-1}, X_0\} \quad n = 1,2,\ldots8 \tag{21}$$

The output index 1 is defined by

$$1 = \begin{cases} 0 \text{ (flat)} & \text{if } a_8 = X_0 = b_8 \\ 1 \text{ (non-maximum)} & \text{if } a_8 = X_0 < b_8 \\ 2 \text{ (non-minimum)} & \text{if } a_8 < X_0 = b_8 \\ 3 \text{ (transition)} & \text{if } a_8 < X_0 < b_8 \end{cases} \tag{22}$$

13. RELATIVE EXTREMA OPERATOR

The extrema operators consist of the relative maximum operator and relative minimum operator. They are recursive operators which have a numeric data domain. They require an input image which needs to be accessed but is not changed and an output image which is successively modified. Initially, the output image is a copy of the input image. The operator must be successively applied in a top-

down left-right scan and then a bottom-up right-left scan until no
changes are made. Each pixel on the output image contains the value
of the highest extrema which can reach it by a monotonic path.
Pixels on the output image which have the same value as those on
the input image are the relative extrema pixels.

The way the relative maxima operator works is as follows.
Values are gathered from those pixels on the output image which corresponds to pixels on the input image which neighbor the given pixel
and which have input values greater than or equal to the input
values of the given pixel. The maximum of these gathered values
are propagated to the given output pixel. The relative minima operator works in an analogous fashion. Figure 7 illustrates the pixel
designations for the normal and reverse scans.

The relative maxima operator uses two primitive functions h
and max. The four argument function h selects the maximum of its
last two arguments if its second argument is greater than or equal
to its first argument. Otherwise, it selects the third argument.
Hence,

$$h(b,c,d,e) = \begin{cases} \max\{d,e\} \text{ if } c \geq b \\ d \text{ if } c < b \end{cases} \tag{23}$$

The primitive function max selects the maximum of its arguments.
In the 8-connected mode, the relative maxima operator lets

$$a_0 = 1_0 \text{ and } a_n = h(X_0, X_n, a_{n-1}, 1_n), = 1,2,3, \text{ and } 4$$

The output value 1 is defined by $1 = a_4$.

In the 4-connected mode, the operator is

$$a_0 = 1_0 \text{ and } a_n = h(X_0, X_n, a_{n-1}, 1_n), n = 1,2,3, \text{ and } 4.$$

The output value 1 is defined by $1 = a_2$.

The relative minima operator is defined similar to the relative
maxima operator with the max function replaced by the min function
and all inequalities changed. Hence, for the relative minima operator, h is defined by:

$$h(b,c,d,e) = \begin{cases} \min\{d,e\} \text{ if } c \leq b \\ d \text{ if } c > b \end{cases} \tag{24}$$

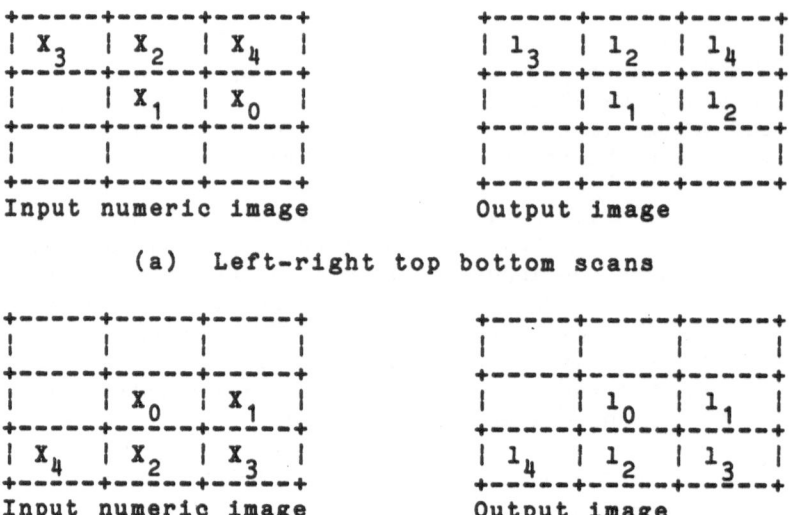

(a)   Left-right top bottom scans

(b)   Right-left bottom top scans

Fig. 7.   Illustration of the pixel designations for the recursive
operators which require forward and reverse scan and a numeric
image, and which recursively produce an output image.

In the 8-connected mode, the relative minima operator lets $a_0 = l_0$ and $a_n = h(X_0, X_n, a_{n-1}, l_n)$, $n = 1, 2, 3$, and 4.

The output value 1 is defined by $1 = a_4$.

In the 4-connected mode, the operator lets $a_0 = l_0$ and

$$a_n = h(X_2, X_4, a_{n-1}, l_n), \quad n = 1 \text{ and } 2.$$

An alternative kind of relative extrema operator can be defined using the symbolic image created by the non minima-maxima
operator in combination with the original numeric image. Such an
operator is a recursive operator and is based on the fact that by
appropriate label propagation, all flat regions can be relabeled
as transition, minima, or maximum regions and that the pixels originally labeled as non-minima or non-maxima can be relabeled as transition regions or true relative minima or maxima.

The initial output image is taken to be the image created by
the minima-maxima operator. Recursive propagation of the labels
from one pixel to its neighbor on the output image is performed
only if the two labels are not the same and the corresponding two
gray tones on the original numeric image are equal.

Let any two neighboring pixels be x and y having respective labels L_x and L_y on the output image. As shown in Figure 8, we have three cases to examine when $L_x \neq L_y$ and x = y:

(a) Either L_x or L_y is a flat (0) and the other one is not (1,2,3). In this case, we propagate the non-zero label into the zero label, thereby eliminating pixels marked as flats.

(b) Either L_x is a minimum (1) and L_y is a maximum (2) or vice versa. In this case, since a region constant in tone cannot simultaneously be a minimum and a maximum, mark both pixels as transitions (3).

(c) Either L_x or L_y is a transition (3). In this case, since a region which is constant in tone and has one pixel marked as transition must be a transition region, mark the non-transition region pixel as a transition region, thereby propagating the transition label.

This propagation rule requires one 4-argument primitive function h defined by:

$$h(x,y,a,b) = \begin{cases} a \text{ if } x \neq y \\ 3 \text{ (transition) if } x = y \text{ and} \\ \qquad (a = 3) \text{ or } (b = 3) \text{ or} \\ \qquad (a = 1 \text{ and } b = 0) \text{ or} \\ \qquad (a = 0 \text{ and } b = 1) \\ 2 \text{ (flat) if } x = y \text{ and } (a = 2 \text{ and } b = 2) \\ 1 \text{ (non-minima) if } x = y \text{ and} \\ \qquad (a = 1 \text{ and } b = 2) \text{ or} \\ \qquad (a = 2 \text{ and } b = 1) \\ \qquad \text{or } (a = 1 \text{ and } b = 1) \\ 0 \text{ (non-maxima) if } x = y \text{ and} \\ \qquad (a = 0 \text{ and } b = 0) \text{ or} \\ \qquad (a = 0 \text{ and } b = 2) \\ \qquad \text{or } (a = 2 \text{ and } b = 0) \end{cases} \qquad (25)$$

Values of pixels in the original numeric input image are denoted by X_n. Values of pixels in the non-minima-maxima labeled image are denoted by l_n. For the operator using 4-connectedness and the standard 3x3 neighborhood designations, let

$$a_0 = l_0$$

and define

$$a_n = h(X_0, X_n, a_{n-1}, l_n), \quad n = 1, 2, 3, \text{ and } 4.$$

```
                            +-+-+
Numeric Neighbor    |x|y|
                            +-+-+
```

```
                            +-+-+    Propagation when       +-+-+  Propagation
Symbolic Neighbor|a|b|    ------------------->  |c|c|  Result
                            +-+-+         x = y              +-+-+
```

b labels a labels	non-mixima	non-minima	flat	transition
non-maxima	non-maxima	transition	non-maxima	transition
non-minima	transition	non-minima	non-minima	transition
flat	non-maxima	non-minima	flat	transition
transition	transition	transition	transition	transition

Fig. 8. Illustration of the propagation table for the recursive relative extrema operator. The table gives the propagation label C for any pair, a,b of labels of neighboring pixels.

The output 1 is a_4.

For the operator using 8-connectedness, let

$$a_0 = 1_0$$

and define

$$a_n = h(X_0, X_n, a_{n-1}, 1_n), \quad n = 1,\ldots,8.$$

The output 1 is a_8.

Propagation can also be achieved by using the forward and reverse scan technique. In this case, the left-right top-bottom scan
is alternated with the right-left bottom-top scan. Using the pixel
designations in Figure 3 for the operator using 4 -connectedness,
let

$$a_0 = l_0$$

and define

$$a_n = h(X_0, X_n, a_{n-1}, l_n), \quad n = 1, 2, 3, \text{ and } 4.$$

The output l is a_4.

14. CONNECTED COMPONENTS OPERATOR

The connected components operator is recursive and has a symbolic data domain. Its purpose is to assign with a unique label
all pixels belonging to the same maximally connected component.
There are a variety of ways of determining connected components including some two-pass algorithms which may require memory for large
internal tables. Discussion of such connected component operators
can be found in Rosenfeld and Pfaltz (1966). The connected components operator described here requires only a small amount of
memory and must repeatedly scan the image until it makes no changes.
It also differs from other neighborhood operators in that it requires an internal state which must be remembered and which can
change or stay the same after each application of the operator. It
also is an operator that can be applied by alternating between top-
down, left-right scans (forward scans) and bottom-up, right-left
scans (reverse scans). Figure 7 illustrates the positioning of the
pixels for the 2x3 windows in each of these types of scans.

We will assume that the symbolic input image has each pixel
labeled with a label from a set L. Corresponding to each label m
$\in$ L is a set S_m of possible labels for the individual connected
components of all regions whose pixels are labeled "1." For nota-
tional convenience we will write $S_m(k)$ to denote the kth label from
the set S_m. For each label m, the connected components operator
must assign a unique label to all maximally connected sets whose
pixels originally had the label m.

We assume that for each label m $\in$ L $S_m = \phi$, that $S_m \cap S_n = \phi$
when m $\neq$ n, and that each of the S_m sets is linearly ordered. The
memory required by the operator is a function f which for each
m $\in$ L specifies a label from the set S_m. The specified label from
S_m is the next label which can be used to label a connected region

of type "m," that is, a connected region whose pixels have the input label "m."

The connected components operator works by propagating labels.
Let d be the label in a pixel and let c be the label in one of its
neighboring pixels. Under certain conditions the label c replaces
the label d. There are five cases governing this process. The case
and action are listed in Figure 9. Notice that for propagation condition (2), the next not yet used label from S_d must be generated.
This requires that the operator have access to a function $f(m)$ which
is an index to the next not yet used label from S_m.

The operator is based on a 2-argument primitive function h
which propagates labels in one of two ways as indicated by the case
analysis. If a pixel and its neighbor have a label from the same
set S_m, then h propagates the minimum of the labels. If a pixel
has not been labeled, then h propagates a label if appropriate,
from a neighboring pixel l. If the neighboring pixel also has not
been labeled it starts a new label. The function h is defined by:

$$
h(c,d) = \begin{cases}
c & \text{if } d \in L, c \in S_d \\
\text{next unused label for region type d} \\
\quad \text{if } d \in L, c \notin S_d \\
d & \text{if } d \in S_m \text{ for some } m \in L \text{ and } c \in S_m \\
\min\{c,d\} \\
\quad \text{if } d \in S_m \text{ and } c \in S_m \text{ for some } m \in L
\end{cases}
\qquad (26)
$$

The operator works the following way. If for some m, $X_0 \in S_m$,
then set $a_0 = X_0$ and leave $f(m)$ unchanged. If $X_0 \in L$, set $a_0 = S_x(m(X_0))$ and $m(X_0)$, set $m(X_0) + 1$.

For 4-connectedness, define $a_n = h(X_{n-1}, a_{n-1})$, n = 1 and 2.
The output label is given by a_2.

For 8-connectedness define $a_n = h(X_{n-1}, a_{n-1})$, = 1,2,3, and 4.
The output label is given by a_4.

```
+--+--+
|c | d|          Two Neighboring Pixels
+--+--+
```

Condition	Propagation
$d \in L$, $c \in S_d$	$d \leftarrow c$
$d \in L$, $c \notin S_d$	$d \leftarrow$ next unused label for region type d
$d \in S_m$ for some $m \in L$, $c \in S_m$	$d \leftarrow d$
$d \in S_m$ and $c \in S_m$ for some $m \in L$	$d \leftarrow \min\{c,d\}$

Fig. 9. The care analysis of what label gets propagated into pixel d by the connected components operator.

15. REACHABILITY OPERATOR

The reachability operators consist of the descending reachability operator and the ascending reachability operator. The operators are recursive and require a numeric input and a symbolic image used for both input and output. Initially the symbolic image has all relative extrema pixels marked with unique labels (relative maxima for the descending reachability case and relative minima for the ascending reachability case). The unique labeling of extrema can be obtained by the connected components operator operating on the relative extrema image. Pixels which are not relative extrema must be labeled with the background symbol "g." The reachability operator, like the connected component operator, must be iteratively and alternately applied in a top-down, left-right scan followed by a bottom-up, right-left scan until no change is produced. The resulting output image has each pixel labeled with the unique label of the relative extrema region that can reach it by a monotonic path if it can only be reached by one extrema. If more than one extrema can reach it by a monotonic path, then the pixel is labeled "c" for common region.

The operator works by successively propagating labels from all neighboring pixels which can reach the given pixel by monotonic paths. In case of conflicts, the label "c" is propagated. Figure 7 illustrates the pixel designations for the reachability operator. To do this, the descending reachability operator employs the four-argument primitive function h. Its first two arguments are labels from the output image and its last two arguments are pixel values from the input image. It is defined by:

$$h(a,b,x,y) = \begin{cases} a \text{ if } (b = g \text{ or } a = b) \text{ and } x < y \\ b \text{ if } a = g \text{ and } x < y \\ \\ c \text{ if } a \neq g \text{ and } b \neq g \text{ and } a \neq b \text{ and } x < y \\ a \text{ if } x > y \end{cases} \qquad (27)$$

The operator uses the primitive function h in the 8-connected mode by letting $a_0 = l_0$ and defining $a_n = h(a_{n-1}, l_n, X_0; X_n)$. n = 1,2,3, and 4. The output label l is defined by $l = a_4$.

The 4-connected mode sets $a_0 = l_0$ and defines $a_n = h(a_{n-1}, l_n, X_0, X_n)$, n = 1 and 2. The output label l is defined by $l = a_2$.

The ascending reachability operator is defined just as the descending reachability operator except that the inequalities are changed. Hence, for the ascending reachability operator, the primitive function h is defined by:

$$h(a,b,x,y) = \begin{cases} a \text{ if } (b = g \text{ or } a = b) \text{ and } x > y \\ b \text{ if } a = g \text{ and } x > y \\ \\ c \text{ if } a \neq g \text{ and } b \neq g \text{ and } a \neq b \text{ and } x > y \\ a \text{ if } x < y \end{cases} \qquad (28)$$

16. CONCLUSION

We have reviewed a variety of neighborhood operators from the point of view of the basic primitive functions whose composition generates the required operator. Many of the primitive functions can be implemented as table-lookup functions. The remainder can be implemented with only a small amount of sequential calculation using the standard logical, comparison, or arithmetic functions on a VLSI processor. This suggests the timely appropriateness of considering VLSI implementations of neighborhood operators for image processing.

17. REFERENCES

Arcelli, C., and Levialdi, S., "Parallel Shrinking by Three Dimensions," Computer Graphics and Image Processing 1:21-30 (1972).

Arcelli, C., and Sanniti di Baja, G., "On the Sequential Approach to Medial Line Transformation," IEEE Trans. Systems, Man and Cybernetics SMC-8(2):139-144 (1978).

Deutsch, E., "Comments on a Line Thinning Scheme," Computer Journal
    12:412 (Nov. 1969).

Deutsch, E., "Thinning Algorithms on Rectangular, Hexagonal and Tri-
    angular Arrays," Communication of Association for Computing
    Machinery 15(9):827-837 (Sept. 1972).

Fraser, J., "Further Comments on a Line Thinning Scheme," Computer
    Journal 13:221-222 (May 1970).

Hilditch, C., "Linear Skeletons from Square Cupboards," Machine
    Intelligence (Meltzer and Michie, eds.), University Press,
    Edinburgh (1969), pp. 403-420.

Levialdi, S., "On Shrinking of Binary Picture Patterns," Communica-
    tion of Association for Computing Machinery 15:7-10 (1972).

Lobregt, S., Verbeck, P.W., and Groen, F.C.A., "Three Dimensional
    Skeletonizations: Principle and Algorithm," IEEE Trans. Pattern
    Analysis and Machine Intelligence PAMI-2(1):75-77 (1980).

Rosenfeld, A., "A Characterization of Parallel Thinning Algorithms,"
    Information and Control 29:286-291 (Nov. 1975).

Rosenfeld, A., "Connectivity by Digital Pictures," Journal of the
    Association for Computing Machinery 17(1):146-160 (Jan. 1970).

Rosenfeld, A., and Pfaltz, J., "Distance Function on Digital Pic-
    tures," Pattern Recognition 1(1):33-61 (July 1968).

Rosenfeld, A., and Davis, L., "A Note on Thinning," IEEE Trans.
    Systems, Man and Cybernetics SMC-6(3):226-228 (March 1976).

Rosenfeld, A., and Pfaltz, J., "Sequential Operation in Digital
    Picture Processing," Journal of the Association for Computing
    Machinery 12(4):471-494 (Oct. 1966).

Rutovitz, D., "Pattern Recognition," J. Royal Statistics Soc. 129(A):
    504-530 (1966).

Stefanelli, R., and Rosenfeld, A., "Some Parallel Thinning Algo-
    rithms for Digital Pictures," Journal of the Association for
    Computing Machinery 18(2):255-264

Tamura, H., "A Comparison of Line Thinning Algorithms from Digital
    Geometry Viewpoint," 4th IJCPR, Tokyo, Japan, 1978.

Yokoi, S., Toriwaki, J., and Fukumura, T., "An Analysis of Topo-
    logical Properties of Digitized Binary Pictures Using Local
    Features," Computer Graphics and Image Processing 4:63-73
    (1975).

REGION RELAXATION IN A PARALLEL HIERARCHICAL ARCHITECTURE

P. A. Nagin, A. R. Hanson*, and M. Riseman**

Tufts New England Medical Center

Boston, Massachusetts 02111 USA

1. INTRODUCTION

Over the past decade the range of image analysis applications has greatly broadened. Usually the first steps of processing involve the transformation of a large spatial array of pixels into a more compact description of the image in terms of visually distinct syntactic elements and their characteristics. These "low-level" processes may achieve the desired goal directly or may serve as the input to a further set of interpretation processes.

1.1 The Need for Parallelism

Given the long range goals of scene analysis systems, one must consider the computational architectures that can facilitate the variety of forms of processing which probably will be required. In almost any application of image analysis, a characteristic which cannot be ignored is the massive amount of visual data which must be processed. For a full-color image of reasonable spatial resolution (512x512) and color resolution (3 colors, 6 bits/color), close to 5 million bits of information must be processed, often repeatedly. Faced with this computational overload, we have made a commitment to parallel processing at the very beginning of our research effort as

---

*School of Language and Communication, Hampshire College, Amherst, Massachusetts 01002 USA
**Computer and Information Science Department, University of Massachusetts, Amherst, Massachusetts 01003 USA

discussed in Hanson and Riseman (1974) and Riseman and Hanson (1974).
If such large amounts of sensory data are eventually to be processed
by a machine in real time, then the use of large parallel array computers appears to be necessary. It is relevant to note that developments in technology imply that such devices are becoming economically feasible.

     A second critical consideration is the need to extract informa-
tion from local areas of the image. Many interesting features are
not a function of individual pixels, but rather a function across
the set of points in a local "window" of the image. The optimum
size of this window varies with the features being sought, the location within the image, the application, etc. For example the variance of intensity over a square neighborhood centered on a pixel can
serve as a texture descriptor of that neighborhood. Given that the
size of this neighborhood must vary with the function employed, the
environment being imaged, and the resolution of detail (which is dependent upon the distance to the imaged object), the extraction of
such features would be facilitated by a hierarchical organization of
layers of decreasing image (and processing) resolution.

2.   A PARALLEL HIERARCHICAL ARCHITECTURE

     These design considerations have led us to simulate a general,
parallel computational structure, called a "processing cone," for
manipulating large arrays of visual data. This parallel array computer is hierarchically organized into layers of decreasing spatial
resolution so that information extracted from increasing sizes of
receptive fields can be stored and further processed. The structure of the processing cone is described in detail in Hanson and
Riseman (1974, 1980A) and applied in Hanson et al. (1975) and Nagin
et al. (1977). This type of architecture has proven very attractive
to other researchers in the field and our work bears a relation to
that of many authors (Uhr, 1972; Klinger and Dyer, 1976; Levine,
1978; Rosenfeld and Thurston, 1971; Hayes et al., 1974; Kelly, 1971;
Price and Reddy, 1977; Ballard et al., 1978). A survey of some of
these uses appears in Tanimoto (1978).

     The function of our processing cone, as shown in Figure 1, is
the transformation and reduction of the massive amount of image data.
The hierarchy of computational processing provides a structure in
which information at higher levels can direct more detailed processing at lower levels of the cone.

     The processing cone is general-purpose in that it may be pro-
grammed by defining a prototype computation to be performed on a
local window (i.e., subarray) of data. In the cone, this prototype
function is applied simultaneously--and in parallel--to all local
windows across the entire array. The user need only specify the

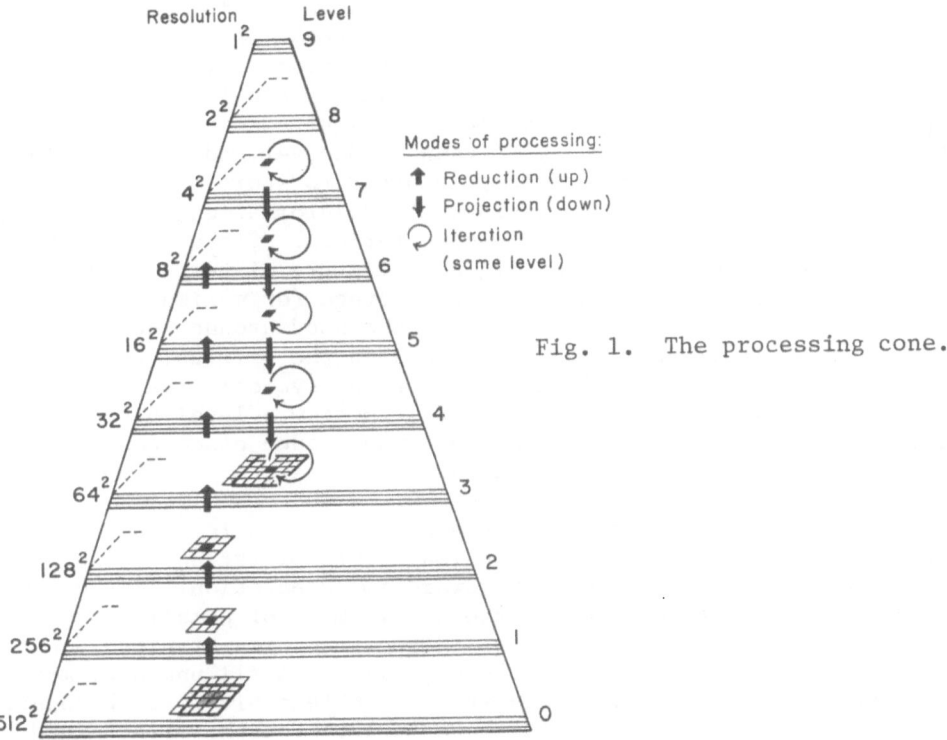

Fig. 1. The processing cone.

definition of the function, the location of the source(s) of the
data within the cone, a description of the size and shape of the
local window, and the destination of the result(s) within the cone.
The cone's operating system simulates lockstep computation just as
if there were parallel arrays of synchronous microprocessors computing on each window; each microprocessor executes a copy of the
prototype computation.

An important characteristic of the processing cone structure
is its hierarchical organization into layers of decreasing resolution. Figure 1 depicts a cone with an initial image resolution at
level 0 of 512x512 pixels; the next layer has one half the resolution on a side (256x256 pixels), etc. Thus, levels 0 through 9
have resolutions of 512x512, 256x256, 128x128, 64x64, . . . , 2x2,
1x1, respectively. Since the prototype function associated with a
cell at level k is applied to a window of data from level k-1 below
it, this type of transformation allows data to be "reduced" up the
cone. The decrease in spatial resolution is achieved by requiring
adjacent cells at level k to receive data from level k-1 windows
which have non-overlapping, but adjacent, 2x2 centers. The effect
of this hierarchical reduction in spatial resolution is that cells
at successively higher layers store and process information extracted
from increasingly larger receptive fields on the image below.

The algorithms which could be implemented within the cone would
be severely limited if only a single value could be stored at each
cell in a layer. Therefore, each cell at a given layer is capable
of storing a vector of information, not just a single scalar value.
For example, the initial data at level 9 might consist of three
color components (red, blue, and green). However, additional memory
at all levels is usually necessary in order to store both intermediate and final results of processing. This information may be used
as the input to further prototype computations. For example, it
might be useful to store at a cell some feature of texture and at
the same time also have available the average (or maximum or variance) of this feature over a local neighborhood around the cell. In
another case, it might be desirable to keep the original color data
as well as a region-label image denoting the symbolic region assignment of each pixel. Thus, it is useful to view a level of the cone
as a collection, or packet, of planes, where each plane can store
either a numeric or a symbolic value.

The design for the cone structure presented here, and which is
being simulated in our research on image interpretation, is not intended to be a blueprint for a hardware implementation. Rather, we
view it as a tool to be used in the development of parallel processing algorithms for image analysis. Since the forms of processing
that are sufficient to achieve these goals are still unknown, maximum flexibility was desired consistent with feasible network connections for a single processor in the network. This has led to a design with connectivity from a given cell to a variable size local
neighborhood of cells; the size of the neighborhood is defined under
program control up to some fixed maximum value.

2.1 Modes of Processing and Algorithm Specification

There are three basic modes of processing available within the
cone: (a) reduction operations upward through the layers of the
cone, (b) horizontal (or lateral) operations with processing restricted to a single level of the cone, and (c) projection operations where information in upper levels of the cone influence computation at lower levels. This flow of information up, laterally, and
down the cone is shown in Figure 1.

2.2 Reductions: Upward Processing within the Cone

During reduction, each function has as its input the data from
a local window of cells at level k and outputs one or more values
into a single cell at level $k+1$. Each window is of size $n \times n$, where
n is greater than or equal to 4 and is even; each such window is
placed over a unique 2×2 set of cells at level k. Since the centers
are non-overlapping, each cell at level $k+1$ is associated with a

particular 2x2 subarray at level k, and this provides the decrease
in spatial resolution at each higher level. In Figure 2a we have
shown the case for n=4 so that each neighborhood overlaps by one
cell the neighborhood adjacent to it in the north, south, east, and
west directions. The overlap of adjacent nxn neighborhoods avoids
such difficulties as, for example, a spatial differentiation operator missing an edge because it lies on the border of adjacent 2x2
centers. The source plane(s) at level k and the destination plane(s)
at level k+1 are specified under program control as the input and
output of the local function. The window size may also be set under
program control.

Averaging represents a simple example of the reduction process.
It is trivial to average an image up the cone using a single function AVERAGE-UP to obtain a sequence of reduced resolution images.
The value found at the 1x1 level represents the average intensity
of the entire image. One merely has to write a single function AVE
which averages the four points in the 2x2 center; AVERAGE-UP is composed from AVE by applying AVE at level k-1 at time k (for k=1, 2,
. . . ,9).

In the algorithms that will be presented in the following sections, the reduction process will be used to compute, for a cell at
level k, a histogram of the values in the 2^k by 2^k cells in the subarray below. This can be done in several ways. The simplest, but
least economical in terms of storage, is to compute a histogram from
level 0 to level 1, and then in all succeeding reductions to add the
four histograms from the 2x2 window at the level below. If the histogram is one-dimensional and there are 64 gray levels, this would
require the storage of 64 values at each cell (i.e., 64 planes of
data). At the lowest levels of the cone, all of this storage is not
necessary because only the actual values need be stored at levels 1
and 2 (i.e., 4 and 16 values, respectively).

A more efficient method is to store histograms only at higher
levels in the cone, say at the 32x32 or 16x16 levels. The histograms could be formed by carrying upward a few sampled values at a
time and sequentially adding them to the partially formed histogram.
This removes all storage of intermediate histograms at the expense
of taking longer to form the histogram. To form a histogram for a
cell covering a 32x32 subarray, it would require 1000 reductions
across 4 levels if only a single value were carried upward at a
time. Thus, histograms can be formed fairly quickly without exor-
bitant amounts of storage.

(a) Reduction processing: $f_{k,k+1}^{t}(I_1,\ldots,I_D;O_1,\ldots O_R)$

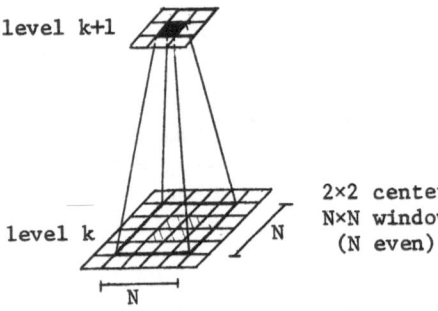

level k+1

level k

2×2 center
N×N window
(N even)

Every cell at level k+1 is
associated with a unique
2×2 window of cells at level k.

(b) Horizontal processing: $f_{k,k}^{t}(I_1,\ldots,I_D;O_1,\ldots O_R)$

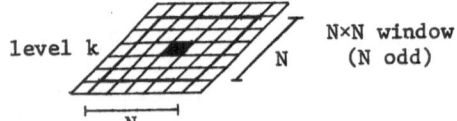

level k

N×N window
(N odd)

Every cell at level k is treated as
the central cell of an N×N window
of cells at level k.

(c) Projection processing: $f_{(k+1,k+2,\ldots,\ell-1,\ell;k)}^{t}(I_1,\ldots,I_{\ell-k};O_1,\ldots O_R)$

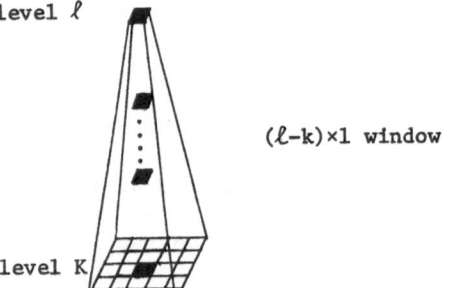

level ℓ

level K

$(\ell-k)\times1$ window

Every cell at level k is associated
with a window of ancestral cells
from level k+1 through the top of
the cone, one from each level. The
particular cell is determined by
the sequence of reduction windows.

Fig. 2. Processing modes in the cone.

2.3 Horizontal (Lateral) Processing: at a Fixed Level in the Cone

During a horizontal processing operation, as illustrated in Figure 2b, each local function receives data at level k from an nxn window (of odd size, n=1, 3, 5, . . .) and places the resultant value into the central cell of that neighborhood. Note that the overlap of adjacent neighborhoods is significant when n is greater than 1. Since every cell at level k has its own unique window placed over it, there is no decrease in the resolution of the data. Again, neighborhood size and the memory planes for source and destination values are all under program control.

Many interesting algorithms require iterative applications of a function. Typical examples are region algorithms and edge algorithms based on relaxation techniques described in Hanson and Riseman (1978A, 1980B). In these algorithms, information in a local neighborhood is used to update a set of likelihoods or confidences of hypotheses at the central cell in the neighborhood. During each iteration the prototype local function is applied simultaneously to every cell at the specified level in a true parallel manner.

The number of iterations of a local function can be fixed prior to computation or it can be determined dynamically as execution proceeds. In the latter case, any of the local copies of the prototype function, as it executes on its local window, can set a global flag which specifies that either another iteration is to be performed or that processing is to be terminated. This implies that termination of a relaxation updating process can be under the control of a global decision mechanism based on the existence (or absence) of a local property across the image.

2.4 Projection: Processing Information Downward in the Cone

In the previous two sections, we have shown how information can flow up and laterally within the cone. These operations have the effect of allowing a local computation to participate in the formation of more global values at levels higher in the cone. However, it is also quite useful to allow global information to influence local computation, for example in planning mechanisms which allow greatly increased efficiency in computation as well as wider contexts of information to direct the processing (see Kelly, 1971; Hanson and Riseman, 1974; Price and Reddy, 1977). This is achieved by a projection process which makes available global information from cells at higher levels in the cone to cells at lower levels. A simple example of the utility of this type of processing is found in thresholding operations. In this application, a threshold value is computed at some high level in the cone, perhaps the 1x1 level (level 9). This information is projected down to levek k (k<9)

where it could be used to set flags on or off, or to set below-
threshold values to zero. The threshold might have been obtained
by computing the mean and variance of the values at level 0 via reduction operations.

Let us examine the projection process a bit more carefully.
Each cell at any level of the cone is a member of a unique 2x2 center of a neighborhood defined by the reduction operator; that is,
it has a parent cell which this neighborhood would map into by reduction. For a given cell at some level, successive reductions denote a set of ancestral cells, with exactly one cell on each higher
level (Figure 2c). If values throughout a cone have already been
computed, then information can be passed down the cone in parallel
by making available all the ancestral information of a given cell
during a projection pass.

In the relaxation algorithms presented later, the probabilities
of pixel labels must be computed from peaks (i.e., cluster centers)
in a histogram of the values of a feature. This can be performed
via projection of the locations of the peaks down to the lowest
level cells which store the raw pixel data (see Section 4.2).

3.   THE PROCESSING CONE AS AN OPERATING SYSTEM FOR IMAGE ANALYSIS

By utilizing the parallel processing cone system as the core
of a low-level image segmentation system, a number of advantages
are realized as discussed by Kohler (1980). First, the cone struc-
ure represents a flexible tool for the development and testing of
image analysis algorithms. It allows algorithms to be constructed,
debugged, documented, and tested relatively easily on a variety of
images.

Second, when the facilities exist for defining algorithms as
sequences of local parallel operations, then it is only necessary
to write a series of functions, each in terms of a prototypical
local window of data. All of the overhead involved in actually
applying the function is handled by the system. Once a local function has been written it can be applied at any level in the cone
without modification. Algorithms then become automatically structured into unit modules of computation, consistent with the tenets
of good program design.

A third important characteristic of the cone involves data display and data movement. The hierarchy of resolution levels permits
rapid access to intermediate results of computations while algorithms are being developed. These levels also provide a type of
focusing mechanism for applying the local function to arbitrarily
smaller areas of the image. This can be achieved by placing a portion of the original data (a $2^k x 2^k$ subimage) at the appropriate

higher level in the cone and beginning the processing at this level.
It is particularly useful in facilitating program development by
allowing economical execution of a function at a high level in the
cone, where there are small numbers of pixels, and then being certain of its proper application at lower levels with their massive
amounts of data. Whenever focus of computation or display of information is desired, then movement of subimages to higher levels
is easily performed.

Finally, it should be emphasized that the cone is not restricted
to the processing of only visual data. For example, a two-dimensional histogram, perhaps of image properties, can be inserted into
the cone at the appropriate level and processed as if it were an
image. All of the algorithms developed for processing images, such
as region growing or thresholding, then become immediately available.

A new general low-level image segmentation system based on the
processing cone was recently implemented on a VAX 11/780 in our
laboratory and is given in Kohler (1980). This system is configured
as shown in Figure 3. It provides a high level of user control over
low-level functions through the use of LISP as the control language.
This choice was dictated somewhat by the requirement that the low-

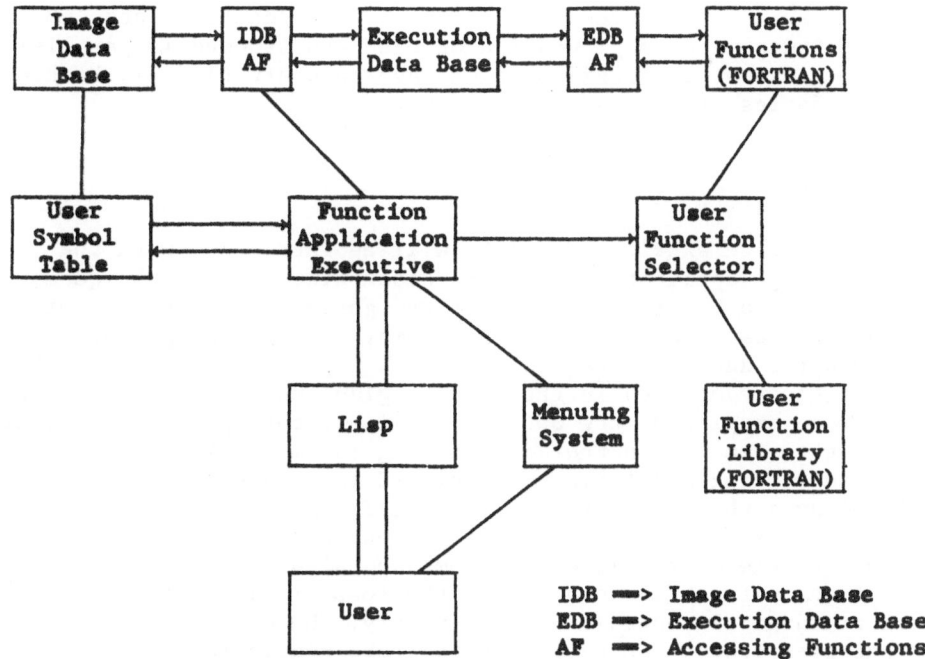

Fig. 3. Overview of low-level image operating system.

level system interface directly into a high-level interpretation
system (see Hanson and Riseman, 1978B). LISP provides the kind of
interactive environment which is well-suited to the development of
complex control strategies for segmentation algorithms.

4.   SEGMENTATION ALGORITHMS AND APPLICATIONS

Let us now focus upon a comprehensive segmentation algorithm
which has been implemented within the framework of the processing
cone. We will first consider an application of image processing to
the relatively constrained biomedical domain of glaucoma detection.
From the ophthalmologist's point of view, glaucomatous tissue damage
gives rise to visually recognizable features that have clinical sig-
nificance. To detect and measure these features objectively, we
present an algorithm which consists of a global histogram-based
pixel classification followed by a local relaxation analysis for
refinement of the pixel label probabilities.

In a later section, the algorithm will be applied to a much
larger domain, that of outdoor scenes. Here, the complexities of
texture, reflection, and shadowing combine to create a greater level
of problem difficulty, requiring a different formulation of the algorithm. In order to overcome weaknesses of the global histogram
analysis, the cluster peak detection can be applied on local receptive fields at a lower level of the cone. Decomposition into subimages and independent segmentation of each can proceed in parallel.
After segmentation, the regions which have been artificially broken
at the boundaries of subimages are sewn together.

4.1 Glaucoma

The region relaxation algorithm will be introduced with the
biomedical image application. Before discussing these techniques,
we will provide a brief background of the goals in processing reti-
nal images for early detection of glaucoma. Atrophy or pallor of
the optic nerve due to increased intraocular pressure is one of the
earliest signs recognized in the disease glaucoma. Associated with
the pallor is cupping of the optic disc, i.e., the optic nerve head.
Pallor is the area within the disc showing maximum color contrast
or the area lacking small blood vessels (Color Plate 1). Cupping
is defined as a three-dimensional depression of the optic disc surface. Both these characteristics of the optic disc have been the
mainstays for the diagnosis of glaucoma and also for following the
course of the patient in relation to the loss of visual field and
medical or surgical control. Unfortunately, the evaluation of these
signs has been mostly subjective and differs considerably from observer to observer.

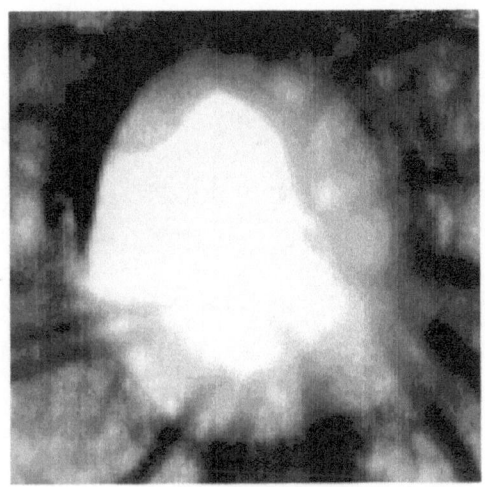

Plate 1
Optic disc image
(Nagin, Hanson, Riseman)

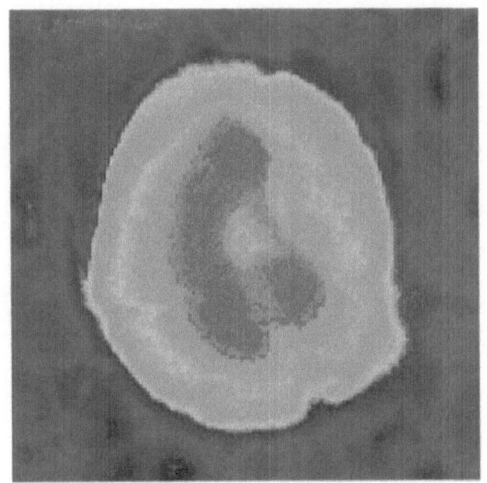

Plate 2
Initial probability labeling
(Nagin, Hanson, Riseman)

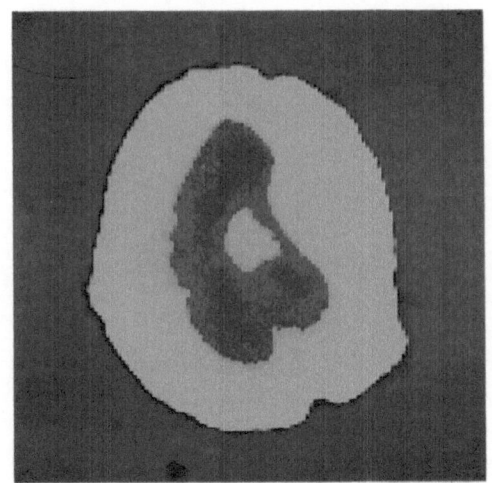

Plate 3
Intermediate relaxation probabilities
(Nagin, Hanson, Riseman)

Plate 4
House scene in U*V*W*
(Nagin, Hanson, Riseman)

Plate 5
U* initial probabilities
(Nagin, Hanson, Riseman)

Plate 6
Intermediate relaxation probabilities
(Nagin, Hanson, Riseman)

Plate 7
Global segmentation as edges
(Nagin, Hanson, Riseman)

Plate 8
Localized segmentation result
(Nagin, Hanson, Riseman)

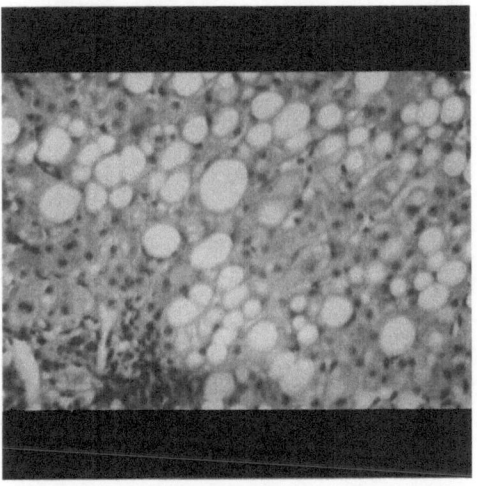

Plate 9
Liver tissue image
(Preston)
(Photography courtesy NASA Ames)

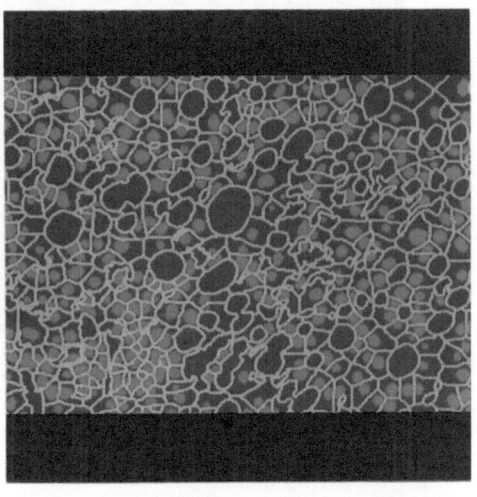

Plate 10
Cell nuclei exoskeleton
(Preston)
(Photography courtesy NASA Ames)

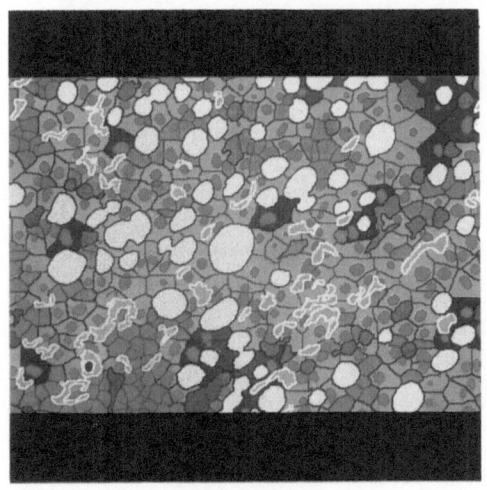

Plate 11
Liver tissue cartoon
(Preston)
(Photography courtesy NASA Ames)

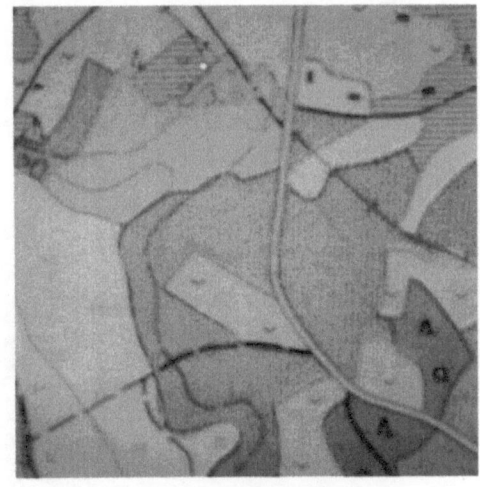

Plate 12
Land use map
(Fukada)

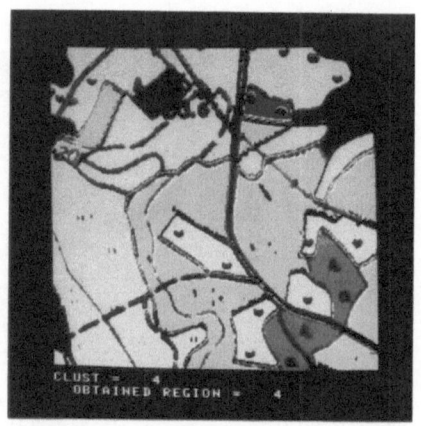

Plate 13
Map analysis with C = 4
(Fukada)

Plate 14
Map analysis with C = 5
(Fukada)

Plate 15
Aerial photograph analysis (C = 6)
(Fukada)

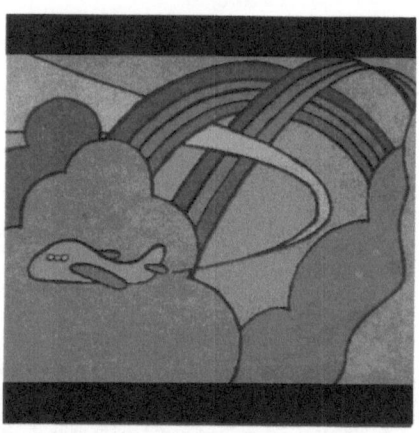

Plate 16
Color-coded image display
(Kamae)

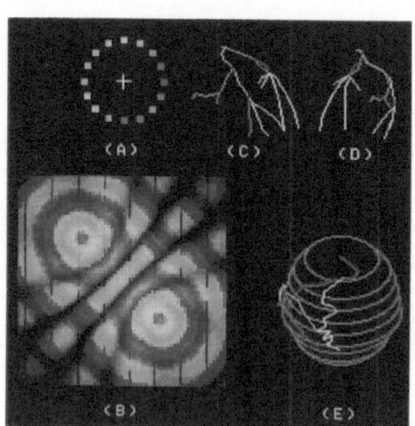

Plate 17
(A) Directional color code
(B) Interference of two cylindrical
    waves
(C,D) Three-dimensional display of
    arteries of the heart
(E) Coronary arteries mapped on
    heart surface
        (Takagi and Onoe)

Ophthalmologists estimate the degree of pallor and cupping by examining the optic fundus via an ophthalmoscope. Although it is fairly straightforward for a trained observer to make a rough evaluation of the condition of the disc, it is extremely difficult and error-prone to detect subtle changes in a single disc over time. It seems that structural changes to the disc precede visual field loss as presented by Schwartz (1976) and therefore accurate detection of disc change is an important factor in initiating preventive treatment.

Our current research is focused on the problem of detecting pallor, with measurement of the 3D cup being left for future investigation. The algorithm used to detect pallor integrates both global and local information with the intention of overcoming the limitations characteristic of each domain.

4.2 Image Analysis Using Histogram Cluster Labels

Histograms are used to obtain a rough initial labeling of pixels and then a spatial relaxation operation is applied to obtain a refined and more consistent labeling. Figure 4 is an overview of the algorithm explored in the following sections.

A simple histogram classification analysis involves two steps: (1) slicing the overall distribution of some image feature at the valleys between the peaks and then (2) labeling each pixel according to the curve component (cluster) into which the pixel falls. In Figure 5a-d the histogram of the digitized, red-filtered image is divided into three components. In the image, these components correspond to retina-choroid (dark background area), optic disc, and central pallor (central whitish area). Figure 5c shows each pixel labeled by cluster type and displayed as a distinct gray level. Figure 5d shows the boundaries between the classified regions in white, superimposed over the raw red image.

Notice that the overlap in the distribution of clusters introduces unavoidable labeling errors in this process. From a pattern recognition viewpoint, an assumption about the form of the distributions allows statistical decision theoretic methods to be used for minimizing errors in the classification process. The difference here is that we have "spatial" information as well, which typically is not available in pattern recognition problems. One views the coordinates of the image as two more features in an N-dimensional space, but this does not acknowledge the spatial characteristics of those particular features. Instead, we use a simple clustering approach followed by a relaxation process. Weaknesses are overcome as described in Section 5 by localization of the algorithm.

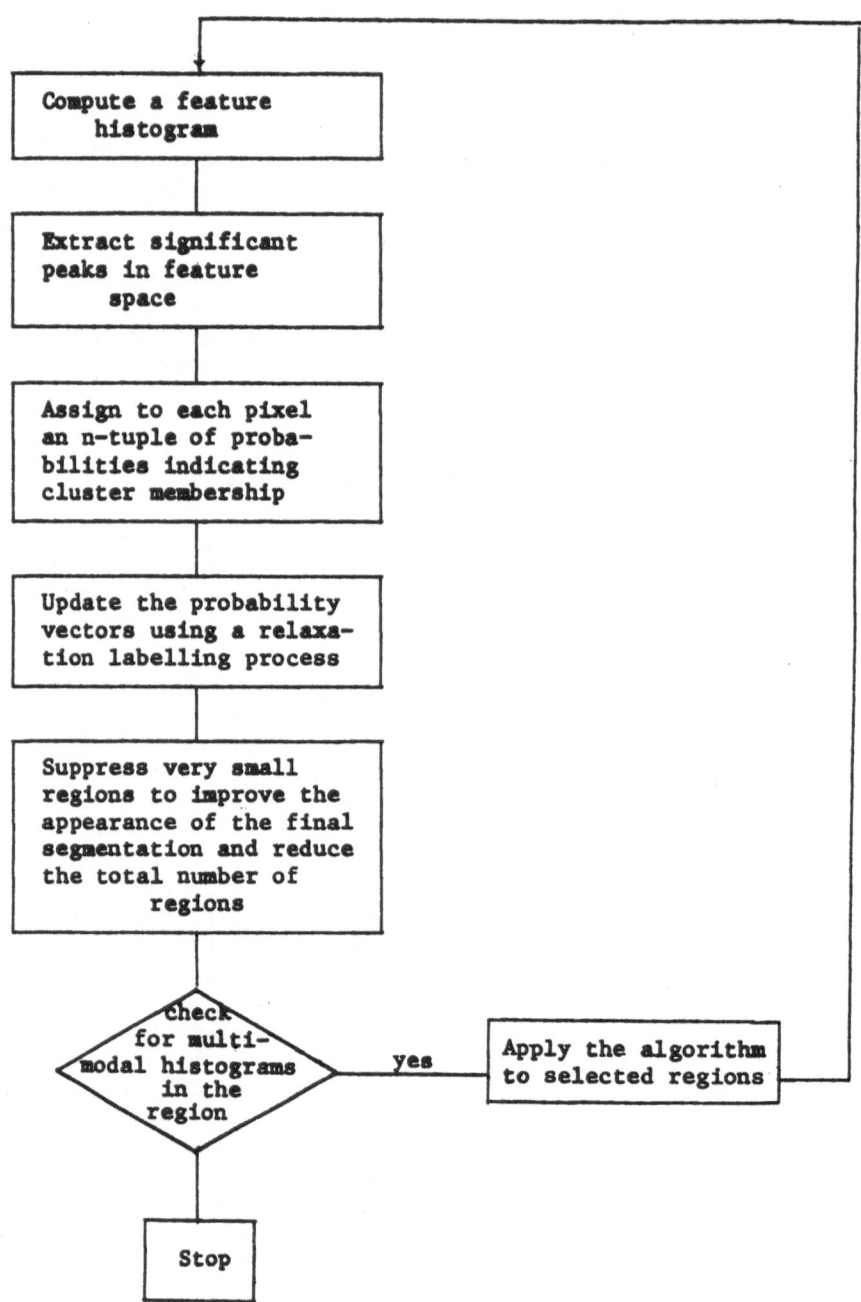

Fig. 4. Summary of the global segmentation algorithm.

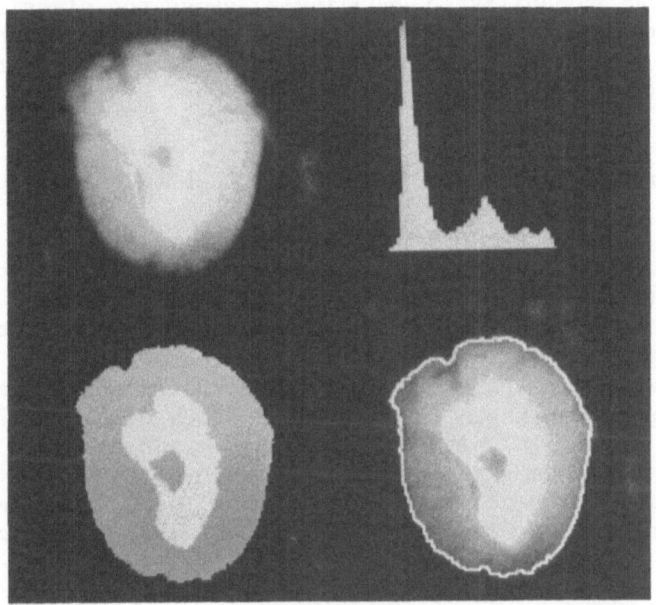

Fig. 5. Discrete classification: (a) raw red filtered image,
(b) histogram of red intensities, (c) spatial classification as
regions, (d) spatial classification as edges.

A more powerful use of the histogram can be derived by performing a probabilistic pixel encoding. Here, the gray level of
each pixel is transformed into a probability vector that indicates
its degree of membership in each of the histogram clusters. We can
assume a very simple model for determining the probability of cluster membership by computing the difference $d_{i\alpha}$ between gray level i
and the gray level of the α cluster mean:

$$P(i,\alpha) = (1/d_{i\alpha}) \, / \, (\sum_{\alpha} 1/d_{i\alpha}) \tag{1}$$

Note that a d value of 0 must be changed to some small value ε, say
0.1. This measure is a monotonically decreasing function of the
Euclidean distance between i and α. Other measures which take into
account the peak/valley structure of the histogram are also convenient to use as given in Kohler (1980).

Color Plate 2 shows the probabilistic encoding of each of the
three clusters from the red histogram. The display was generated
by interpreting the vector of probabilities associated with each
pixel as an (RGB) triple for a color monitor; therefore a saturated
red, green, or blue color implies that a single class is highly
probable while a mixed color implies uncertainty of class membership.

Notice that the probabilistic encoding suspends the classification of the areas of the image that are unclear, i.e., at the choroid-disc boundary and, more significantly for this research, at the disc-pallor boundary.

4.3   Refinement of Initial Labeling Using Spatial Relaxation

The next step is to determine the best label for each pixel. It seems reasonable to assume that the best label should be a function of both the initial (global) label at a pixel as well as the labels of the neighboring pixels. One approach is to use the so-called relaxation algorithms to update the vector of label probabilities at each pixel based upon the surrounding contextual information as proposed in Nagin (1979A) and Eklundh et al. (1979). Relaxation is a parallel, iterative technique and has been implemented within the cone structure as follows. Given $p_i^k(\alpha)$, the probability of label α at pixel i and time k, compute the new probability at time k+1 by first computing the neighborhood support for label α as follows:

$$q_i^k(\alpha) = \sum_j \sum_\alpha r_{ij}(\alpha,\alpha')\, p_j^k(\alpha')$$

where the r_{ij} are precomputed (see below) from the initial labeling and measure the compatibility of α at the central pixel with the labels in the surrounding neighborhood. The outer sum is computed over the neighborhood of i and the inner sum is computed over the labels. Note that the geometry of the neighborhood of a pixel is in general arbitrary. We have chosen the smallest, most compact neighborhood, i.e., the 3x3 area surrounding each pixel.

The initial values at each pixel can be used to estimate globally the likelihood that a pair of labels co-occur in each of four orientations. The joint probability of a pair of labels and the unary label probabilities allow orientation-dependent conditional probabilities to be computed. These values are then used as compatibility coefficients, r_{ij}, in the relaxation labeling scheme for updating probabilities. This approach has been examined by Hanson and Riseman (1978A) and by Nagin (1979A). Other work related to this appears in Rosenfeld et al. (1976), Peleg (1980), and Eklundh et al. (1979A).

The probabilistic labeling at time k+1 is computed as follows:

$$p_i^{k+1}(\alpha) = p^k(\alpha)(1+q^k(\alpha)) \,/\, \sum_{\alpha'} p_i^k(\alpha')(1+q_i^k(\alpha')) \qquad (2)$$

Color Plate 3 shows the effect on the probabilities of applying three iterations of relaxation. Notice that the boundary areas are the slowest to converge. Figure 6 shows this data in a different format by displaying the edges between labeled regions, superimposed over the raw red-filtered image. Notice that the relaxation process has not converged at this point and may not converge for several more interations. Moreover, it has been shown by Nagin (1979A) and Peleg (1980) to diverge in certain cases. Work on monitoring the course of the algorithm is in progress by Eklundh and Rosenfeld (1978) and Peleg (1979).

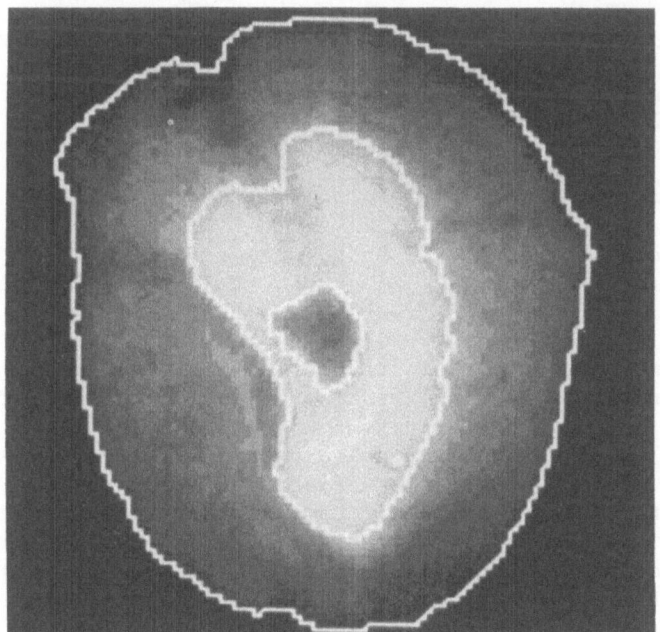

Fig. 6. Edges between labeled regions after 5 iterations of relaxation.

4.4   Segmentation Algorithm Applied to a Natural Image

We now turn to a more difficult image domain, i.e., that of outdoor scenes. The scene depicted in Figure 7 presents a difficult image processing problem for a number of reasons. First, the physical scene has undergone a number of stages of information

Fig. 7. Color components of house scene: (a) red, (b) green,
(c) blue, (d) intensity.

degradation including the photographic and digitization process and
a spatial averaging (blurring) process to reduce the amount of data
to manageable levels. The effect of these processes is to introduce noise, blur edges, and to create hybrid pixel values--mixed
pixels--which are not easily classifiable. Moreover, the image
displays inherent visual complexities such as irregular texturing,
object occlusion, and irregular changes in gradients. Finally,
the image is complex because in 3-color space there is a large,
unknown number of object classes to be discriminated, most of which
overlap to varying degrees.

4.4.1 *The Data*

The raw image used in these experiments consists of a 512x512
array of pixels in which each pixel is represented as a triple of
six-bit numbers. The components of a pixel correspond to its light
intensity when scanned through red, green, and blue (RGB) filters.
To lessen the computational overload, the original data has been
reduced in the cone by independently blurring each component via a
2x2 spatial averaging process, yielding a resolution of 256x256
pixels (Figure 7). After data reduction, the image was transformed
from the raw RGB components into the U*V*W* (Color Plate 4) opponent

color space after Pratt (1978), Nagin (1979A), and Kender (1976).
This transformation appears to enhance the contrast between objects
and is therefore sometimes useful for segmentation.

4.4.2 *Relaxation Results*

The following segmentation experiments were performed using
the U* or "red-green" component of the image, since the corresponding image appeared to give the greatest degree of object discrimination. The peaks and valleys for the U* distribution were detected
by a simple heuristic process given in Nagin (1979A) and are marked
on the histogram shown in Figure 8. Color Plate 5 shows the initial
labeling of each pixel with respect to the 3 peaks in the U* component histogram. As in Color Plate 2 the display was generated by
interpreting the vector of probabilities associated with each pixel
as an (RGB) triple for a color monitor.

How can the initial segmentation be evaluated? Since there is
no ground truth data available with which to generate an error rate,
the evaluation must be subjective. First, notice that the roof of
the house, the tree crown on the right, and the sky appear as a
single region. Cluster B, which is relatively wide, apparently contains the distributions of all of these objects. Since they happen
to lie adjacent to each other in the image, they receive the same
region label and appear as a single region in the segmentation.
This situation is referred to as *overmerging* and results from widely
overlapped clusters in feature space.

Other areas of concern are (1) "fragmentation" of the left
tree due to partial cluster overlap between A and B, (2) fragmentation of the thin roof gutter (clusters A and B), (3) fragmentation
of the shutters (clusters A and B), and (4) overmerging of the
bushes and lawn regions.

Next consider the behavior of the parallel relaxation scheme
after it has been applied to the initial labeling of pixels. Color
Plate 6 shows the probabilities after 5 iterations of relaxation.
Color Plate 7 shows the U*V*W* image with edges superimposed wher-
ever the highest probability label of a pixel (at iteration 5) differs from that of any of its neighbors.

The impact of the relaxation process is difficult to evaluate
since the image itself is so complex (see Nagin et al., 1979B).
However, subjectively, very little change appears to have occurred
here beyond clean-up of small, scattered region fragments. The process does not, for instance, bind together nearby region fragments
to form "whole" regions. In the case of the thin roof gutter, the
algorithm tends to suppress the gutter fragments into the roof

Fig. 8. U* component ("red-green") and corresponding histogram.

instead of leading to the formation of a single gutter line. In a
careful study by Nagin (1979A) using test images, we showed that
this problem could be traced in part to the impact of the compatibility coefficients on the updating of label probabilities. Recall
that these coefficients are computed across the entire image and
therefore reflect the average image structure for pairs of neighboring pixels. In a heterogeneous image such as the house scene
presented here, the average structure (embedded in the coefficients)
is too diluted to adequately represent and promote any particular
structure that might occur locally.

5.   LOCALIZATION

The previous section demonstrated two fundamental problems with
a segmentation algorithm that is based on global image statistics.
These problems--cluster overlap and "insensitive" compatibility co-
efficients--were shown to yield a relatively poor segmentation of
an image with reasonably well defined regions. Since both of these
problems stem from too wide a view of the data, we propose a solution via moderate localization.

In this scheme, which is summarized in Figure 9, the image is
arbitrarily partitioned into small (e.g., 16x16 or 32x32 pixel)
subimages. The algorithm presented in the previous section is then
applied independently and in parallel to each subimage using the
cone structure. In this manner, local activity can be exploited to
yield improved cluster detection as well as yielding compatibility
coefficients that are representative of local dependencies in the
data.

It is important to note that it is necessary to extend or overlap each subimage into the adjoining subimages in order to obtain
effective results. By overlapping, there is an increased likelihood that the cluster detection process will yield more consistent
clusters in going from one subimage to the next and it is less
likely that small portions of a region extending into an adjacent
subimage will be lost.

Clearly, there is a problem with this approach since some regions will be artificially broken along the boundaries of the subimages. Therefore, a region merging process is in order. Specifically, given a pair of regions lying along a subimage boundary, it
is necessary to detect whether the regions have approximately the
same distribution of gray levels. When this situation is detected,
the boundary between the regions should be eliminated.

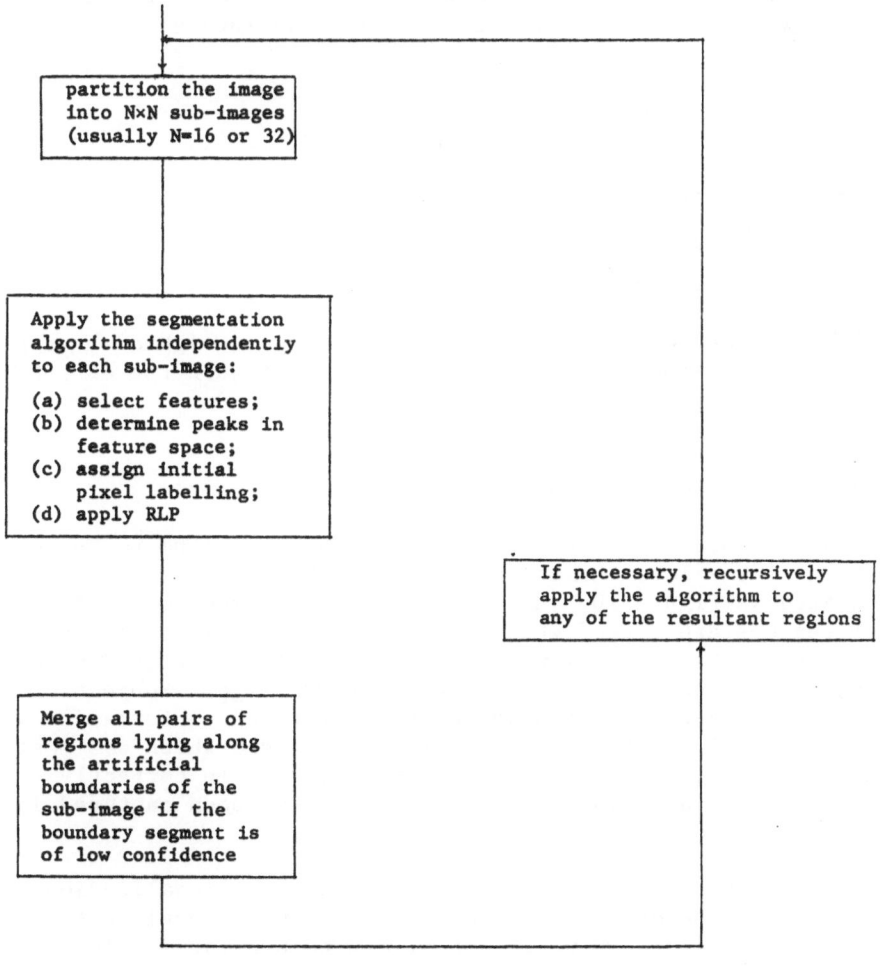

Fig. 9. Summary of the localized segmentation algorithm.

5.1 Region Merging

The region merging process can be accomplished via the same procedure that is used to cluster the data originally. That is, one can compute the histogram across the two regions and then check for modality. If the combined histogram is unimodal, then the regions should be merged, otherwise the boundary should be left intact. This procedure requires the extraction of a histogram for each region and an adjacency flag so that the decision of merging or not can take place at a cell in the cone whose receptive field covers both regions.

A procedure which demands smaller amounts of storage involves a statistical decision based on the variances of each separate region relative to the variance of the combined distribution as given in Yakamovsky (1979) and Prager (1980). The confidence that regions i and j are separate regions can be defined as:

$$Cij = Vij/V_i V_j \qquad\qquad (3)$$

where

    Vij   the variance across both regions i and j

    Vi   the variance across region i

    Vj   the variance across region j.

For this application, the computation of V_{ij} was weighted by region size using the method of Nagin (1979A) to counteract the effects of differences in area between the regions.

5.2 Results of Localization

Now consider the result of applying the algorithm to the outdoor scene. The local segmentation algorithm uses 32x32 partitions as depicted in Figure 10. Each partition is segmented and then the border regions are merged when the merging statistic specifies that it is appropriate. In Color Plate 8 the final result is displayed as region edges superimposed over the U*V*W* image. Subjectively, the results appear to correspond quite well to the information in the raw data. In particular, the right-hand tree which was lost in the global segmentation (Color Plate 7) has been correctly segmented here. In addition, much more detail has been captured in the bushes and house regions.

One should note that the desirable degree of localization is dependent upon the data in the image. One cannot localize to the point where there are insufficient pixels to provide meaningful

Fig. 10.   U* image artificially sectored in 32x32 subimages.

clusters, yet one cannot be so global that clusters will be hidden
by information in other areas of the image.   The cone provides an
excellent architecture for varying the size of the subimage, because it only requires variation of the level of the cone at which
cluster labels are extracted and projected.   The region merging at
the boundaries requires the extraction of region adjacency information and can also be performed in the cone in somewhat parallel
fashion.

6.   CONCLUSION

This chapter presents a hierarchical architecture for parallel
computation in image analysis algorithms.   The storage and processing is organized into layers of decreasing resolution and information flows upward, laterally, and downward through the cone
structure during computation.

The algorithms described were shown to be effective in a range
of domains from general outdoor scenes to biomedical images.   The
first algorithm presented was based primarily on global feature information derived from a series of parallel operations applied to
the raw image data.   Pixel labels are associated with detectable

peaks in a histogram. Initial assignment of the probability of each label is based upon distances in feature space. A relaxation process is used to update these values based upon the spatial context of each pixel.

This algorithm was applied successfully to a relatively constrained image, i.e., an image with few clusters in feature space and with a homogeneous image structure. However, when confronted with a more complex image involving many types of visual activity, the global algorithm tended to leave regions grossly overmerged and fragmented. Recovery from these problems, via recursive segmentation after Ohlander (1975) and region merging after Nagin (1979A) is sometimes possible, but computationally expensive.

In response to these problems, a localized segmentation algorithm was formulated. By decomposing the image into small subimages, the segmentation algorithm showed tremendously improved results. Clusters which were hidden in the global histogram appeared in the localized histograms. Furthermore, the compatibility coefficients were better tuned to the local image structures. The localization algorithm has an additional facet which was not explored here. By decomposing the image, it is possible to select a different feature for each subimage. However, the criterion for sewing the image back together is not as obvious.

7.   ACKNOWLEDGEMENTS

This research was supported by the Office of Naval Research under Contract ONR N00014-75-C-0459, by the National Institutes of Health under Biomedical Research Support Grant RR07048, and by the National Eye Institute under Grant NIH 2 R01 EY00936.

8.   REFERENCES

Ballard, D. H., Brown, C. M., and Feldman, J. A., "An Approach to Knowledge-Directed Image Analysis," in Computer Vision Systems (Hanson, A., and Riseman, E., eds.), New York, Academic Press (1978).

Eklundh, J., and Rosenfeld, A., "Convergence Properties of Relaxation," TR 701, Computer Science Center, Univ. of Maryland (Oct. 1978).

Eklundh, J. O., Yamamoto, H., and Rosenfeld, A., "A Relaxation Method for Multispectral Pixel Classification," IEEE Trans. Pat. Anal. Mach. Intel. $\underline{2}$:72-75 (1979).

Hanson, A., and Riseman, E., "Preprocessing Cones: A Computational Structure for Scene Analysis," COINS Tech. Rpt. 74C-7, Univ. Massachusetts (Sept. 1974).

Hanson, A. R., Riseman, E. M., and Nagin, P., "Region Growing in Textured Outdoor Scenes," Proc. of 3rd Milwaukee Symp. Auto. Comput. Control (1975), pp. 407-417.

Hanson, A. R., and Riseman, E. M., "Segmentation of Natural Scenes," in Computer Vision Systems (Hanson, A., and Riseman, E., eds.), New York, Academic Press (1978).

Hanson, A. R., and Riseman, E. M., "VISIONS: A Computer System for Interpreting Scenes," in Computer Vision Systems (Hanson, A., and Riseman, E., eds.), New York, Academic Press (1978).

Hanson, A. R., and Riseman, E. M., "Processing Cones: A Computational Structure for Image Analysis," in Structured Computer Vision (Tanimoto, S., and Klinger, A., eds.), New York, Academic Press (1980).

Hanson, A. R., Riseman, E. M., and Glazer, F. C., "Edge Relaxation and Boundary Continuity," in Consistent Labeling Problems in Pattern Recognition (Haralick, R. M., ed.), Plenum Press, 1980.

Hayes, K. C., Jr., Shah, A. N., and Rosenfeld, A., "Texture Coarseness: Further Experiments," IEEE Trans. Sys. Man. Cyber. 4: 467-472 (1974).

Kelly, M. D., "Edge Detection in Pictures by Computer Using Planning," Mach. Intel. 6:379-409 (1971).

Kender, J. R., "Saturation, Hue and Normalized Color: Calculation, Digitization Effects, and Use," Tech. Rpt., Dept. Computer Science, Carnegie-Mellon University (Nov. 1976).

Klinger, A., and Dyer, C. R., "Experiments on Picture Processing Using Regular Decomposition," Computer Graphics and Image Processing 5:68-105 (March 1976).

Kohler, R. R., Ph.D. Dissertation, COINS Department, University of Massachusetts (in preparation).

Levine, M. D., "A Knowledge Based Computer Vision System," in Computer Vision Systems (Hanson, A., and Riseman, E., eds.), New York, Academic Press (1978).

Nagin, P. A., Hanson, A. R., and Riseman, E. M., "Region Extraction and Description Through Planning," COINS Tech. Rpt. 77-8, Univ. Massachusetts (May 1977).

Nagin, P. A., "Studies in Image Segmentation Algorithms Based on Histogram Clustering and Relaxation," COINS Tech. Rpt. 79-15 and Ph.D. Thesis, Univ. Massachusetts (Sept. 1979).

Nagin, P., Kohler, R., Hanson, A., and Riseman, E., "Segmentation, Evaluation, and Natural Scenes," Proc. IEEE Conf. Pat. Recog. Image Proc., Chicago (Aug. 1979).

Ohlander, R., "Analysis of Natural Scenes," Ph.D. Thesis, Carnegie-Mellon Univ. (Apr. 1975).

Peleg, S., "Monitoring Relaxation Algorithms Using Labeling Evaluation," TR 842, Computer Science Center, Univ. Maryland (Dec. 1979).

Peleg, S., "A New Probabilistic Relaxation Scheme," IEEE Trans. Pat. Anal. Mach. Intel. 2(4):362-369 (1980).

Pratt, W. K., Digital Image Processing, New York, John Wiley and Sons (1978).

Prager, J. M., "Extracting and Labeling Boundary Segments in Natural Scenes," IEEE Trans. Pat. Anal. Mach. Intel. PAMI-2:16-27 (1980).

Price, K., and Reddy, R., "Change Detection and Analysis in Multi-spectral Images," Proc. 5th Internat'l Joint Conf. Artif. Intel., Cambridge (1977), pp. 619-625.

Riseman, E. M., and Hanson, A. R., "The Design of a Semantically Directed Vision Processor," COINS Tech. Rpt. 74C-1, Univ. Massachusetts (Jan. 1974).

Rosenfeld, A., Hummel, R., and Zucker, S., "Scene Labelling by Relaxation Operations," IEEE Trans. Sys. Man Cyber SMC-6:420-433 (1976).

Rosenfeld, A., and Thurston, M., "Edge and Curve Detection for Visual Scene Analysis," IEEE Trans. Comput. (1971), pp. 562-569.

Schwartz, B., "Primary Open-Angle Glaucoma," in Clinical Ophthalmology (Duane, T. D., ed.), Harper and Row (1976).

Tanimoto, S. L., and Pavlidis, T., "A Hierarchical Data Structure for Picture Processing," Comput. Graph. Image Proc. 2:104-119 (1975).

Tanimoto, S. L., "Regular Hierarchical Image and Processing Struc-
    tures in Machine Vision," in Computer Vision Systems (Hanson,
    A., and Riseman, E., eds.), New York, Academic Press (1978).

Uhr, L., "Layered 'Recognition Cone' Networks that Preprocess,
    Classify, and Describe," IEEE Trans. Comput. (1972), pp. 758-
    768.

Yakamovsky, Y., "Boundary and Object Detection in Real World Images,"
    J. ACM 23:599-618 (1979).

GENERALIZED CELLULAR AUTOMATA

A. Rosenfeld

Computer Vision Laboratory, Computer Science Center

University of Maryland, College Park, MD 20742 USA

1.   INTRODUCTION

Cellular automata are arrays of processing elements which operate in a synchronous, parallel fashion and can be used to process or classify their arrays of input data. This chapter summarizes work done over the past few years at the University of Maryland on generalizations of the basic cellular automaton concept. It shows how the speed and power of cellular automata can be increased by extending the array into an appropriately connected larger structure or by allowing the individual processors to have larger amounts of internal memory. It also extends the cellular automaton concept to networks of processors in which each node has bounded degree and discusses the graph processing and recognition capabilities of such networks.

2.   ONE-DIMENSIONAL CELLULAR AUTOMATA

In one dimension, a *bounded cellular automaton* (BCA) B is a string of identical processing elements called "cells" which operate in a synchronous, parallel manner at a succession of discrete time steps. At any given step, each cell is in a specified "state"; we shall assume that the state of a cell is defined by the contents of some part of its memory, e.g., a particular register. Each cell can also read the states of its left and right neighbors in the string; if it is at an end of the string, it knows this by sensing a special "state" in one of its neighbors, which never changes. A step of B's operation is defined by every cell simultaneously entering a new state, which is a function of its

63

own and its neighbors' old states. The "transition function" which
maps these triples of old states into new states is the same for
every cell. Information to be processed is input to B as an
initial string of states, or initial "configuration." The configurations at subsequent time steps thus constitute processed
versions of the initial string. Recognition of an input string as
belonging to a specified class is defined by a "distinguished cell"
c_0 of B, which we shall take to be the leftmost cell, entering a
special "accepting" state. It is well known that BCA's have the
same acceptance power as linear-bounded automata.

     As a simple illustration of how BCA's operate, suppose that
the initial string consists of 0's and 1's, and we want to decide
whether the set of 1's is connected, i.e., consists of a single
"run." Informally, we do this by initiating a signal at the rightmost cell (which can identify itself since only this cell has no
right neighbor) and shifting the signal one cell leftward at each
time step. Specifically, if the rightmost cell is initially in
state 0, after the first step it goes into the signal state s_0,
and if it is initially in state 1, it goes into s_1. If a cell in
state 0 sees s_0 on its right, it goes into s_0; if a cell in state 1
sees s_0 or s_1 on its right, it goes into s_1; and if a cell in
state 0 sees s_1 on its right, it goes into s_2. Finally, if a cell
in state 0 sees s_2 on its right, it goes into s_2; if a cell in
state 1 sees s_2 on its right, it goes into s_3; and if a cell in
state 0 or 1 sees s_3 on its right, it goes into s_3. Thus eventually some s_i reaches the leftmost cell c_0. It is easily seen
that s_0 reaches c_0 iff the initial string σ has no 1's; s_1 or s_2
reaches c_0 iff σ has exactly one run of 1's; and s_3 reaches c_0 iff
σ has more than one run of 1's. Thus c_0 can classify σ based on
which signal reaches it.

     In the illustration just given, the number of time steps re-
quired until c_0 can make its decision is equal to the length $|\sigma|$
of the input string. In general, this is a lower bound on recog-
nition time; in fact, if c_0 makes a decision about σ in $t < |\sigma|$
time steps, the decision cannot depend on all of σ, since the
state of c_0 at time t is independent of the initial states of the
cells at distances greater than t from c_0. Thus B is inherently
no faster than a sequential processor A which scans a tape containing σ, since for some tasks A too could be designed to require
only $|\sigma|$ steps. (There do exist tasks for which A's are inherently
slower than B's, e.g., any A might require at least $O(|\sigma|^2)$ steps
for a task that some B can do in $O(|\sigma|)$ steps; but the absolute
lower bound of $|\sigma|$ steps is the same for both A's and B's.) In
Section 4 we shall show how, by building an expontentially tapering
"triangle" of BCA's, we can reduce the lower bound on BCA recogni-
tion time to $\log |\sigma|$, which is a substantial improvement.

BCA's are usually defined in such a way that the amount of memory in each cell is a constant, independent of the number $|\sigma|$ of cells. Thus a cell cannot in general know its position in the string, since this position is specified by a $\log_2|\sigma|$-bit number; and similarly, the distinguished cell c_0 cannot know how many cells there are, since this number too has $\log_2|\sigma|$ bits. As we shall see in Section 5, allowing each cell to have $O(\log|\sigma|)$ memory substantially increases the versatility and usefulness of BCA's.

3.   MULTI-DIMENSIONAL CELLULAR AUTOMATA

A two-dimensional BCA is a rectangular array of cells which operates exactly as in the one-dimensional case, except that each cell (other than those in the array borders) has four neighbors—upper, lower, right and left. We take the distinguished cell to be the cell in the upper left corner of the array; this cell is identifiable as the only one having no upper and left neighbors. The lower bound on acceptance time for 2-D BCA's is the city block diameter (=height + width) of the array; note that this is inherently faster than acceptance by a sequential processor, which in general requires O(array area = height x width) steps. As we shall see in Section 4, further speedup to O(log diameter) can be achieved by building an exponentially tapering "pyramid" of 2-D BCA's; and in Section 5 we shall discuss the advantages of having O(log array area) memory in each cell of a 2-D BCA.

BCA's based on other types of arrays of processors (hexagonal, three-dimensional, etc.) can also be defined. More generally, we can define a *graph cellular automaton* (GCA) as a graph of bounded degree having processors (cells) at its nodes, where the neighbors of each cell are defined by the graph structure. GCA's and their recognition capabilities will be treated in Section 6.

It should be pointed out that the cellular automata defined in this section have all been *deterministic*, i.e., the new state of a cell is uniquely determined by its own and its neighbors' old states. One can also define nondeterministic BCA's (or GCA's) in which the transition function maps the old state into a set of new states. Each cell of a nondeterministic CA can thus be regarded, at any given step, as being in a set of possible states, and we say that c_0 has recognized the input if its set of states contains an accepting state. We will occasionally be concerned with nondeterministic CA's in the subsequent sections.

The literature on one- and two-dimensional BCA's is not large; much of it is reviewed in a recent book by Rosenfeld (1979), which also has a chapter dealing with the material in Section 4. For further details on the Section 5 material see Dyer and Rosenfeld

(1980-in press). Sections 2 and 3 are based on Chapters 1-4 of
the Ph.D. dissertation of Dyer (1979). GCA's were introduced in
the Ph.D. dissertation of Wu (1978) and are treated in slightly
abridged form in Wu and Rosenfeld (1979, 1980).

4.   CELLULAR PYRAMIDS

A *triangle cellular automaton* (TCA) is a "stack" of one-
dimensional BCA's of exponentially tapering lengths, where adjacent
layers in the stack are connected to form a complete binary tree.
Thus if the longest BCA, constituting the base of the triangle,
has length $|\sigma|=2^n$, the others have lengths $2^{n-1},2^{n-2},...,2,1$, and
the height of the tree is $n=\log|\sigma|$. Each cell has two "brother"
neighbors in its own layer (or one, if it is at the left or right
end); two "sons" in the layer below it (unless it is in the base);
and one "father" in the layer above it (unless it is at the apex).
We take the apex cell to be the distinguished cell c_0. A string
input to the base of the TCA is said to be recognized ("accepted")
if c_0 eventually enters an accepting state.

A *pyramid cellular automaton* (PCA) is defined analogously in
two dimensions as a stack of two-dimensional BCA's of exponentially
tapering sizes, with adjacent layers connected to form a complete
quadtree. If the base of the pyramid is $2^n x 2^n$, the successive
layers are $2^{n-1}x2^{n-1}$, $2^{n-2}x2^{n-2},...,2x2$, 1x1, so that the height is
n. Each cell has (in general) four brothers in its own layer, four
sons in the layer below it, and one father in the layer above it.
Arrays to be accepted are input to the base, and the apex cell is
the distinguished cell.

It can be shown that TCA's are equivalent in recognition
power to BCA's, even if information is allowed to pass only upward
and downward, but not sideways (i.e., each cell can sense the
states of its father and sons, but not of its brothers). Allowing
information to pass only upward makes them strictly weaker than
BCA's if they are deterministic, but not if they are nondeterministic. Analogous results hold for PCA's.

On the other hand, TCA's and PCA's are faster than BCA's for
many tasks. To illustrate their operation, we show how a TCA can
recognize the connectedness of the 1's in an input string σ of 0's
and 1's in $\log|\sigma|$ time steps. Initially, the base cells are in
state 0 or 1, and all other cells are in a "quiescent" state. A
quiescent cell whose sons are nonquiescent enters a new state as
defined by the table below. Thus in $\log|\sigma|$ steps (the tree
height), c_0 enters a nonquiescent state. It is easily seen that
this state is 0 iff σ has no 1's; it is 1, ℓ, r, or s iff σ has
exactly one run of 1's; and it is f iff σ has more than one run of
1's. Note that in this TCA, information passes upward only.

(State of left son, state of right son)	New state of cell
(0,0)	0
(1,1)	1
(0,1), (0,ℓ), or (ℓ,1)	ℓ
(1,0), (1,r), or (r,0)	r
(0,r), (ℓ,0), or (ℓ,r)	s
(1,ℓ), (ℓ,ℓ), (r,1), (r,r), or (r,ℓ)	f
(x,f) or (f,x) for x = 0,1,r,ℓ,s	f
(0,s) or (s,0)	s
(x,s) or (s,x) for x = 1,r,ℓ,s	f

The following are some other tasks that a TCA can perform in $O(\log|\sigma|)$ steps; we assume for concreteness that σ is a string of 0's and 1's:

(1)   Recognize whether the number of 1's in σ is odd or even (parity).

(2)   Detect the presence of a 1, or of a specified substring, in σ.

(3)   Count the number of 1's, or of occurrences of a specified substring, in σ, by outputting the successive bits of this number, least significant bit first.

(4)   Recognize whether the number of 1's is equal to, greater than, or less than the number of 0's (equality, majority).

In fact, these tasks can be done even if information is allowed to pass upward only.

Not all tasks that can be performed by a TCA can be carried out in $\log|\sigma|$ time. A TCA can recognize whether the 0's and 1's in σ form a balanced parenthesis string in $O(\log^2|\sigma|)$ time, and can recognize whether σ is symmetric (a palindrome) in $O(|\sigma|)$ time, allowing only upward transmission of information, and these results seem to be the best possible.

In general, a TCA can accept any set of strings accepted by a finite-state automaton, and this process requires only $O(\log|\sigma|)$ time. On the comparison between TCA's (or PCA's), frontier-to-root tree acceptors, and diameter-limited perceptrons see Rosenfeld (1979) and Dyer (1979).

Many of these results also hold for PCA's. Specifically, a PCA can recognize parity, equality, and majority in an input array Σ of 0's and 1's in $O(\log|\Sigma|)$ time, and it can detect the presence

of a 1 or count the number of 1's, even if information is allowed
to pass upward only; but to detect or count specified subarrays at
that speed, sideways transmission of information in the base is required. It can also determine whether the 1's constitute a rectangle or square, or more generally, whether each row and column
contains at most one run of 1's, and the runs in adjacent rows or
columns overlap (row and column convexity), in $O(\log|\Sigma|)$ time,
using upward transmission only; but it requires $O(|\Sigma|)$ time to recog-
nize symmetry (about the vertical axis, e.g.) under this restriction.

5.   MEMORY-AUGMENTED CELLULAR AUTOMATA

     In this section we consider the advantages of allowing the
memory in each cell of a BCA to grow, e.g., logarithmically, as the
number of cells increases. On the relationship between such memory-
augmented BCA's and tape-bounded Turing acceptors see Dyer and Rosenfeld (1980-in press) and Dyer (1979). The "log memory" assumption
is especially attractive because it is easily satisfied in practice,
i.e., for any realistic number of cells, the required amount of
memory is very modest. It turns out that the class of input sets
accepted by log-memory BCA's is the same as that accepted by deterministic, nonerasing stack automata. In the discussion that follows,
the input will be regarded as a digitized wave-form or image.

     A BCA can "count" the number of its cells, or the number of
occurrences of a specified symbol or subpattern in its input, by
using a subset of its cells as an adder. These counting processes
become much simpler for log-memory BCA's, in which a single cell
can hold numbers of the required size. Thus if the input is an
image, we can compute histograms of gray levels or local property
values, or scatter plots of pairs of values (e.g., gray level co-
occurrence matrices), simply and efficiently. We can also compute
moments of the input, and in particular, locate its centroid and
compute central moments. All of these processes take O(input
diameter) time. Computation of the autocorrelation can also be
performed, but takes longer.

     In one dimension, the connected components of a string of 0's
and 1's are just the runs of 1's, and these can easily be labeled
and counted in O(diameter) time. Connected component counting in
O(diameter) time is also possible in two dimensions, using a shrinking process. With log memory, connected component labeling can also
be done in O(diameter) time.

     The run length code of the input can be computed by a log-mem-
ory BCA in O(diameter) time and output in O(number of runs) time;
and similarly the input can be reconstructed from the code. The
chain codes of region borders in a 2-D input of 0's and 1's, together with coordinates of their starting points, can be computed

and output in O(diameter) + O(border length) time, and the input can be reconstructed from these codes. The distance transform and medial axis of such an input can be computed in O(diameter) time; and the input can be reconstructed similarly. The quadtree representation can also be computed, and the input reconstructed from it, in O(diameter) time.

In two dimensions, log-memory BCA's can be used to compute a variety of region properties. The area and perimeter of a region R can be computed in O(diameter of R) time, as can the region's compactness (area/perimeter^2) and elongatedness (area/width^2), where the width of R is the number of shrinking steps required to annihilate it completely. The (extrinsic) diameter of R can be computed similarly; in fact, this diameter is equal to the length of the longest side of R's framing rectangle, which can be constructed in O(diameter of R) time. Similar remarks apply to the height and width of R. Computing intrinsic diameter is more difficult; it can be done in O(intrinsic diameter)^2 time. Another difficult problem is that of determining whether R is digitally convex; recent results show that this can be done in O(perimeter of R) time.

Memory augmentation is also useful in TCA's and PCA's. (In fact, it often suffices to give each cell an amount of memory that grows with its height in the triangle or pyramid; since the number of cells drops exponentially with height, this keeps the total amount of memory small.) It makes possible more natural algorithms for various tasks, such as local pattern counting, and also allows measurement of region dimensions, etc. in O(log diameter) or (log^{2}diameter) time. The quadtree representation can also be constructed by an augmented PCA in O(log diameter) time.

6.   CELLULAR GRAPHS

In all of the cellular automata considered above, each cell had a bounded number of distinguishable neighbors--2 or 4, in a 1- or 2-D BCA; 5 or 9, in a TCA or PCA. This section briefly discusses a generalized class of graph cellular automata (GCA's) in which each cell has at most d distinguishable neighbors, for some fixed d, but the neighbors are otherwise arbitrary. Thus a GCA is defined by a graph γ having the cells as nodes and in which each node has degree $\leq$ d. As previously, the input to a GCA is defined by the pattern of initial node states, and recognition takes place by a distinguished node c_0 (which must be specially marked, so that it can identify itself) entering an accepting state. It has been shown by Wu and Rosenfeld (1980) that such GCA's are equivalent in acceptance power to sequential automata that have graph-structured "tapes."

GCA's can serve as formal models for networks of processors, and it is of interest to study their ability to recognize or compute various graph properties. We assume here that the graph structure remains fixed throughout a computation; more recently, GCA's that can reconfigure themselves in the course of a computation have also been studied.

As in the case of BCA's, the state of c_0 at time t cannot depend on the initial states of the nodes farther than t from c_0; thus in general, recognition by a GCA will require O(graph diameter) time. Many basic properties of the given graph γ can in fact be computed by the GCA in O(diameter) time, by constructing a breadth-first spanning tree of the graph rooted at c_0 to serve as a communication network. This construction itself takes O(diameter) time; when it has been completed, the GCA can compute its radius (from c_0 as center), count its nodes (or detect and count nodes having specific labels), etc., in O(diameter) time. In these counting processes, the cells on a longest path from the root to the frontier of the tree can be used as a counter. Constructing a depth-first spanning tree takes longer (O(number of nodes)), as does finding the "center" of γ (=a node from which γ has minimum radius). Individual cut nodes and bridges can be identified in O(diameter) time, but it is harder to identify all of them simultaneously; this process can be speeded up if we use a log-memory GCA (each node has memory that grows with the log of the number of nodes).

The classes of graphs accepted by GCA's include all the finite sets of graphs, and are closed under finite union and intersection. We also have closure under concatenation operations, and under the operation of constructing the line graph.

Given any graph γ_0, we can define a GCA that accepts the graph γ iff it has a subgraph isomorphic to γ_0. For many classes of graphs, including arrays, trees, etc., this acceptance requires only O(diameter) time. Thus GCA's provide an efficient parallel method of checking subgraph isomorphism.

GCA's can also be defined that accept many standard classes of graphs in O(diameter) time, including strings, cycles, rectangular and square arrays, bipartite graphs, and Eulerian graphs. Recognition of such properties as planarity, on the other hand, appears to be much more difficult.

7.   CONCLUDING REMARKS

Thanks to the rapidly declining size and cost of processors, cellular machines of significant size are beginning to be built. Cellular algorithms are ideal candidates for VLSI implementation

and deserve to be better known to hardware designers. Studies such
as those described in this chapter give us insights into the computational capabilities of such machines, and should be useful as aids
in their design and utilization. It is hoped that this chapter will
help draw attention to the potential usefulness of generalized cellular automata studies to the designer and user of highly parallel
arrays and networks of processors.

8.   ACKNOWLEDGEMENTS

The support of the U.S. Air Force Office of Scientific Research
under Grant AFOSR-77-3271 is gratefully acknowledged, as is the help
of Kathryn Riley in preparing this chapter. This chapter is based
on the Ph.D. dissertations of Charles R. Dyer and Angela Y. Wu.

9.   REFERENCES

Dyer, C. R., "Augmented Cellular Automata for Image Analysis," Ph.D.
    Dissertation, University of Maryland, College Park (1979).

Dyer, C. R., and Rosenfeld, A., "Parallel Image Processing by Memory
    Augmented Cellular Automata," IEEETPAMI (1980) (in press).

Rosenfeld, A., Picture Languages, New York, Academic Press (1979).

Wu, A. Y., "Cellular Graph Automata," Ph.D. Dissertation, University
    of Maryland, College Park (1978).

Wu, A., and Rosenfeld, A., "Cellular Graph Automata (I and II),"
    Info. Control 42:305-353 (1979).

Wu, A., and Rosenfeld, A., "Sequential and Cellular Graph Automata,"
    Info. Sciences 20:57-68 (1980).

THEORETICAL CONSIDERATIONS ON A FAMILY OF DISTANCE TRANSFORMATIONS AND THEIR APPLICATIONS

S. Yokoi, J-i. Toriwaki*, and T. Fukumura**

Faculty of Engineering, Mie University

Kamihama-cho, Tsu, 514 JAPAN

1. INTRODUCTION

The distance transformation (DT) was defined for processing binary pictures by Rosenfeld and Pfaltz (1967). It was also pointed out that the skeleton is an important concept in relation to the DT. The DT and the skeleton have been used for a variety of purposes, for example, description, data compression, thinning or smoothing of binary pictures. Levi and Montanari (1970) defined a grey weighted distance transformation (GWDT) and a grey weighted skeleton by introducing grey value information into the DT and skeleton of binary pictures. These transforms have been also used effectively for feature extraction, thinning or preprocessing of grey pictures.

In this chapter, we extend these algorithms in two ways. First, we define the distance transformation by variable operators (VODT) as an extension of the DT. The VODT is considered to be in a suitable form to be implemented for parallel processing. It includes various types of distance transformations of binary pictures. The DTLP (distance transformation of line pattern) which was defined by Toriwaki et al. (1978) is one of them. Next, this chapter defines another important type of VODT. It is defined using local minimum filters and is called the distance transformation by local

---

*Department of Electrical Engineering, Nagoya University, Furo-cho, Chikusa-ku, Nagoya 464 JAPAN
**Department of Information Science, Nagoya University, Furo-cho, Chikusa-ku, Nagoya 464 JAPAN

minimum filters (LMDT). It is considered to include many important
algorithms according to the definition of the local minimum filters.
We show some examples of these. Corresponding to the LMDT, a skeleton is defined in a natural way. It is proved that resotrability
from the skeleton always holds for every type of LMDT. Therefore,
VODT provides a new method for feature extraction or data compression.

    In the latter half of the chapter, we present a generalized
grey weighted distance transformation (GGWDT). It is an extension
of the GWDT in the sense that positive initial values assigned in
the initial-value picture are substituted into corresponding elements in O-pixels in the input picture. This extension is important in the sense that it leads to successive decomposition and
generation of a picture. By the successive decomposition of a picture, we can obtain a sequence of initial-value pictures and decomposed pictures. They are considered to express a kind of structural
information about the grey-value distribution of the original picture. We discuss these points. Finally, we study the properties
of successive applications of the GGWDT (GWDT) to a picture, corresponding to the above successive decomposition. In the process,
the sequence of pictures converges to a kind of limit state. It has
almost the same grey-value distribution structure as a binary picture. Hence it is known that the GWDT (GGWDT) has the effect of
losing grey-value information. Relationships between the various
distance transformations discussed in this chapter are shown in
Figure 1.

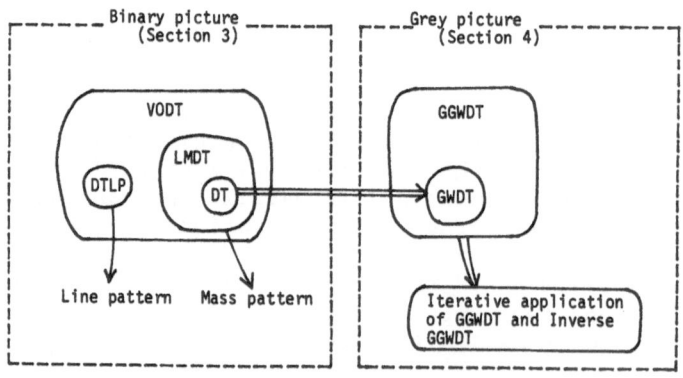

DT : Distance transformation
DTLP : Distance transformation on a line pattern
LMDT : Distance transformation by local minimum filters
VODT : Distance transformation by variable operators
GWDT : Grey weighted distance transformation
GGWDT : Generalized grey weighted distance transformation

Fig. 1. Relational structure of the family of distance transforms.

2.   BASIC DEFINITIONS

All considerations in this paper are made on the basis of the algebraic formulation of picture operations proposed in Yokoi et al. (1976). We often use notations, definitions and results described in this reference without explicit description, although some definitions are presented in this section.

(*DEFINITION* 1)   (Picture) Let I be the set of all integers and R be the set of all real numbers. A picture is defined as a mapping from I x I into R. Each element $(i,j) \in$ I x I is called a "picture element," a "pixel," or a "point." The image f_{ij} of an element (i,j) is called its grey value. A picture which has a grey value f_{ij} at each element (i,j) is denoted by $F = \{f_{ij}\}$. F is called a positive-valued picture if $f_{ij} \geqq 0$ for $\forall$ $(i,j) \in$ I x I, and a binary picture if $f_{ij} = 0$ or 1 for $\forall$ $(i,j) \in$ I x I. For a positive-valued picture F, two sets of elements are defined as follows:

$$R_+(F) = \{ (i,j) ; f_{ij} > 0 \} \tag{1}$$

$$R_0(F) = \{ (i,j) ; f_{ij} = 0 \} \tag{2}$$

The set of all pictures is called a picture space (P). The set of binary pictures is called a binary picture space (P_B).

(*DEFINITION* 2)   (Operator) An operator (unary operator) O is defined as a mapping from the picture space P_1 into P_2 , where P_1 and P_2 are subsets of P. The picture which is obtained by applying an operator O to a picture F is expressed by O (F). P_1 is called the domain of O.

(*DEFINITION* 3)   (Binary operator) A procedure which obtains a picture $H = \{h_{ij}\}$ from two pictures $F = \{f_{ij}\}$ and $G = \{g_{ij}\}$ is called a binary operator. If the following equation holds among F, G and H, it is called a binary point operator (or pointwise operator).

$$h_{ij} = \gamma(f_{ij}, g_{ij}) \quad \text{for} \forall (i,j) \in \text{I x I} \tag{3}$$

where γ is an arbitrary real function of two variables independent of i and j.

(*DEFINITION* 4)   (Serial composition of operators) Given two operators O_1 and O_2, a serial composition of O_1 and O_2 $(O_1 \cdot O_2)$ is defined by

$$(O_1 \cdot O_2) \ (F) \stackrel{\Delta}{=} O_1(O_2(F)) \tag{4}$$

Furthermore, the following notations are used,

$$(O_N \cdot O_{N-1} \cdot \ldots \cdot O_1)\ (F) = \left[\prod_{i=1}^{N} O_1\right]\ (F) \tag{5}$$

$$(O \cdot O \cdot \ldots \cdot O)\ (F) = O^N\ (F)$$

(*DEFINITION* 5) (Parallel composition of operators) Given two operators O_1 and O_2 and a binary point operator *, a parallel composition of operators O_1 and O_2 by * is defined by

$$(O_1 * O_2)\ (F) \overset{\Delta}{=} O_1\ (F) * O_2(F) \tag{6}$$

Furthermore, $\displaystyle\sum_{i=1}^{k} * \cdot O_i \overset{\Delta}{=} O_1 * O_2 * \ldots * O_n$

(*DEFINITION* 6) (Equality and order relation) For two pictures $F = \{f_{ij}\}$ and $G = \{g_{ij}\}$,

$$F = G \leftrightarrow f_{ij} = g_{ij} \quad \text{for } \forall (i,j)\ \varepsilon\ I \times I \tag{7}$$

$$F \overset{S}{=} G \leftrightarrow f_{ij} = g_{ij} \quad \text{for } \forall (i,j)\ \varepsilon\ S\ \varepsilon\ I \times I \tag{8}$$

$$F \overset{\geq}{=} G \leftrightarrow f_{ij} \overset{\geq}{=} g_{ij} \quad \text{for } \forall (i,j)\ \varepsilon\ I \times I \tag{9}$$

For two operators O_1 and O_2,

$$O_1 = O_2 \leftrightarrow O_1(F) = O_2(F) \quad \text{for } \forall F\ \varepsilon\ P_1 \tag{10}$$

$$O_1 \overset{\geq}{=} O_2 \leftrightarrow O_1(F) = O_2(F) \quad \text{for } \forall F\ \varepsilon\ P_1 \tag{11}$$

(*DEFINITION* 7) (Neighborhood) For a pixel (i,j), a set of pixels given by equation (12) is called a neighborhood of (i,j).

$$N(i,j) \overset{\Delta}{=} \{\ (p,q)\ ;\ (p-i, q-j)\ \varepsilon\ A\ \} \tag{12}$$

where A is an arbitrary subset of I x I.

Especially,

$$F(i,j) = \{\ (i+1,j),(i-1,j),(i,j),(i,j-1),(i,j+1)\ \} \tag{13}$$

is called the 4-neighborhood of (i,j).

$$E(i,j) = F(i,j) \cup \{\ (i-1,j-1),(i-1,j+1),(i+1,j-1), \atop (i+1,j+1)\ \} \tag{14}$$

is called the 8-neighborhood of (i,j).

(*DEFINITION* 8)   (Classification of pixels) A pixel $P = (i,j)$ of a positive-valued picture $F = \{f_{ij}\}$ is classified as follows.

Border point $\leftrightarrow f_{ij} > 0$ $\exists (\alpha,\beta) \in N(i,j)$, $f_{\alpha\beta} = 0$ $\qquad$ (15)

Interior point $\leftrightarrow f_{ij} > 0$, $f_{\alpha\beta} > 0$ for $\forall (\alpha,\beta) \in N(i,j)$ $\quad$ (16)

Local minimum point $\leftrightarrow f_{ij} \overset{\leq}{=} f_{\alpha\beta}$ for $\forall (\alpha,\beta) \in N(i,j)$ $\qquad$ (17)

Local maximum point $\leftrightarrow f_{ij} \overset{\geq}{=} f_{\alpha\beta}$ for $\forall (\alpha,\beta) \in N(i,j)$ $\qquad$ (18)

Zero point $\leftrightarrow f_{ij} = 0$ $\qquad\qquad\qquad\qquad\qquad\qquad$ (19)

where $N(i,j)$ is the 4-neighborhood or 8-neighborhood of P.

(*DEFINITION* 9)   (Local minimum filter and local maximum filter) An operator $\phi(\mu)$ which is defined by equation 20 (equation 21)) is called a local minimum filter (local maximum filter), $(G = \{g_{ij}\}$, $F = \{f_{ij}\})$.

$$G = \phi(F), \quad g_{ij} = \min \{ f_{pq} ; (p,q) \in N(i,j) \}$$

$$\text{for } \forall (i,j) \in I \times I \qquad (20)$$

$$G = \mu(F), \quad g_{ij} = \max \{ f_{pq} ; (p,q) \in N(i,j)\}$$

$$\text{for } \forall (i,j) \in I \times I \qquad (21)$$

where $N(i,j)$ is an arbitrary neighborhood of (i,j) and independent of (i,j).

3.   DISTANCE TRANSFORMATION BY VARIABLE OPERATORS (VODT)

In this section we will define a generalization of the distance transformation for binary pictures. The distance transformation (DT) was defined by Rosenfeld and Pfaltz (1967). Toriwaki et al. (1978) showed that the DT is expressed by compositions of a type of local minimum filter. If we use a wider class of operators instead of these, a wide variety of distance transformations can be defined. They are called distance transformations by variable operators (VODT). The distance transformation of line pattern (DTLP) defined by Toriwaki et al. (1978) is one of the important algorithms included among these. It is described in Section 3.2. Another important type of VODT (LMDT) is discussed in Section 3.3 This type of distance transformation is considered to be a direct extension of the

DT in the sense that the LMDT is defined by extending the local minimum filters used in the DT with the result that its skeleton can be defined. It is shown that the skeleton produced by the LMDT also has the important property of restorability of the original picture.

3.1   Distance Transformation by Variable Operators (VODT)

First let us define a thinning operator which plays an important role in the definition of VODT.

(*DEFINITION* 10)   (Thinning operator) If a position invariant operator O, as described in Yokoi et al. (1975), has the following property, it is called a thinning operator.

$$O(F) \overset{\leq}{=} F \quad \text{for} \forall F \in P_B$$

We now define the distance transformation by variable operators (VODT), which is one of the main objectives of this chapter.

(*DEFINITION* 11)   (Distance transformation by variable operators (VODT)) For an arbitrary binary picture $F = \{f_{ij}\}$, an operation which obtains a picture $D = \{d_{ij}\}$ according to equation (22) is called a distance transformation by variable operators (VODT). D itself is also called the VODT of F and each d_{ij} is called the distance value of (i,j).

$$d_{ij} = \begin{cases} o & \text{if } f_{ij} = 0 \\ \\ n & \text{if } g_{ij}^{(n)} = o \text{ and } g_{ij}^{(n-1)} = 1 \text{ and } f_{ij} = 1 \end{cases} \tag{22}$$

where

$$G^{(n)} = \{g_{ij}^{(n)}\} = \left(\prod_{i=1}^{n} \phi_i \right)(F)$$

and $\{\phi_k\}$ is a sequence of prespecified thinning operators. An illustration of VODT is shown in Figure 2.

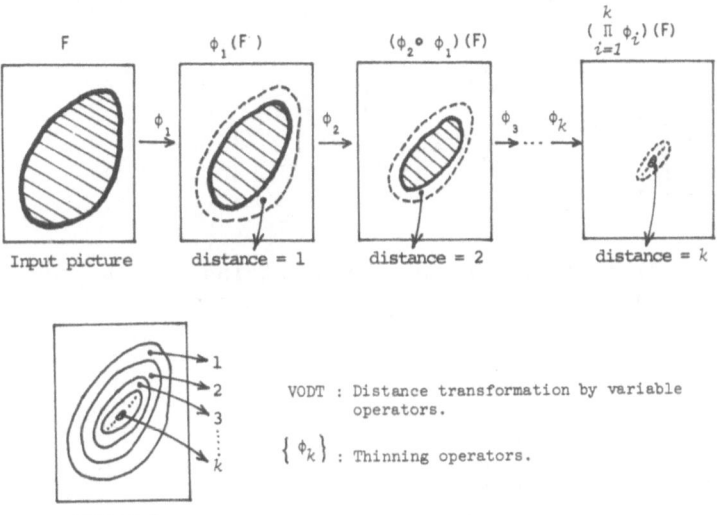

Fig. 2.   Illustration of the variable operator distance transform.

Depending on the definition of the thinning operators, various types of VODT can be considered. We discuss them in the subsequent sections. As is known from the definition, the VODT is an algorithm which is suitable for parallel processing. This will be further clarified by the following operator expressions describing funda- mental properties of the VODT.

(*PROPERTY* 1) Operators which extract pixels with certain dis- tance values are expressed as follows.

(1) An operator DL[n] which extracts all pixels (i,j) where $d_{ij} \geq n$.

$$
DL[n] = \begin{cases} \prod\limits_{i=1}^{n-1} \phi_i & n \geq 2 \\[2mm] I \text{ (identity operator)} & n = 1 \end{cases} \tag{23}
$$

where $\prod\limits_{i=1}^{k} O_i \overset{\Delta}{=} O_k\, O_{k-1} \cdot \ \cdot \ \cdot \ \cdot O_1$ (a serial composition of operators)

(2) An operator DIST[n] which extracts all pixels (i,j) where d_{ij} = n.

$$DIST[n] = \begin{cases} (I - \phi_n) \cdot \left(\begin{array}{c} n-1 \\ \Pi \\ i=1 \end{array} \phi_i \right) & n \overset{\geq}{=} 2 \\ \\ I - \phi_1 & n = 1 \end{cases}$$
(24)

(*PROPERTY* 2) An operation DC which obtains VODT D from a binary picture F is given by

$$DC = \sum_{i=1}^{N}{}_+ \; (M[i] \cdot DIST[i])$$

$$= \sum_{i=1}^{N}{}_+ \; P_i, \; P_i = \begin{cases} \begin{array}{c} i-1 \\ \Pi \, \phi_k \\ k=1 \end{array} & i \overset{\geq}{=} 2 \\ \\ I & i = 1 \end{cases}$$
(25)

where $\sum_{i=1}^{n}{}_+ \; 0_i \overset{\Delta}{=} 0_1 + 0_2 + \ldots + 0_n$ (a parallel composition of operators) and M[c] is a constant multiplication operator, i.e., $M[c](F) = \{cf_{ij}\}$, $(F = \{f_{ij}\})$. An operator P which extracts pixels satisfying a certain condition C is defined as follows:

$$G = P(F) \; (G = \{g_{ij}\}, \; F = \{f_{ij}\})$$
(26)

$$g_{ij} = \begin{cases} 1 \text{ if } (i,j) \text{ satisfies } C \text{ in } F \\ \\ 0 \text{ else} \end{cases}$$

3.2 DTLP (Distance Transformation of Line Pattern)

The DTLP was originally defined in Toriwaki et al. (1978). It was shown there that the DTLP is an effective tool for feature extraction in line pattern analysis. We show here that the DTLP is included in the definition of the VODT. The following property will be useful in understanding the function of the DTLP.

(*PROPERTY* 3) THE VODT with the following operator used for the thinning operators $\{\phi_k\}$ agrees with the distance transformation of line pattern (DTLP) of type I. The operator is given by

$$\phi_k = I \ominus EP \quad \text{(independent of k)}$$
(27)

where EP is an operator for extracting pixels which satisfies condition A in Toriwaki et al. (1978). That is,

$$G = EP(F) \quad (G = \{g_{ij}\}, \; F = \{f_{ij}\}) \tag{28}$$

$$g_{ij} = \begin{cases} 1 \text{ if } (i,j) \text{ satisfies condition } A \text{ in Toriwaki et al.} \\ \quad (1978) \\ 0 \text{ else} \end{cases}$$

There are three kinds of definitions for EP depending on the definition of connectivity (4-connectivity, 8-connectivity, or mixed connectivity). The sets of pixels which satisfy condition A agrees approximately with the set of endpoints (the pixels whose connectivity number is 1).

We show here only that the DTLP is a kind of VODT. A detailed discussion of the DTLP is given in Toriwaki et al. (1978). An illustration of the DTLP is given in Figure 3.

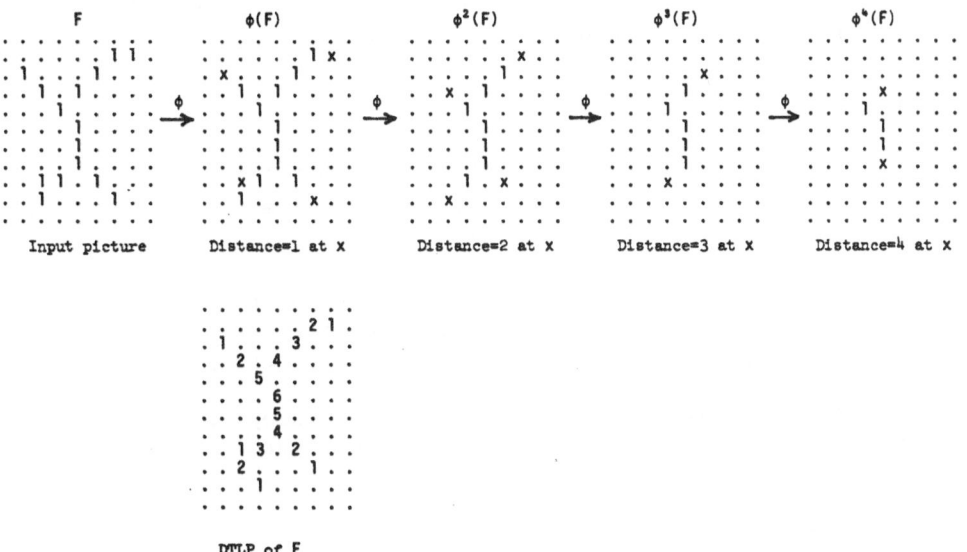

Fig. 3.   Illustration of the distance transform for line pattern.

3.3 The LMDT and Its Skeleton

In Section 3.1 we defined the VODT and showed that the DTLP is an important member of the set of VODTs. In this section, we define another important member. It is defined using local minimum filters. This type of VODT is considered to be a natural extension of the original definition of the DT. Its skeleton is also defined. Furthermore, in this case, similar properties as those of the original DT hold in extended forms.

(*DEFINITION* 12)   (LMDT) The type of VODT in which local minimum filters are used for thinning operators $\{\phi_k\}$ in *DEFINITION* 11 is called a distance transformation by local minimum filters (LMDT). The transformed picture is also called the LMDT. The ϕ_k's are given as follows.

$$G = \phi_k(F) \qquad (G) = \{g_{ij}\}, \; F = \{f_{ij}\} \tag{29}$$

$$\leftrightarrow g_{ij} = \min \{ f_{pq} \; ; \; (p,q) \; \varepsilon \; N^k(i,j) \; \}$$

where $N^k(i,j)$ is a neighborhood of (i,j). Notice here that different neighborhoods can be used for the variable k. The concept of the minimal path is important for the geometrical interpretation of the DT. In the case of LMDT, it changes from the original concept as follows.

(*PROPERTY* 4) Let $D = \{d_{ij}\}$ be the LMDT of a picture $F = \{f_{ij}\}$. Equation (30) holds at each pixel (i,j) in D.

$$d_{ij} = \min \{|\pi|\} \tag{30}$$

where $\pi = (p_0, p_1, \cdots, p_n)$ such that $p_0 \; \varepsilon \; R_0(F)$, $p_n = (i,j)$ $p_{i-1} \; \varepsilon \; N^{mi}(p_i)$ for $1 \leqq i \leqq n$ ($m_{i-1} \leqq m_i$ for $2 \leqq i \leqq n$) and $|\pi| = m_n$. The path which determines d_{ij} is called the minimal path to (i,j).

It is known from this property that the concept of the minimal path is extended from that given in the case of the DT, in the manner in which different neighborhoods are used for the p_i.

Given the definition of the LMDT and the concept of the minimal path, the LMDT can be interpreted as shown in Figure 4 by analogy

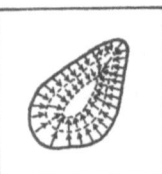
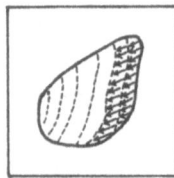
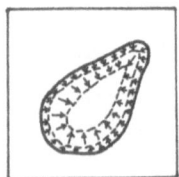

The grass burns from the outside in a constant speed. (4 or 8-neighbor distance)

If the wind blows hard in a certain direction, the grass burns from one side. (directional distance[7])

If the humidity changes depending of time, the speed at which the grass burns changes. (octagonal distance[6])

Fig. 4.   Analogy between the LMDT and the grass burning process.

to the "grass-burning" process in which the grass, corresponding to the set of pixels of value 1, burns from the outside, corresponding to the set of pixels of value 0.

Next, we present the algorithm for performing the LMDT.

(*PROPERTY* 5) LMDT is obtained by the following parallel algorithm.

ALGORITHM 1   Calculate a sequence $\{G^{(k)}\}$ according to equation (31).

$$G^{(k)} = \phi_k(G^{(k-1)}) + F, \ k = 1, 2, \ldots \tag{31}$$

where $G^{(o)} = \{g_{ij}^{(o)}\}$, $g_{ij}^{(o)} = \begin{cases} M \ (M: \ \text{a sufficiently large number}) \\ \text{if} \quad k = 1 \\ 0 \ \text{else} \end{cases}$

and $\{\phi_k\}$ are the local minimum filters in *DEFINITION* 12.

When $G^{(n)} = G^{(n-1)}$ holds for a particular value of n, the algorithm is terminated. The picture $G^{(n)}$ is the LMDT of F. Notice that here the operator (filter) ϕ_k is varied for each k. On the contrary, in the case of the DT, the same filter (4-neighbor or 8-neighbor local minimum filter) is used constantly for each k.

The skeleton is also defined for the LMDT and its important property--restorability of the original picture--holds.

(*DEFINITION* 13)   (Skeleton of LMDT) Let $F = \{f_{ij}\}$ be a binary picture and $D = \{d_{ij}\}$ be its LMDT. The set of pixels which satisfies equation (32) is called the skeleton (of the LMDT) of F. Let

$$d_{ij} \overset{\geq}{=} \max \ \{ \ d_{pq} \ ; \ (i,j) \ \varepsilon \ N^k \ (p,q) \ , \ k = d_{ij} \ \} \tag{32}$$

The condition expressed by equation (32) is considered to be a kind of local maximality in the picture D. It should be noted here that various neighborhoods are used corresponding to the value of d_{ij}. Next we show how the skeleton of the LMDT may be extracted using local minimum filters.

(*PROPERTY* 6) The subset which is the skeleton of the LMDT in a picture F whose distance values are equal to k is extracted by the following operator SK[k].

$$SK[k] = \begin{cases} \{I \ominus (\mu_k \cdot \phi_k)\} \cdot \left(\prod_{i=1}^{k-1} \phi_i \right) & k \geq 2 \\ I \ominus (\mu_1 \cdot \phi_1) & k = 1 \end{cases} \qquad (33)$$

where $\{\mu_k\}$ is a sequence of local maximum filters determined corresponding to the $\{\phi_k\}$ using equation (34).

$$A = \mu_k(B) \qquad (A = \{a_{ij}\}, \ B = \{b_{ij}\}) \qquad (34)$$

$$a_{ij} = \max \{b_{pq} \ ; \ (i,j) \ \varepsilon \ N^k(p,q)\}$$

Notice the set of neighborhoods $\{N^k\}$ which is the same as that of the $\{\phi_k\}$ is also used here.

(*PROPERTY* 7) The skeleton of the LMDT of a picture F is extracted by the following operator.

$$SK = \sum_{i=1}^{N} \mathbf{0} \ SK[i] \qquad (35)$$

where $\sum_{i=1}^{N} \mathbf{0} \ O_i = O_1 \ \mathbf{0} \ O_2 \ \mathbf{0} \ . \ . \ . \ \mathbf{0} \ O_N$, $\left(\prod_{i=1}^{N} \phi_i \right) (F) = O$ (O-value

picture)

From the LMDT skeleton, the original picture can be restored in the same way as from that of the DT. This can be proven rigorously by the algebraic transformation of the operator expression for the skeleton in *PROPERTY* 6.

(*PROPERTY* 8) The original picture can be restored from its skeleton with respective distance values as shown in equation (36).

$$\sum_{i=1}^{N} \mathbf{0} \left\{ (\prod_{k=1}^{i-1} \mu_k) \cdot SK[i] \right\} (F) = F \qquad (36)$$

(PROOF)   Let $A_\ell = \left(\prod_{k=1}^{\ell-1} \mu_R \right) \cdot SK[\ell] \ (F)$

$$A_N = \left[\prod_{k=1}^{N-1} \mu_k\right] \cdot (I \ominus \mu_N \cdot \phi_N) \cdot \left[\prod_{k=1}^{N-1} \phi_k\right] (F)$$

$$= \left[\prod_{k=1}^{N-1} \mu_k\right] \cdot \left\{\left[\prod_{k=1}^{N-1} \phi_k\right] \ominus \mu_N \cdot \left[\prod_{k=1}^{N} \phi_k\right]\right\} (F)$$

$$= \left[\prod_{k=1}^{N-1} \mu_k\right] \cdot \left[\prod_{k=1}^{N-1} \phi_k\right] (F)$$

$$A_{N-1} \oplus A_N = \left[\prod_{k=1}^{N-1} \mu_k\right] \cdot \left[\prod_{k=1}^{N-1} \phi_k\right] (F) \oplus \left[\prod_{k=1}^{N-2} \mu_k\right] \cdot (I \ominus \mu_{N-1} \cdot \phi_{N-1})$$

$$\cdot \left[\prod_{k=1}^{N-2} \phi_k\right] (F)$$

$$= \left[\prod_{k=1}^{N-2} \mu_k\right] \cdot \left\{ (\mu_{N-1} \cdot \phi_{N-1}) \oplus (I \ominus \mu_{N-1} \cdot \phi_{N-1}) \right\}$$

$$\cdot \left[\prod_{k=1}^{N-2} \phi_k\right] (F)$$

Since $\mu_{N-1} \cdot \phi_{N-1} \oplus \{ I \ominus (\mu_{N-1} \cdot \phi_{N-1}) \} = I \oplus (\mu_{N-1} \cdot \phi_{N-1}) = I,$

$$A_{N-1} \oplus A_N = \left[\prod_{k=1}^{N=2} \mu_k\right] \cdot \left[\prod_{k-1}^{N=2} \phi_k\right] (F)$$

Similarly,

$$A_M \oplus A_{M+1} \oplus \ldots \oplus A_N = \left[\prod_{k=1}^{M-1} \mu_k\right] \cdot \left[\prod_{k=1}^{M-1} \phi_k\right] (F)$$

Hence $\sum_{i=1}^{N} \oplus A_i = (\mu_1 \cdot \phi_1) (F) \oplus (I \ominus \mu_1 \cdot \phi_1) (F) = F$

Obviously the LMDT is determined by assigning the sequence of local minimum filters which are specified by the neighborhoods in *DEFINITION* 12. Examples of the LMDT and the resultant skeletons for a sample picture are shown in Figure 5.

4.   GENERALIZED GREY WEIGHTED DISTANCE TRANSFORMATION (GGWDT)

The grey weighted distance transformation (GWDT) is the extension of the distance transformation for binary pictures given in
Rosenfeld and Pfaltz (1967) to grey pictures so that information on
density values as well as the shape of a figure can be taken into
consideration during the transformation (see Levi and Montanari,
1970).

In this section we investigate the effects of the GWDT on the
structure of the grey level distributions of the input picture. To
do this, the generalized GWDT (GGWDT) is defined, in which an arbitrary initial-value picture can be assumed. It is then shown that
any positive-valued picture can be decomposed into a sequence of
initial-value pictures and an elemental picture and is inversely
generated from the elemental picture using the set of initial-value
pictures by applying the GGWDT successively. A kind of convergence
property of the grey weighted skeleton is also derived. These results reveal that GWDT or GGWDT has a remarkably close relationship
to the structure of a grey picture.

4.1 GGWDT and IGGWDT

(*DEFINITION* 14) (GGWDT) An operator which obtains a picture
$G = \{g_{ij}\}$ from a positive picture $F = \{f_{ij}\}$ according to the following equation is called the generalized grey weighted distance transformation (GGWDT) and denoted by J (i.e., $G = J(F)$).

$$d_{ij} = \min_{\pi \in \Pi} \{ |\pi| \} \tag{37}$$

where Π is the set of all paths $\pi = (p_1, p_2, \ldots p_n)$ such that
$p_1 \in R_0(F)$, $p_n = (i,j)$ and $p_{i-1} \in F(p_i)$ or $E(p_i)$ on the picture
F' defined as follows.

$$F' = \{ f'_{ij} \} , \quad f'_{ij} = f_{ij} \quad \text{if } (i,j) \in R_+ (F) ,$$

$$= a_{ij} \quad \text{if } (i,j) \in R_0 (F) .$$

$A = \{a_{ij}\}$ is a given picture which is called the initial-value
picture such that $a_{ij} \geq 0$ if $(i,j) \in R_0(F)$ and $a_{ij} = -1$, otherwise.
The path π^* which determines g_{ij} is called the minimal path to
(i,j). $|\pi| = f_{p_1} + f_{p_2} + \ldots + f_{p_n}$.

(*DEFINITION* 15) (IGGWDT) Given the GGWDT $J(F)$ and $R_0(F)$ of F,
an operator which obtains the original picture F and the initial-
value picture A is called the inverse generalized grey weighted distance transformation (IGGWDT) and denoted by J^{-1}.

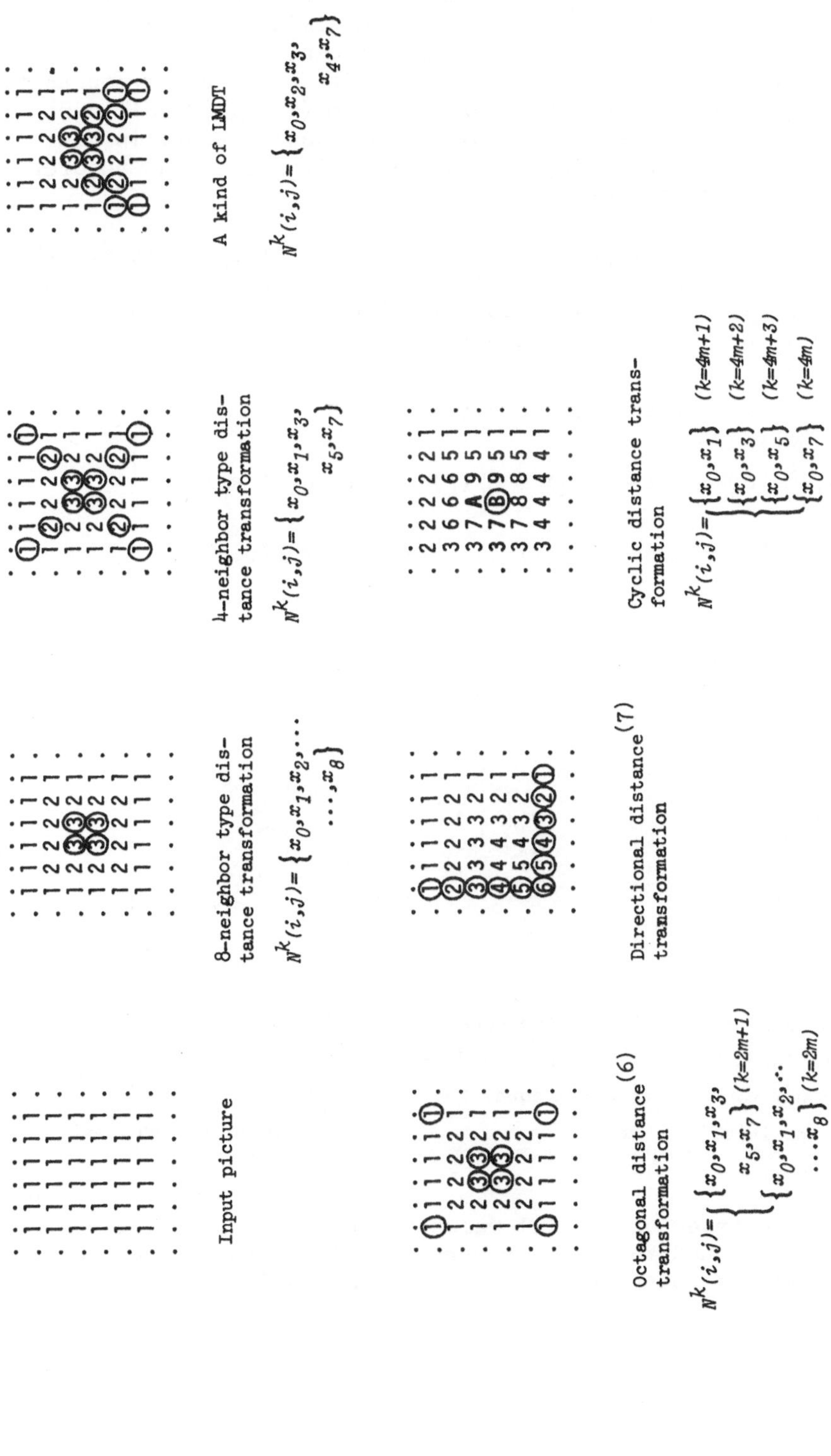

A kind of LMDT

$$N^k(i,j) = \{x_0, x_2, x_3, x_4, x_7\}$$

4-neighbor type distance transformation

$$N^k(i,j) = \{x_0, x_1, x_3, x_5, x_7\}$$

Cyclic distance transformation

$$N^k(i,j) = \begin{cases} \{x_0, x_1\} & (k=4m+1) \\ \{x_0, x_3\} & (k=4m+2) \\ \{x_0, x_5\} & (k=4m+3) \\ \{x_0, x_7\} & (k=4m) \end{cases}$$

Input picture

8-neighbor type distance transformation

$$N^k(i,j) = \{x_0, x_1, x_2, \ldots, x_8\}$$

Directional distance[7] transformation

Octagonal distance[6] transformation

$$N^k(i,j) = \begin{cases} \{x_0, x_1, x_3, x_5, x_7\} & (k=2m+1) \\ \{x_0, x_1, x_2, \ldots x_8\} & (k=2m) \end{cases}$$

Fig. 5. Various types of LMDT and the resulting skeletons (circled numbers).

As is known from the above definition, g_{ij} in $R_+(F)$ is the integral of the density of the input picture F along the path $\pi*$. The minimal path $\pi*$ is selected so as to minimize this integral over all paths to (i,j) from any of the element in $R_0(F)$. The picture $A = \{a_{ij}\}$ defines the initial value (or the integral constant) of the integral along the path starting from (i,j). Significant subsets of the GGWDT are derived by assuming appropriate initial picture types. The following are important from both the theoretical and practical viewpoints.

(*DEFINITION* 16)   (Special types of GGWDT) If the initial-value picture satisfies the condition (i) or (ii) below, the corresponding GGWDT is called a special type of GGWDT (Type 1 or Type 2) and denoted by J_1 or J_2, respectively (Figure 6).

(i) $a_{ij} = 0$ for $\forall$ (i,j) ε $R_0(F)$

(ii)   ① $a_{ij} = a_{pq}$ for $\forall$ (i,j) ε $R_0(F)$, $\forall$ (p,q) ε $R_0(F)$

        such that (i,j) ε E(p,q) or F(p,q)

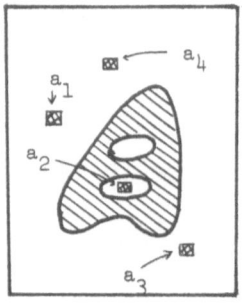 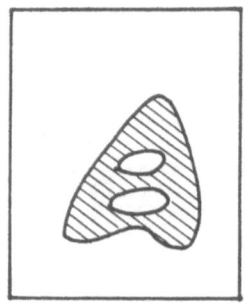 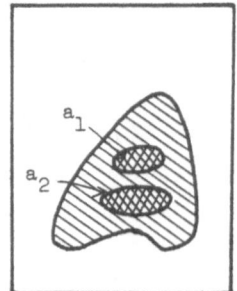

(a) In the case of J (b) In the case of J_1 (c) In the case of J_2

all points in $R_0(F)$ all points in $R_0(F)$ all points in a connected

are set equal to   are set equal to   component of $R_0(F)$

arbitrary positive   zeros   are given an identical

values   value

Fig. 6.   Three types of initial-value pictures.   Shaded regions are $R_+(F)$, cross-hatched regions are where the initial-value picture is set to a positive constant, other regions are $R_0(F)$.

  ⊘ (i,j) is a local minimum point of GGWDT G

$$\text{for} \quad (i,j) \; \varepsilon \; R_0(F) \cap R_+(A)$$

The first type of GGWDT (J_1) is exactly the same as the grey
weighted distance transformation (GGWDT) introduced by Levi and
Montanari (1970). The initial-value picture A is not necessary in
this case, because $A \overset{R_0(F)}{=} F$. In other words, the GWDT is extended
to the GGWDT in the manner in which positive initial values given
in the initial-value picture A are substituted into corresponding
elements of $R_0(F)$ before the search for the minimal path. The difference between J and J_1 might appear to be small, but it is
actually very significant because this extension leads to the successive decomposition and generation of a picture by the GGWDT.

  In the second type of GGWDT (J_2), some constraints are imposed
on the form of the initial-value picture A. Suppose that all elements belonging to $R_0(F)$ are divided into connected components.
Then all elements in the same component must have the same value in
A. Furthermore, positive initial values have to be given so that
elements with these values may become local minimum points in the
obtained GGWDT $J_2(F)$. It will be seen later that these limitations
are introduced quite reasonably.

  The following property is of critical importance because it
provides a practical procedure to determine the GGWDT.

  (*PROPERTY* 9) Between an input picture F and its GGWDT G, the
following equation holds.

$$G \overset{R_+(F)}{=} \phi(G) + F, \; G \overset{R_0(F)}{=} A \tag{38}$$

where ϕ is a local minimum filter which obtains a picture X = $\{x_{ij}\}$
from an input picture Y = $\{y_{ij}\}$ such that x_{ij} = min $\{ y_{p,q} ; (p,q)$
ε E(i,j) or F(i,j) $\}$. The proof is omitted here.

  (*COROLLARY*) For the Type I GGWDT, $G = \phi(G) + F$. The practical procedures for performing the GGWDT and the IGGWDT are derived from *PROPERTY* (9).

  (*PROPERTY* 10) Equation (38) can be solved by the following algorithm.

  ALGORITHM 2 Calculate a sequence of pictures $\{G^{(k)}\}$ from an
input picture F according to equation (5).

$$G^{(k)} \overset{R_+(F)}{=} \phi(G^{(k-1)}) + F \, , \ G^{(k)} \overset{R_0(F)}{=} A \tag{39}$$

where

$$G^{(0)} \overset{R_0(F)}{=} A \, , \ G^{(0)} \overset{R_+(F)}{=} M$$

Here M is a picture in which all grey values are equal to a sufficiently large number M. If the condition $G^{(n)} = G^{(n-1)}$ holds for a certain value of k (k = n), the algorithm is terminated and $G^{(n)}$ gives GGWDT of F. Note that *ALGORITHM* 2 has a form similar to *ALGORITHM* 1. Hence it is known that the GGWDT and the LMDT are obtained by the same type of algorithm. They are different in that different local minimum filters $\{\phi_k\}$ are used for k in the case of the LMDT. *ALGORITHM* 2 was first given by Shikano et al. (1972) but only for the Type 1 GGWDT J_1 and in a somewhat different form. It was also presented in Rosenfeld and Kak (1976). From the above consideration (*PROPERTY* 10), we can see that this algorithm is naturally derived as a procedure for solving picture equation (38) iteratively.

(*PROPERTY* 11) When the GGWDT of a picture F is given with the definition of $R_0(F)$, F can be restored from G by the operator I-ϕ; in other words,

$$F \overset{R_+(F)}{=} (I-\phi) \ (G) \tag{40}$$

A is also restored by $A \overset{R_0(F)}{=} G$.

4.2 Decomposition of Pictures by the GGWDT and Iterative Application of the GWDT

In this section, we analyze the relationship between the GGWDT and the structure of a grey picture or the effects of the GGWDT on both the geometrical and grey-value distribution structure of pictures. The major concern here is how a grey picture is changed by GGWDT.

(*DEFINITION* 17) (Iterative application of GGWDT) An iterative application of the GGWDT J to a picture F is defined as follows.

$$J^{(n)} (F) = J(J^{(n-1)} (F)) \text{ for n = 2,3,..., } J^{(1)}(F) = J(F) \tag{41}$$

(*PROPERTY* 12) Let $H = \{h_{ij}\}$ denote IGGWDT of a positive-valued

picture $F = \{f_{ij}\}$, that is, $H = J^{-1}(F)$. The following properties
hold: (1) $h_{ij} = f_{ij}$, if (i,j) is a zero point or a border point of
F, (2) $h_{ij} = 0$, if (i,j) is an interior point and a local minimum
point of F, and otherwise (3) $0 \leq h_{ij} \leq f_{ij}$.

(PROPERTY 13) Let $H^{(n)}$ and $A^{(n)}$ $(n = 1,2, \ldots)$ be sequences
of pictures and initial-value pictures which are determined by the
iterative application of J_2^{-1} to a given picture F. That is, $H^{(n)}$
$= J_2^{-1}(H^{(n-1)})$. Then the sequence of pictures $\{H^{(n)}\}$ converges.
In other words, there exists a definite integer k such that $H^{(n)} =$
$H*$ for $\forall n \geq k$, where $H*$ is a picture independent of n which we call
the "elemental picture." Moreover, if the grey values of all elements of F are integers, then $k \leq$ max $\{ f_{ij} ; (i,j)$ is an interior
point of $F \}$. Thus the limit picture is reached after some finite
number of iterations n.

(PROPERTY 14) The elemental picture $H*$ consists of only border
points and zero points. It does not include interior points.

(PROPERTY 15) Any two initial-value pictures $A^{(p)}$ and $A^{(q)}$
$(p \neq q)$ have no elements with positive values in common, i.e.,
$R_+(A^{(p)}) \cap R_+(A^{(q)}) = \Theta$ (Θ is an empty set) for arbitrary integers
p and q $(p \neq q)$.   (42)

(PROPERTY 16) For an arbitrary picture F, the sequence
$\{ H^{(n)} , A^{(n)}, n = 1, \ldots , k \}$ is determined uniquely, where k
is the minimum possible number of iterations in which the picture
$H*$ is obtained.

All these properties reveal new aspects of the GGWDT. First,
an arbitrary positive-valued picture can be generated from an
elemental picture $H*$ by iterative application of the Type 2 GGWDT
with the sequence of initial-value pictures. The elemental picture
$H*$ is considered to be a kind of line figure carrying grey values
in the sense that it consists only of border points and zero points.
It represents a kind of border of the area having almost homogeneous
grey-value distributions. On the other hand, the initial-value
picture is considered to express the region in which the grey-value
structure is homogeneous. Hence an arbitrary grey picture F can be
generated by iterating the procedure of transferring positive
values assigned by the initial-value picture into the input picture
and applying the GGWDT J_2 to it. Since any two pictures $A^{(p)}$ and
$A^{(q)}$ $(p \neq q)$ in a sequence of initial-value pictures $\{A^{(n)}\}$ have no
common elements with positive values, it is known that each of the
initial-value pictures carries information inherent to it. For
detailed discussion on these properties, see Toriwaki et al. (1978).
Examples of an elemental picture and the initial-value picture are
shown in Figure 7.

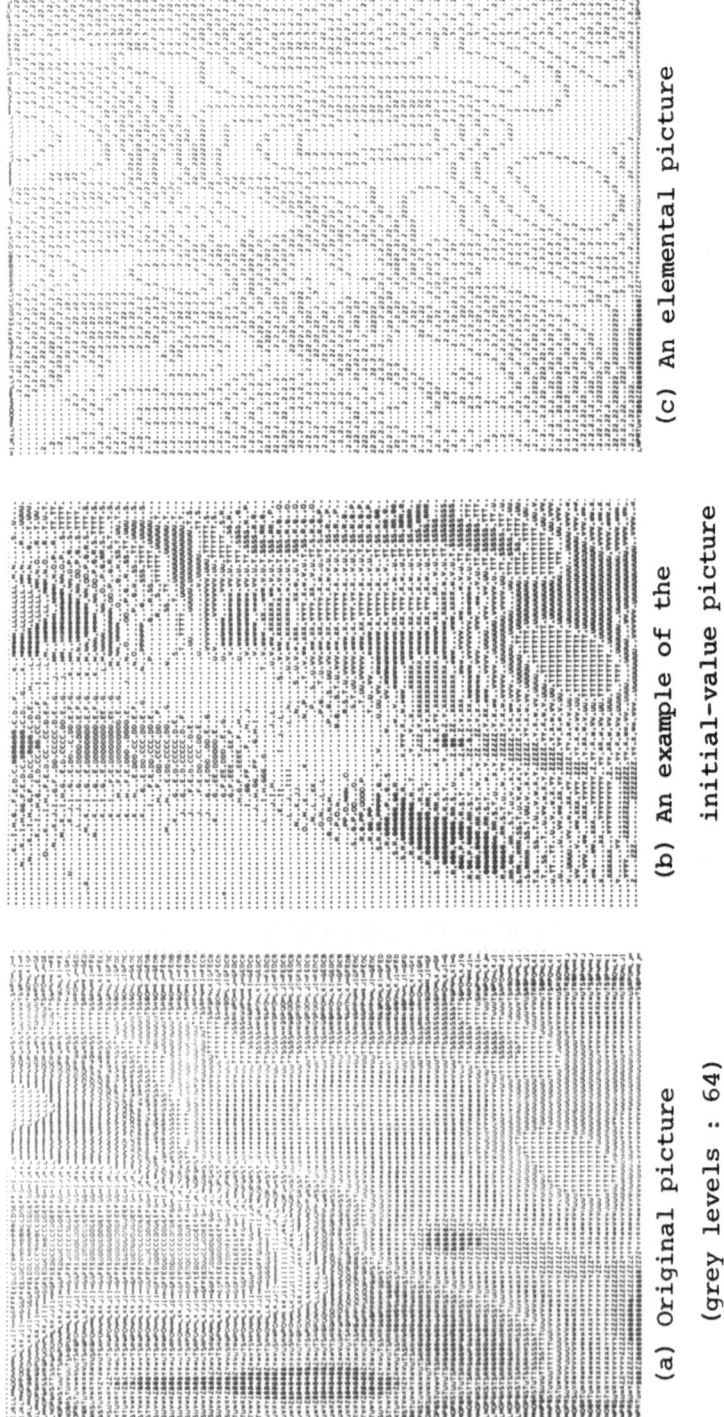

(a) Original picture
(grey levels : 64)

(b) An example of the
initial-value picture

(c) An elemental picture

Fig. 7. An example of the use of the generalized grey-weighted distance transform.

Next we consider the results of iterative application of the GWDT to a given picture. Generally, the GWDT $J_1(F)$ of an arbitrary positive-valued picture F reflects both information about the geometric shape of the borders of $R_+(F)$ and the grey-value distribution of the elements in $R_+(F)$. However, significant features of the grey-value distribution of the original picture may be lost in the application of the GWDT. Since the GWDT is frequently employed in preprocessing for thinning grey pictures, the effects of the GWDT on the structure of the grey-value distribution should be elucidated. Concerning this problem, we can present the following properties.

(*PROPERTY* 17) The sequence of pictures $\{J_1{}^{(n)}(F)\}$ converges in the sense of the invariant GWS (grey weighted skeleton), that is,

$$\exists_{N > 0} , \; S(J_1{}^{(p)}(F)) = S(J_1{}^{(q)}(F)) \quad \text{for} \; \forall \, p \overset{\geq}{=} N, \forall \, q \overset{\geq}{=} N \quad (43)$$

where $S(F)$ represents the GWS of the picture.

(*PROPERTY* 18) Let $S^{(\infty)}(F)$ denote the limit of the GWS given by equation (43) and F_B denote the binarized picture of F, i.e., $F_B = \{f_{Bij}\}$, $f_{Bij} = 1$ if $f_{ij} > 0$, 0 if $f_{ij} = 0$.

Then

$$S^{(\infty)}(F) \supset S(F_B) \qquad\qquad\qquad (44)$$

If F itself is a binary picture, $S(J_1{}^{(n)}(F)) = S^{(\infty)}(F)$ for $\forall n \overset{\geq}{=} 1$.

5.   CONCLUSION

In this chapter, we have discussed generalizations of the DT and the GWDT. First we defined the VODT as an extension of the DT for binary pictures. The VODT includes various kinds of algorithms. The DTLP, which was defined for feature extraction of line patterns, is one of the important algorithms included in these. Next, we defined the LMDT using local minimum filters. In this case the skeleton was also defined, and it was demonstrated that the original picture could be restored from the skeleton. Several examples were given. These considerations offer new methods for feature extraction; and it was shown that various types of distance transformations can be arranged in a systematic way.

Then we discussed a generalization of the GWDT (GGWDT), decomposition of pictures by the GGWDT, and the asymptotic property of the iterative application of the GGWDT. These considerations revealed new characteristics of the GWDT such that all positive grey

pictures are decomposed and generated by successive applications of
the generalized GWDT. Thus it was shown that the GGWDT relates to
the structure of grey pictures more closely than was known previously. The application of the VODT (LMDT) and the GGWDT to
feature extraction and regeneration of actual pictures as well as a
more precise estimation of the effects caused by the GWDT remain to
be studied in the future.

6.   ACKNOWLEDGEMENT

    The authors thank Professor Namio Honda of Nagoya University
for valuable advice and encouragement as well as their colleagues
at Nagoya University for useful discussions.

7. REFERENCES

Levi, G. and Montanari, U., "A Grey Weighted Skeleton," In. Control
    17:63 (1970).

Rosenfeld, A. and Pfaltz, J. L., "Sequential Operations in Digital
    Picture Processing," J. Assoc. Comput. Mach. 13:471 (1967).

Rosenfeld, A. and Kak, A. C., Digital Picture Processing, New York,
    Academic Press (1976).

Shikano, K., Toriwaki, J., and Fukumura, T., "A Wave Propagation
    Method for Conversion of Grey Pictures into Line Figures,"
    Trans. Inst. Electronics Comm. Engrs. Japan 55(10):669 (1972)
    [available in English, Sys. Comput. Controls 3(5):58 (1972)].

Toriwaki, J., Kato, N., and Fukumura, T., "Parallel Local Operations
    for a New Distance Transformation of a Line Pattern and Their
    Applications," Proc. 4th Internat. Joint Conf. Pattern Recog.
    (1978), p. 649.

Toriwaki, J., Naruse, T., and Fukumura, T., "Fundamental Properties
    of the Grey Weighted Distance Transformation," Trans. Inst.
    Electronics Comm. Engrs. Japan 60-D(12):1101 (1978) (in
    Japanese).

Yokoi, S., Toriwaki, J., and Fukumura, T., "Theoretical Analysis of
    Parallel Processing of Pictures using Algebraic Properties of
    Picture Operations," Proc. 3rd Internat. Joint Conf. Pattern
    Recog. (1976), p. 723.

IMAGE DATA MODELING AND LANGUAGE FOR PARALLEL PROCESSING

H. Enomoto, T. Katayama, N. Yonezaki and I. Miyamura

Tokyo Institute of Technology

Ookayama, Tokyo, JAPAN

1.   INTRODUCTION

In order to process image data, it is necessary to assume a suitable model and apply it to the observed image data by means of some programming language. For this purpose, the concept of a hierarchical structure in image data should be introduced with respect to a model applicable to the image data as a whole.

When images are considered to be curved surfaces in a 3-dimensional Euclidian space, structure lines are useful for image processing as described by Enomoto and Katayama (1976) and in Enomoto et al. (1976). These characteristic lines are generalizations of ridge and valley lines. Division lines carry mode information on the surface and edge lines express edge information in the ordinary sense.

These schemata describe types of models; system descriptions may be considered to be predicates describing the relations between data and parameters under the constraints of the schema. Furthermore, many types of system descriptions specify a learning process. In the case of an image database, the key extraction process must accompany learning. Then a hierarchical class of data semantics relating to the process should be taken into consideration with respect to the model in recognizing key information. These requirements lead to several guidelines for parallel image data processing.

In this chapter, we first describe the requirements for image features and the definitions of structure lines. Next we discuss the image data model using structure lines and propose a language for parallel processing which is useful for image data analysis and pattern recognition.

2.   STRUCTURE LINES

Features of images should be invariant under a parallel displacement and rotation of the coordinates. If $G(x,y)=0$ yields some
features of an image and (x,y) is converted to (ξ,η), $G'(\xi,\eta)=0$
should be obtained as the result of the coordinate transformation
and $G'(\xi,\eta)=0$ should be equivalent to $G(x,y)=0$. This means that
$G(x,y)$ should be invariant.

Let ϕ be the brightness or height function of x and y. Let
ϕ_x, ϕ_y, ϕ_{xx}, ϕ_{xy} and ϕ_{yy} denote the respective partial derivatives
of ϕ. $L(\phi_x,\phi_y)=\phi_x^2+\phi_y^2$ is invariant under the above coordinate transformation and $L(\phi_x,\phi_y)=0$ gives the characteristic points $\phi_x=\phi_y=0$
and is invariant under this transformation. The most fundamental
invariant formula is the Hessian matrix given by

$$H(x,y) = \begin{bmatrix} \phi_{xx} & \phi_{xy} \\ \phi_{xy} & \phi_{yy} \end{bmatrix} \tag{1}$$

Let $\nabla^2\phi=\text{trace}[H(x,y)]$. This quantity and $||H(x,y)||$ are both well-
known invariant formulas. Finally, let X and $X\perp$ denote grad ϕ and
the vector orthogonal to grad ϕ, respectively.

Definitions of three kinds of structure lines are given as
follows:

Definition 1

Characteristic line:

$$C(\phi) = X^t \perp HX = \phi_{xy}(\phi_x^2 - \phi_y^2) + (\phi_{yy} - \phi_{xx})\phi_x\phi_y = 0 \tag{2}$$

Division line:

$$D(\phi) = X^t \perp HX \perp = \phi_{yy}\phi_x^2 + \phi_{xx}\phi_y^2 - 2\phi_{xy}\phi_x\phi_y = 0 \tag{3}$$

Edge line:

$$E(\phi) = X^t HX = \phi_{xx}\phi_x^2 + \phi_{yy}\phi_y^2 + 2\phi_{xy}\phi_x\phi_y = 0 \tag{4}$$

Let p be an arbitrary point in (x,y) space; then a local coordinate system (u,v) with origin p is defined in such a way that
the u-axis is directed along grad ϕ and the v-axis is tangent to
the contour line. Then the structure lines are given, respectively,
by

$$C(\phi)=\phi_{uv}=0 \tag{5}$$

$$D(\phi)=\phi_{vv}=0 \tag{6}$$

$$E(\phi)=\phi_{uu}=0 \tag{7}$$

These definitions characterize the following properties besides the property of invariance.

2.1 Theorems

(C-1) A C-line is a collection of points where the density of the contour lines is an extremal along the contour line.

(C-2) A C-line is a collection of inflection points of lines of force.

(C-3) A C-line is a collection of points where the direction of the principal axis of H coincides with that of grad ϕ or the contour line.

(D-1) A D-line is a collection of inflection points of contour lines.

(D-2) A D-line is a collection of points where the normal curvature along a contour line is zero.

(D-3) A D-line is a collection of points where the angle between an eigenvector of H and the contour line is given by arctan $\sqrt{-\lambda_1/\lambda_2}$ where λ_1 and λ_2 are the eigenvalues of H.

(E-1) An E-line is a collection of points where the density of contour lines is an extremal for the direction of the lines of force.

(E-2) An E-line is a collection of points where the normal curvature along the lines of force is zero.

(E-3) An E-line is a collection of points where the angle between an eigenvector of H and grad ϕ is given by arctan $\sqrt{-\lambda_1/\lambda_2}$.

These theorems show that the characteristic points ($\phi_x=\phi_y=0$) give the primary features of an image and that the structure lines are fundamental features based on characteristic points. These lines suggest the mutual relationships among characteristic points. Figure 1 illustrates the concept of structure lines.

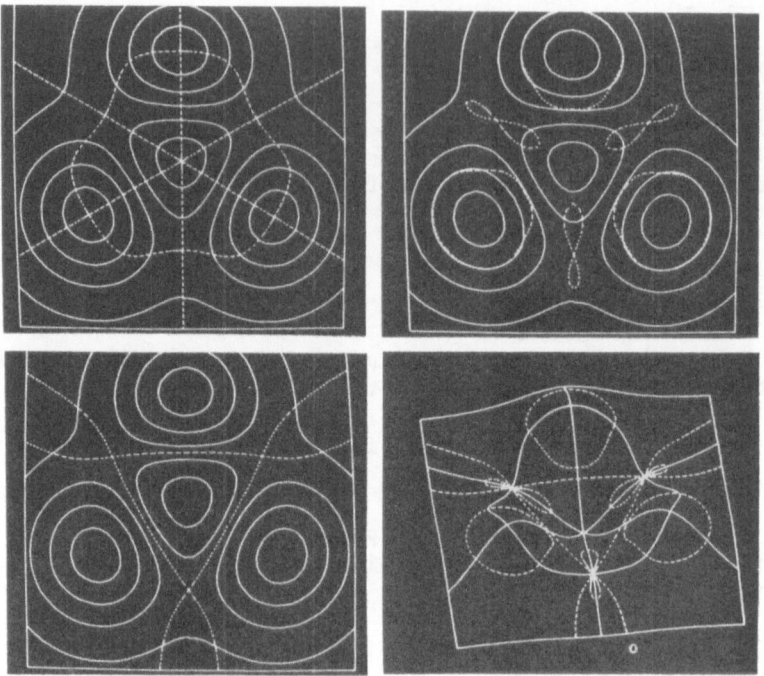

Fig. 1. Structure line concepts: (a) C-line, (b) E-line,
(c) D-line, (d) perspective.

image classification relevant to the collecting method or place or
to data about objects in the images. The more interesting keys
should be concerned with relations between partial features.

As to partial features, there are some characteristic values
or characteristic patterns which depend upon statistical properties
such as texture. In this case an individual point in each image
has a respective characteristic value or pattern corresponding to
a local feature. Here the image data retrieval is carried out
according to the existence of points characterized by the partial
feature within some range of values or patterns.

Structure retrieval should be done using some geometrical
structure described by the partial features. This geometrical
structure has some specified global properties derived from the
particular partial features. This causes the model to be important
to the image database.

It is important to note the following points when extracting partial features.

1.   The image features are specified and computed with respect to operational values of the model and the environment.

2.   Primary features should be statistically orthogonal to each other. Compound features are expressed using these primary features. Accordingly, almost all features can be expressed in a fairly simple and efficient manner.

3.   The usual adaptive operations should be limited to applications to stationary images and are not suitable for use in edge detection. The edge detection operation is used to select some suitable line corresponding to $E(\phi)=0$.

4.   When the target features are so weak that they are masked by strong features, it is very difficult to extract the weak features directly. It is necessary to use the hierarchical method in this case to extract the stronger features in sequence.

5.   When compression schemata or hierarchical feature extraction methods are employed, there is the possibility that some information will be lost and these losses may result in some errors.

Even if image structures are simple, features have geometrical structures depending on the image model. Features can be represented by a limited class of graphs expressing the geometrical structure. Thus the process of image structure retrieval includes both an algorithm for extracting specific graphs from the object image and a partial matching algorithm for comparing the feature structure and the key graph structure. Since the keys are expressed in terms of graphs, each graph structure must correspond with one of the concepts introduced by us. This fact means that a line drawing corresponding to a graph structure must express an exact image of the original picture and lead to the hierarchical relation between the picture, the image and the pattern shown in Figure 2.

We can derive many global properties from the structure lines. This has been found to lead to efficient performance in image analysis and recognition. Also global interconnecting relations expressed by graphs may be useful for managing image data bases.

3.   REQUIREMENTS FOR FEATURE RETRIEVAL

In order to retrieve image features there must be some keys defining specific interests in the images. The keys may relate to

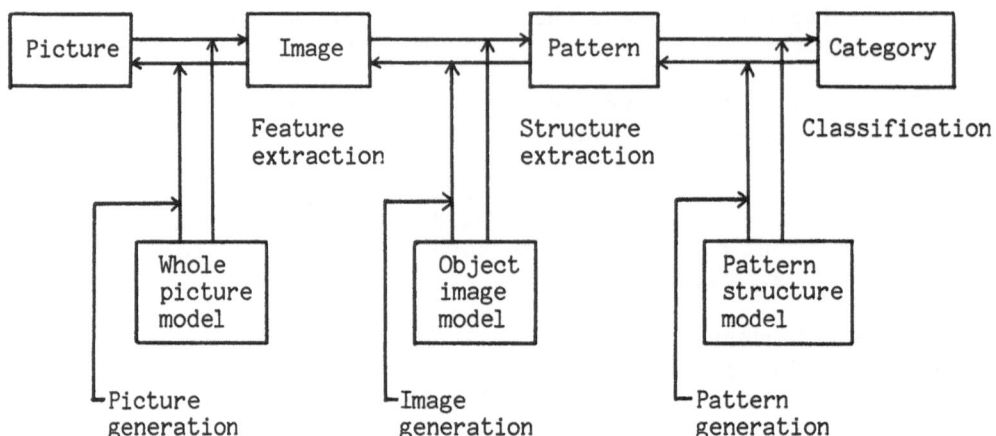

Fig. 2. Picture, image, and pattern relational hierarchy.

4.   LANGUAGE REQUIREMENTS FOR PARALLEL IMAGE PROCESSING

     In general, a programming language should provide facilities
for writing any complex process in a simple form and computing results in an efficient way. For simplicity let us point out several
facts. First, generally used programming languages describe jobs
as procedures and objects whose processing must be embedded in a
procedure. If we divide a job into a definition section and a control section, the definition section can describe the target or recursive definition of the job and the control section can describe
the procedure so that control is efficient. If a recursive definition is included in the definition section, the corresponding control section may be described in a recursive step not using back-
tracking so that efficient computation is possible.

     Secondly, non-deterministic expressions have very interesting
properties which are useful for simple description of both the
definition and control sections. Due to the commonly used non-
deterministic sense in automata theory, the execution of an object
job may lead to a sequence of non-deterministic processes whose
path reaches a termination. Accordingly, there are many cases
where non-deterministic expressions become extremely simple compared with deterministic ones.

     A target or object can usually be expressed in two ways. One
way is to represent the object as a set of points in the space
satisfying the multiple conditions expressed by the equations or
predicates. Examples are linear simultaneous equations or a set
of differential equations satisfying initial conditions on the
structure lines. Definition 1 provides examples.

The other way is to express the object as a set of points which attain the maximum value of a given objective function under some constraints, conditions, or predicates. The definition of structure lines described in the theorems is an example of this. These representation methods have no direct relation to the procedure seeking the final solution. This fact indicates that the separate description of the definition section and the control section may result in a simpler expression and an efficient search for the solution. For example, when we are required to obtain a point satisfying Definition 1 along a given line, the main portion of the definition section consists of Definition 1 and, in the control section, two local variables p1 and p2 indicating two points separated by a given distance must be declared.

Next, the following algorithms are described in the control section, if C(p1) and C(p2) have the same sign, points p1, p2 are extended so as to alter the sign. But if C(p1) and C(p2) have different signs, a binary search algorithm is applied to obtain a point p satisfying the condition $|C(p)| \leq \epsilon$. This example indicates that the separation of a definition section and a control section increases the possibility that they can be expressed by typical descriptions. This fact also shows the usefulness of the abstract data type, and it also contributes to simple descriptions.

The described forms of programs are suitable for correction or modification because revisions of a software system will be restricted to some parts in the program. Similarly this technique will be useful for verification or validation of programs, and the reliability of a software system is expected to increase.

Most image processing involves complex and time-consuming computations. The languages which are used in real-time or efficient image analysis and recognition should have sufficient parallel computation facilities. As described above, a non-deterministic scheme has very interesting properties for parallel processing. We must also consider recursive operations.

In the case of the extraction of structure lines from a given image or a curved surface within a specific boundary, the definition of structure lines, where the tangent to a structure line cannot be defined and multiple structure lines cross, requires that definitions be defined in a hierarchical manner.

In a case where the hierarchical definition becomes simple for a given specification, the parallel processing language should have descriptive facilities for parallel processing with different operands derived subsequently at each hierarchical level. (The FFT is an example.) Besides this parallelism, parallel processing of homogeneous operations in different image positions is also quite efficient for the real-time purposes. It is also important to have

some facility for the mutual synchronization and for avoiding system
deadlocks caused by competition between parallel processes.

Image data bases are becoming more important to image analysis,
recognition, and understanding. From the above discussion we can
state that the nucleus of the image data base should be a retrieval
system of graphs which represent the intrinsic image structures produced from the concept of structure lines. Then the graphical representation method determines the structure of the schemata and the
whole structure of the data base management system.

One method for representing a graph is the utilization of a set
of trees. The languages, Lorel-1 and Lorel-2, have been developed
by Enomoto et al. (1978) to process relational structures. They
have proved useful for this purpose. From this point of view, programming languages convenient for image processing must have some
facilities for manipulating graph structures and have a close relation to data base management systems.

5.   A LANGUAGE FOR PARALLEL PROCESSING WITH NON-DETERMINISTIC AND RECURSIVE FEATURES

There are many proposed programming tools for parallel processing. Fork, join, co-begin, co-end, critical section and path
expression are well-known and given in Hansen (1973) and Haberman
(1977). They are not always convenient for parallel processing as
described in the previous section. Taking the point of view described in the previous section, we propose a language for parallel
processing with non-deterministic and recursive features. We propose a program form in which a program is divided into a definition
section and a control section, and the control section can be constructed hierarchically. After compilation, the program is translated into a procedural form.

In this chapter, descriptions are limited to the translated
procedural form, because the formal expressions and control sections
require a preliminary explanation or knowledge about the abstract
data type, etc. Also it is not necessarily required for the explanation of parallel processing. In order to indicate a parallel
operation in a non-deterministic sense, we use "§:<statement list>;§"
or "§:<label>:<statement list>:§". The symbol "§" is introduced in
a sense that any process parenthesized by the symbol "§" can be
executed in a parallel manner.

The statement list stands for a sequential processing of statements separated by the symbol ";" as usual. For synchronization
purposes, condition statements have two forms: One condition using
"→" means that if the <condition part> doesn't have the value true,

the <execution part> is kept waiting until the value of the <condition part> becomes true. The other condition using "→>" means that if the <condition part> doesn't have the value true at its processing time, the following <execution part> is skipped and the system control transfers to the next statement.

For the purpose of parallel processing of homogeneous operations, the symbol "$" is used and a new cell is created by an allocate statement. The expression "$<variable>" declared in a block is interpreted so that $<variable> composes a set labeled <variable> and an element is selected from the set at the beginning of the block, and the selected element doesn't change to another element in the set until a changing or assignment statement appears. Because there exists a necessity for association with $<variable>, the "recall" statement is introduced. When "recall" is executed, the element associated with $<variable> is reallocated. If there are no associated elements, this statement is skipped.

For the purpose of recursive operation and compiling an interface between the definition section and control sections, the "reenter" statement is used. When "reenter" is met in a block, the process returns to the beginning of the innermost block. Furthermore many commonly used statements can be used and usual scope rules are applied in this language. The main parts of the syntax of this language are shown in Figure 3. As an example, a program for extracting structure lines and obtaining the connection relation between singular points of structure lines is shown in Figure 4.

<statement>:: =<condition statement> | <unit statement>

<condition statement>:: =<condition part> → <execution part> |

                            <condition part> ⇸ <execution part>

<execution part>:: =<term> | DO; <statement list> OD;

<term>:: =<parallel term> | <primary>

<primary>:: =<statement> | begin; <statement list> end;

<statement list>:: =<term>; <statement list> | <term>

<parallel term>:: =§: <statement list> :§ <parallel term> |

              §: <statement list> :§ §: <statement list> :§

    Fig. 3.   Syntax of a language for parallel processing.

```
Procedure A(Q,B) returns {CNCT<$P>};
    dcl (Q, PSET, B) {<real, real>};
    dcl <$P<real, real>, (DR, CNCT) {<real, real>}>;
    PSET= Λ;
    §§ main(Q, B, PSET): begin;
        dcl <($Y, PYD, YD) <real, real>>;
        Q≠Λ ⇒> DO; allocate($Y); $Y= SELECT(Q) OD;
        Q=Λ →> DO; continue; exit(§§main) OD;
        §A1: generate(§§main): A1§;
        §A2: (PYD<$Y>, YD<$Y>)= BEXDL($Y, B);
            §§ track($Y): begin;
                §§ tksing($Y): begin;
                    while not NRB(YD<$Y>, B) ∧ not NRSING(YD<$Y>)
                            DO; (PYD<$Y>, YD<$Y>)= EXTDL(PYD<$Y>, YD<$Y>) OD;
                    not NRB(YD<$Y>, B) ∧ NRSING(YD<$Y>) →> PP= SKSING(YD $Y );
                    NRB(YD<$Y>, B) ⇒> DO; YD<$Y>∈ Q →> DELETE(YD<$Y>, Q);
                                            ADD(YD<$Y>, CNCT<recall($P=$Y)>);
                                            YD<$Y>∉Q →> ADD(YD<$Y>, CNCT<recall($P=$Y)>)   OD;
                §§ tksing: end;
                §§ regcnr(PP, $Y, PSET): begin;
                    PP ∉ PSET ⇒>
                        DO; allocate($P); $P= PP; ADD($P, PSET);
                            ADD($P, PSET): DR<$P>= SKDR($P, $Y) OD;
                    PP ∈ PSET →>
                        DO; recall($P= PP);
                            $Y∈CNCT<$P>→> DO; continue; exit(§§regcnr) OD;
                            $Y∉CNCT<$P>⇒>
                                DO; DELETE(TRACED(DR<$P>, $Y), DR<$P>);
                                    ADD($Y, CNCT<$P>); ADD($P, CNCT<recall($P=$Y)>) OD;
                §§ regcnr: end;
            §§ track: end;
            §§ branch(DR<$P>): begin;
                DR<$P>≠Λ ⇒>
                    DO; allocate($Y); $Y= $P; PYD<$Y>= $P; YD<$Y>= SELECT(DR<$P>) OD;
                DR<$P>=Λ →> DO; continue; exit(§§branch) OD;
                §B1: generate(§§branch): B1§;
                §B2: generate(§§track) : B2§;
            §§ branch: end;
        :A2§
    §§  main: end;
end;
```

Fig. 4. Parallel procedure for extracting structure lines.

6. CONCLUSION

The concept of structure lines in images has been introduced. Structure lines are shown to be useful for describing the global properties of images. Requirements on the feature retrieval for images have been explained.

From considering the properties of structure lines, requirements for a language for parallel processing in image analysis, recognition, and understanding have been described. It has been explained that the non-deterministic and recursive concepts are important for image processing.

A language for parallel processing with these features has been proposed and its usefulness is shown in an example. This language is suitable for a multiple processors system. Implementation of an operating system is considered fairly easy. Furthermore a suitable scheduler can behave as a good organizer by considering the states of the processors.

7. ACKNOWLEDGEMENTS

It is our pleasure to thank our colleagues in the Enomoto-Katayama Laboratory of the Tokyo Institute of Technology.

8. REFERENCES

Enomoto, H., and Katayama, T., "Structure Lines in Images," Proc. 3rd Intern'l Joint Conf. Pattern Recog. (1976).

Enomoto, H., Katayama, T., and Yoshida, T., "Computer Experiments on Global Properties of Structure Lines of Images Using Graphic Display and Its Considerations," Info. Proc. Japan 17(7) (1976).

Enomoto, S., Miyachi, T., Katayama, T., and Enomoto, H., "On the Programming Language Lorel-2," Info. Proc. Japan 19(6): (1978)

Haberman, A. N., "On the Concurrency of Parallel Processes," in Perspective on Computer Science (Jones, A. K., ed), Association for Computing Machinery (1977).

Hansen, P. B., Operating System Principles, New York, Prentice-Hall (1973).

A STUDY ON PARALLEL PARSING OF TREE LANGUAGES AND ITS APPLICATION TO SYNTACTIC PATTERN RECOGNITION

N. S. Chang and K. S. Fu

School of Electrical Engineering, Purdue University

West Lafayette, Indiana 47907 USA

1.   INTRODUCTION

The tree systems approach to pattern recognition has received increasing attention by Fu and Bhargava (1973) and Fu (1974). It has been shown to be an effective method for the classification of bubble chamber events, fingerprint pattern recognition, and the LANDSAT data interpretation as discussed in Fu (1977). Instead of using a string representation of primitives and relations, tree systems use a tree structure to represent a pattern in terms of its primitives and relations. Multidimensional patterns can be described more efficiently and effectively by a tree language than by a string representation. As long as patterns can be described by a tree structures, the associated tree grammars can be easily constructed or inferred. Tree automata can then be used to classify unknown patterns according to the constructed pattern grammars.

Often there are too many minor variations to be included in grammar rules for describing similar patterns. A solution to this problem is to consider only the basic structures for similar patterns during the grammar development stage and to handle variations and noise as errors during the recognition stage. By using minimum-distance or maximum-likelihood criteria, an Error-Correcting Tree Automaton (ECTA), as described by Lu and Fu (1976), can choose a least erroneous pattern, described by grammar rules for a distorted or noisy pattern. With error correcting features, this linguistic approach to pattern recognition is no longer sensitive to noise. It is an effective method to handle real world data. However, one must trade-off the long computation time required by an error-correcting automaton.

Section 2 proposes several software engineering techniques for the implementation of a more efficient ECTA and a parallel parsing algorithm for tree languages is introduced. The procedure applied to the recognition of roads in LANDSAT images are described in Section 3.

2. PARALLEL PARSING OF TREE LANGUAGES

This section first describes tree grammars and gives a lexical analyzer for them. Next methods for parallel parsing are presented and simulation results are given.

2.1 Tree Grammars

Definitions and notations that appear in this paper are those used by Fu and Bhargava (1973), Fu (1977), and Lu and Fu (1976). They are briefly reviewed below.

Definition 2.1

A grammar G_t = (V,r,P,S) over $<\Sigma,r>$ is a tree grammar in expansive form where

V is a set of terminal and nonterminal symbols,
Σ is a set of terminal symbols,
r : $\Sigma \rightarrow$ N where N is the set of non-negative intergers, is the rank associated with symbols in Σ,
S is the starting symbol,
and P is a set of production rules in the form of

$$X_0 \rightarrow \underset{X_1 X_2 \ldots X_{r(x)}}{\overset{x}{\bigwedge}} \quad \text{or } X_0 \rightarrow \chi$$

where x ϵ Σ and $X_0, X_1, \ldots, X_{r(x)}$ ϵ V-Σ.

For a two-dimensional digitized picture, the input data is a two-dimensional array of gray levels. A simple and practical tree-systems approach for two-dimensional pictures is to divide the whole picture into windows. Then single pixels with different gray levels are chosen to be the pattern primitives. Every pixel in a window corresponds to a node label in a tree representation. For implementation, a tree structure can be chosen arbitrarily and is then fixed for grammars written for this structure. A convenient tree structure is shown in Figure 1 where the window size is 5x5.

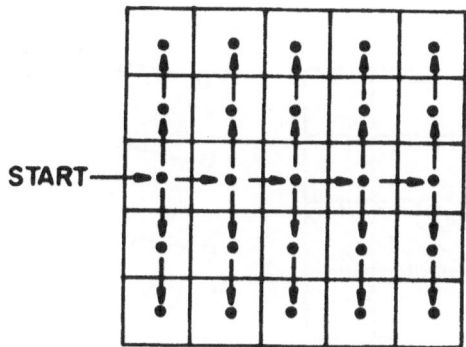

Fig. 1. Tree structure.

Example 2.1

The windowed pattern shown in Figure 2 (left) has the tree representation shown in Figure 2 (right) where only binary value primitives are considered, i.e., a dark pixel corresponds to 1; otherwise, 0.

There are various ways to choose primitives for tree grammars. Windowed patterns might be chosen as primitives for a second-level tree system. These primitives are then connected by another tree structure and characterized by a second-level tree grammar. Therefore multi-level tree system problems can be solved by applying the same tree automata repeatedly at each level.

Example 2.2

The following tree grammar G_0 generates the following four patterns (using the tree structure shown in Figure 1):

1. only the pixels in the third row are 1's. [Figure 2 (left)],

2. only the pixels in the second row are 1's,

3. all the pixels are 1's,

4. all the pixels are 0's.

$$G_0 = (V_0, r, P_0, S) \text{ over } <\Sigma, r>$$

where $V_0 = \{U_1, U_0, X_0, A_0, S, \$, 0, 1\}$

$\Sigma = \{\blacksquare, \square\}$
      1   0

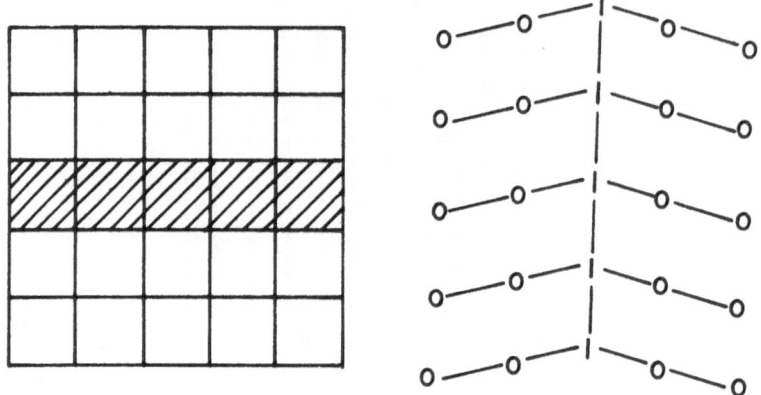

Fig. 2. A windowed pattern (left) and its tree representation (right).

$r = \{0,1,2,3\}$

$P_0 : S \rightarrow$ $\underset{U_1}{\overset{\$}{\big|}}$ (1) ; $\underset{U_0}{\overset{\$}{\big|}}$ (2) ; $\underset{A_0}{\overset{\$}{\big|}}$ (3) ; $\underset{X_0}{\overset{\$}{\big|}}$ (4)

$U_1 \rightarrow$ $\underset{X_0 \; U_1 \; X_1}{\overset{0}{\bigwedge}}$ (5) ; $\underset{X_0 \quad X_1}{\overset{0}{\bigwedge}}$ (6)

$U_0 \rightarrow$ $\underset{X_0 \; U_0 \; X_0}{\overset{1}{\bigwedge}}$ (7) ; $\underset{X_0 \quad X_0}{\overset{1}{\bigwedge}}$ (8)

$X_0 \rightarrow$ $\underset{X_0 \; X_0 \; X_0}{\overset{0}{\bigwedge}}$ (9) ; $\underset{X_0 \quad X_0}{\overset{0}{\bigwedge}}$ (10) ; $\underset{X_0}{\overset{0}{\big|}}$ (11) ; 0 (12)

$A_0 \rightarrow$ $\underset{A_0 \; A_0 \; A_0}{\overset{1}{\bigwedge}}$ (13) ; $\underset{A_0 \quad A_0}{\overset{1}{\bigwedge}}$ (14) ; $\underset{A_0}{\overset{1}{\big|}}$ (15) ; 1 (16)

$X_1 \rightarrow$ $\underset{X_0}{\overset{1}{\big|}}$ (17) ; 1 (18)

2.2 Lexical Analyzer for Tree Grammars

Every production rule of an expansive tree grammar can be written in a string representation as $X_0 \to \chi X_1, X_2 \ldots X_{r(x)}$. With this standard format, production rules can be arranged in a specific order so that efficient algorithms can then be used to search grammar rules in the parsing procedure. Since there is no addition and deletion of grammar rules in the parsing procedure, grammar rules can be stored sequentially to save storage space, and only internal codes for nonterminals need to be stored for parsing. The operation of a lexical analyzer for a tree grammar is proposed to transform grammar rules into internal codes and to arrange them in suitable sequences for the searching algorithm chosen for parsing.

Algorithm 1

A lexical analyzer for tree grammar

Input: An expansive tree grammar $G = (V, r, P, S)$ over $\langle \Sigma, r \rangle$

Output: Sorted tree grammar rules in internal codes.

Method:

   (1) Sort all the nonterminals.

   (2) For each nonterminal, assign its sequence in the sorted symbol table as its internal code.

   (3) For each production rule, translate all the nonterminals into their internal codes.

   (4) Sort the translated grammar rules in the order of the right-hand side nonterminals.

Since error-correcting tree automata are basically bottom-up syntax analyzers, the right-hand side nonterminals are designated as sorting keys in Step 4. As for ordinary (non-error-correcting) tree automaton, the left-hand side nonterminal should be sorted.

Even more efficient searching algorithms such as table look-up, hashing techniques, or scattered storage searching can be used if grammar rules have been preprocessed by a lexical analyzer. Since all the possible sorting keys in Step 4 of *Algorithm 1* are all combinations of nonterminals, for a set of tree grammar rules with m_i nonterminals and n degrees (the maximum number of branches of a rule), there are $k = \sum_{i=0}^{n} m^i$ possible sorting keys. The most efficient table look-up searching method can be applied in parsing when a lexical analyzer is defined as follows. A simple hash function which maps a combination of nonterminals $(X_1, X_2, \ldots, X_n)$ into one of the k consecutive intergers $0, 1, 2, \ldots, k-1$ can be defined as

$$h(X_1, X_2, \ldots, X_n) = \sum_{i=1}^{n} J_i * m^{(n-i)}$$

where J_i is the sequence of X_i in the nonterminals' table. Collision occurs only for the same combination of right-hand side nonterminals which should be linked together. Therefore, this hash function maps into one of the k consecutive locations where the starting addresses for a linked list of terminals and left-hand side nonterminals are stored for this combination of right-hand side nonterminals.

Algorithm 2

A lexical analyzer for tree grammar

Input: An expansive tree grammar $G = (V, r, P, S)$ over $<\Sigma, r>$ which has m nonterminals and the maximum number of branches of a rule is n.

Output: Internal codes of G which are suitable for table look-up searching in parsing.

Method:

   (1)   Define a function CODE which maps each nonterminal into its sequence in the nonterminal symbol table. CODE returns zero for the null symbol.

   (2)   For i=0 to $\sum_{i=0}^{n} m^i - 1$ let Link (i) $\leftarrow$ 0

   (3)   For each production rule $X_0 \rightarrow \chi X_1 X_2 \ldots X_n$
         Let $h \leftarrow \sum_{i=1}^{n} CODE(X_i) * m^{n-i}$. Attach χ and $CODE(X_i)$ to the linked list pointed to by h.

The table look-up method is the most efficient searching method, however, it requires much more storage than sequentially allocated grammar rules as suggested by *Algorithm 1*. A suitable chosen hash function can provide an algorithm whose space and time requirements are between those of the two extreme cases.

2.3 Parallel Parsing of Tree Languages

There are several subtrees connected with each node of a tree, while each subtree is again composed of subtrees and a node. A simple parsing procedure is performed repeatedly on every node of a tree automata. These independent subtrees can, of course, be parsed in parallel. The basic parsing procedure is nothing but searching for matched grammar rules and creating new tasks to be parsed at other levels of the tree structure. All the required basic instructions are just integer comparison, integer addition, and branching operations. Therefore a special purpose multiprocessing system can be easily implemented as a tree automaton. For example, an associative processor having content search and multiwrite capability can be used to reduce the search time greatly as demonstrated by Thurber and Wald (1975).

In order to partition the basic parsing procedure into subtasks to be processed in parallel, the parsing procedure needs to be considered in detail. In this chapter, a parallel parsing algorithm for minimum–distance Structure–Preserved ECTA(SPECTA) is presented as an illustration. The replacement of a node label by another terminal in a tree is considered as a substitution transformation. The distance between two trees α and β with the same structure $d(\alpha,\beta)$ is defined as the smallest number of substitution transformations required to derive β from α. For a given tree grammar G and an input tree β, the minimum-distance SPECTA is formulated to find a tree α such that $d(\alpha,\beta)$ is the minimum among all the trees that can be generated by G and have the same structure as β. The operation of a SPECTA is to construct a transition table from frontiers to the root of β. For each node b of the tree structure of β, there is a corresponding set of triplets denoting t_b in the transition table. Each triplet (X,ℓ,k) is added to t_b if the kth production rule is applied at node b so that the label of node b with its branches are reduced to the left–hand side nonterminal X of the kth rule, and there are ℓ substitution errors involved in subtree b.

Algorithm 3

A parallel parsing algorithm for minimum–distance SPECTA.

Input: G = (V,r,P,S) and an input tree β.

Output: Transition table of β.

Method:

(1) For every frontier b of β, perform (2) in parallel.

(2a) Repeat (2b) for all grammar rules with zero rank.

(2b)   If TERMINAL(k) = LABEL(b)   then add $(X_0,0,k)$ to t_b.

else add $(X_0,1,k)$ to t_b.

where X_0 is the left-hand side nonterminal of the kth rule, TERMINAL(k) is the terminal of the kth rule, and LABEL(b) is the label on node b of tree β.

(2c)   Designate the father node of b as a new frontier.

(3a)   For every nonempty set of the transition table, perform (3b) in parallel.

(3b)   If there are more than one item in a set having the same state, delete the item with a larger number of errors.

 (4)   For every lowest level frontier b, perform (5a) in parallel.

(5a)   Select one item from each set of its son's transition table entries and form all the combinations.

(5b)   For every combination $((X_1,\ell_1,k_1),\ (X_2,\ell_2,k_2),\dots,$ $(X_n,\ell_n,k_n))$ perform (6) in parallel.

(6a)   Repeat (6b) for every grammar rule of the form

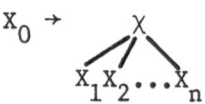

$$X_0 \to$$

(6b)   If X = LABEL(b)   then add $(X_0,\ell_1+_2\dots+\ell_n,k)$ to t_b

else add $(X_0,\ell_1+\ell_2+\dots+\ell_n+1,k)$ to t_b.

(6c)   Same as (2c).

 (7)   Same as (3).

 (8)   Repeat (4) until the root of β is parsed.

 (9)   If (S,ℓ,k) is in the transition table entries for the root, then β is accepted by G with ℓ errors, otherwise there is no tree which has the same structure as β.

   The minimum-distance error correction of β can be easily traced out from the transition table. Although triplets are used in the transition table entries, a doublet containing only the rule number and error counts is sufficient to represent the corresponding trip-let because the corresponding state of the triplet can be obtained from the left-hand side nonterminal of that production rule. Trip-

lets are used only for better illustration. Also an explicit form
of the grammar rules is used, instead of using the internal form
as suggested in *Algorithm 1* or *2*.

Example 2.3

    Given a noisy pattern of size 3x3 as shown in Figure 3 (left),
the tree structure β used is similar to Figure 1. Its tree representation is shown in Figure 3 (center). The transition table of β
by using the minimum-distance SPECTA with respect to grammar G_0 (defined in *Example 2.1*) is shown in Figure 4. In this figure the correction of β is traced out and the corrected triplets are marked
with asterisks. The corrected label of a node is obtained from the
terminal of the corrected rule used at that position. The corrected
pattern is shown in Figure 3 (right), which is one of the patterns
described by G_0.

    If there are n_1, n_2, and n_3 candidate states for a node's
three subtrees, respectively, then there are $n_1 \times n_2 \times n_3$ subtasks
processed in parallel. Therefore not only the independent subtrees are parsed in parallel, but the repeated subtasks within a
basic parsing procedure are also processed in parallel as described
in *Algorithm 3*. Fewer candidate states are kept by an ordinary
(non-error-correcting) tree automaton than its corresponding ECTA
because only no-error states are kept. Therefore ordinary tree
automata are just special cases of ECTA and their parallel parsing
can be obtained by modifying *Algorithm 3*. Instead of keeping error
counts for the minimum-distance criterion, ECTA keeps probabilities
in the transition table by using the maximum-likelihood criterion.
Therefore *Algorithm 3* can also be modified for ECTA using the maximum-likelihood criterion.

    If a lexical analyzer has been used to preprocess the tree
grammar, an efficient searching algorithm can then be used in 6(a)

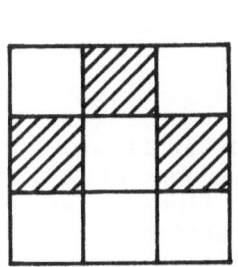 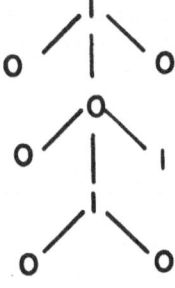 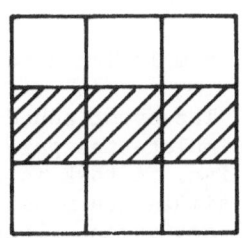

Fig. 3. A noisy pattern (left), its tree representation (center),
and the corrected pattern (right).

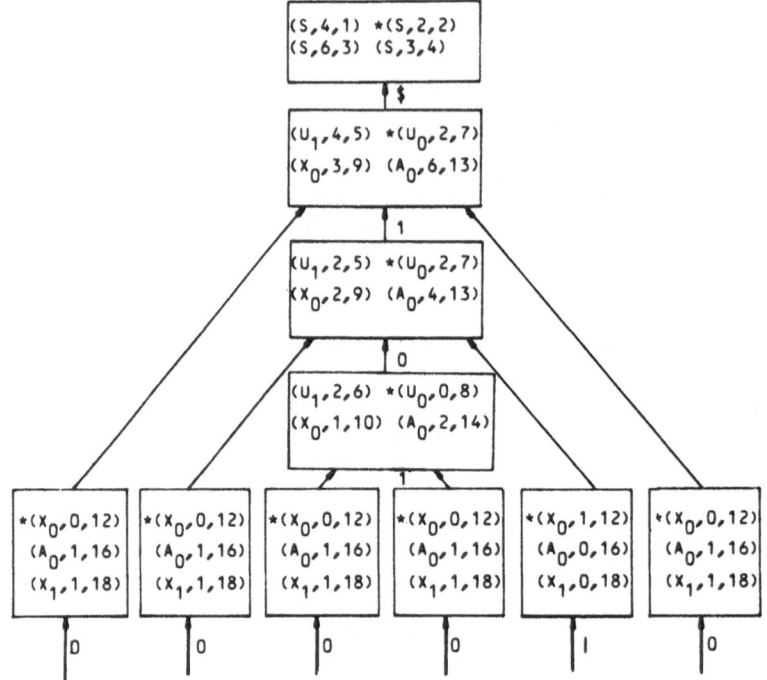

Fig. 4. The transition table corresponding to Figure 3.

of *Algorithm 3*. Since this step is in the inner loop of that procedure, parsing efficiency can be improved greatly by using the
lexical analyzer described in *Algorithm 1* or *2*.

It is assumed that all the parallel subtasks in *Algorithm 3*
can be processed simultaneously, however, if the number of processors are not large enough to handle all the subtasks simultaneously then some queuing, scheduling, and dispatching operations
are required for the implementation of *Algorithm 3*. A simulation
program has been developed to simulate such a multiprocessing system. Results are presented in the next section.

2.4 Simulation Results

The event scheduling approach given by Fishman (1973) is used
to simulate the parallel parsing tree automaton on a sequential
machine. A task is characterized by a task assignment block in
the simulation program. Once a task is created, an associated
task assignment block is prepared and then sent to a task assignment queue. A supervisor dispatches tasks from this queue to non-
busy processors. The progress of this system is marked by the

occurrence of a series of events. For example, the events in which
we are interested are the creation of a new task, the completion of
a task, and the memory interference between processors. Whenever
an event is scheduled, a record identifying the event and the time
at which it is to occur is filed in an event list. After processing
an event, the simulation program searches this event list to find
and perform the event with the earliest scheduled time. Then simulated time is advanced to this scheduled time, thus skipping the
"dead" time between two consecutive events. Simulation results are
obtained through data collected from the simulated time.

     All algorithms and simulations are programmed in the language
"C" described by Ritchie and Thompson (1974) and executed on a PDP
11/45 computer with 32K core memory. Although the simulations are
written in the high-level programming language C, the estimated time
required to execute an equivalent microprogram is used by the simulation program. The Data General ECLIPSE microinstructions format and
capabilities are used as reference (see Chang et al., 1978; Data
General, 1975). Since the execution of every microinstitution requires the same amount of processing time, the number of microinstructions needed for the execution of a procedure is collected.
Since only simple operations are used in tree automata, the translation of these operations into microinstructions is quite obvious.
These time constants can be easily modified for any other chosen
multiprocessing system such as those described by Kuck (1977), Bear
(1973), Rosenfeld (1969), or Sastry and Kain (1975).

     Simulations were made on five sets of tree grammars. Digi-
tized input pictures were first divided into non-overlapped fixed-
size windows. Tree structures similar to Figure 1 were used to
connect pixels within each window as a tree representation. Only
binary primitives were used for these five grammars. Grammar G_1
was used to recognize highway-like patterns in a LANDSAT image.
The recognition procedure is described in detail in Section 3. Figure 5 (left) is a digitized LANDSAT image of the Chicago area and
Figure 5 (right) is a map of the same area. Figure 6 (left) shows
the thresholded results which are used as input for *Algorithm 3*.
A simple postprocessor, which connects broken lines along the edges
of consecutive windows, is applied to the output of *Algorithm 3*.
The corrected highway network is shown in Figure 6 (right).

     Grammars G_{22}, G_{34}, G_{38} and G_{68} were developed by Lu and Fu
(1978) to discriminate different textures. (See Figure 7.) Statistical results are listed in Table 1. Since the same amount of input
data is parsed by each grammar and a binary search method is used in
Algorithm 3, the characteristics of each grammar rule determined the
average number of elementary operations needed to parse a fixed-size
window pattern.

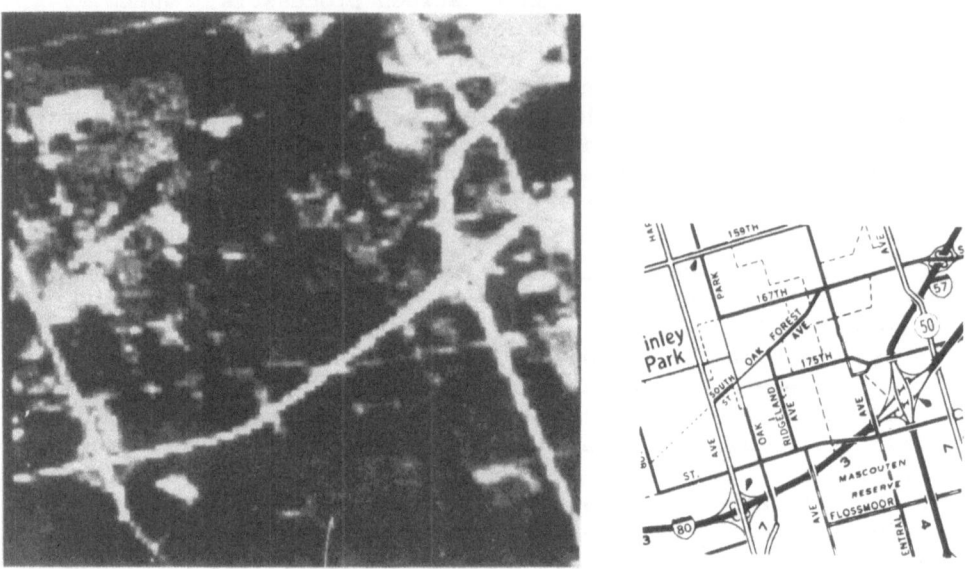

Fig. 5.   LANDSAT image (left) and the corresponding map (right).

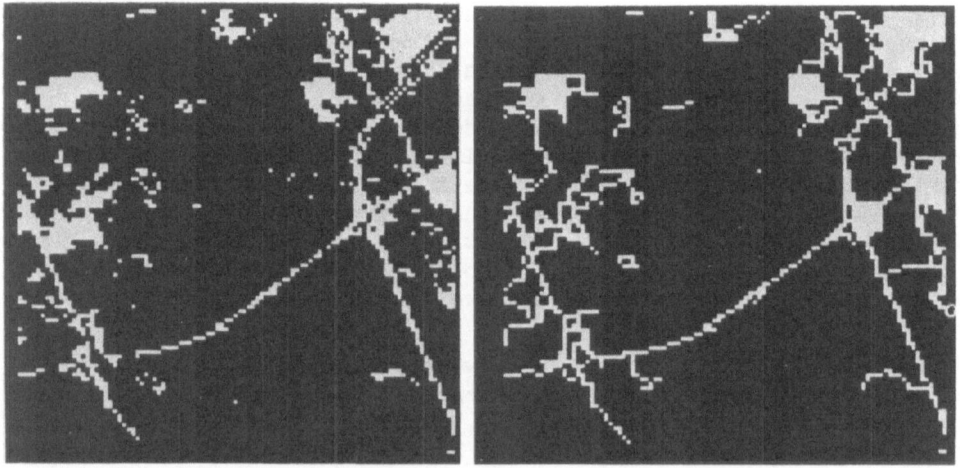

Fig. 6.   A thresholded (binary) image (left) and the recognized
pattern (right).

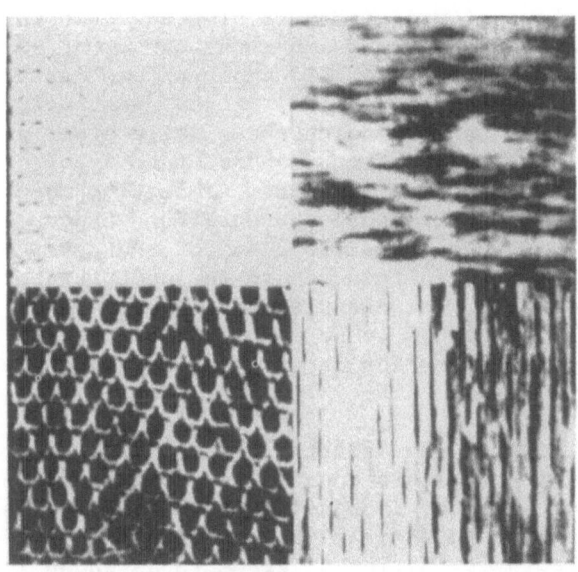

Fig. 7. Digitized textures (left to right, top to bottom): netting, water, reptile skin, woodgrain.

Table 1 - Statistics for Grammars

Grammar Statistics	G_1	G_{22}	G_{34}	G_{38}	G_{68}
No. of rules	84	260	158	248	31
No. of rules (rank 0)	7	3	3	3	2
No. of rules (rank 1)	7	16	11	17	8
No. of rules (rank 2)	10	57	68	39	18
No. of rules (rank 3)	60	184	76	189	31
No. of Nonterminals	15	74	68	56	7
No. of starting rules	10	58	58	40	3
No. of operations to parse a 9x9 window	2×10^5	6×10^6	2×10^6	8×10^6	3×10^4

The first step of each simulation concentrates upon parsing a 9x9 window using all the available processors. The averaged speed-up for various numbers of processors is plotted as shown in Figure 8. Although there are variations among the different grammars, all curves increase linearly for a small number of processors and saturate for a large number of processors. At any instance, there are several active processors working on tasks while other processors are idle; however, there is a fixed number of active processors between any two consecutive events in this simulation model. The accumulated time between consecutive events for various numbers of active processors provides a distribution for that system as shown in Figure 9. The numbers in this figure indicate the total number of available processors. For example, for a system with 10 processors, 10 processors are found to be working together for about 26% of the time and 9 processors for about 4% of the total parsing time, etc.

Since the parsing of the "central branch" is a time consuming task with only a few processors engaged in processing, no significant speed-up is obtained by increasing the number of processors after saturation. Therefore speed improvements are almost saturated for 4 to 16 processors for different grammars as shown in Figure 8. There are two peaks for the distributions shown in Figure 9. One

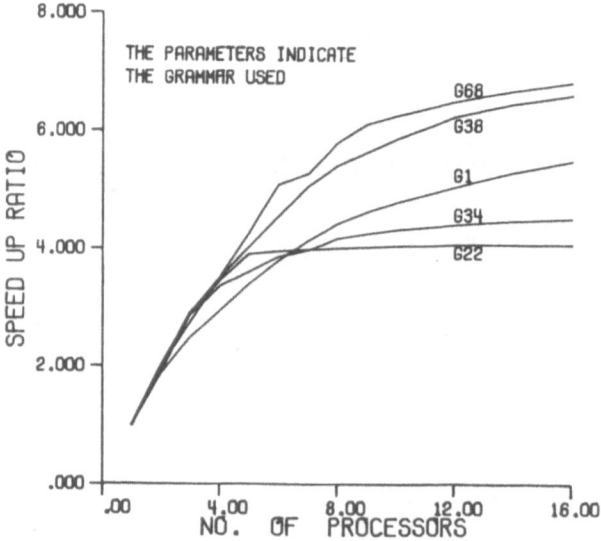

Fig. 8. Graph of speed-up factor for a window processed using several grammars.

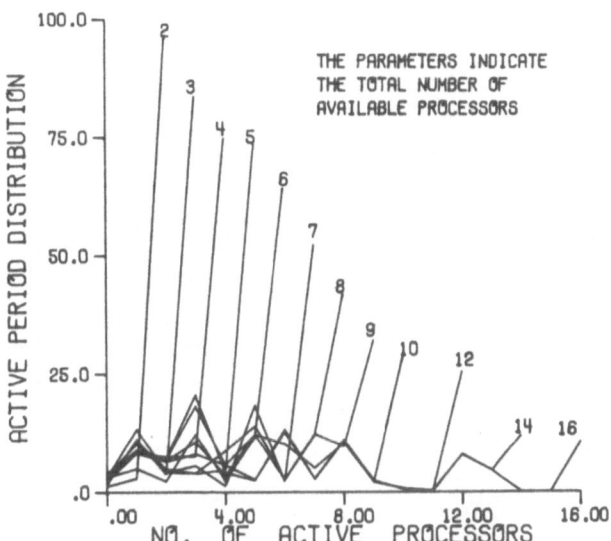

Fig. 9.   Distribution of the number of active processors.

peak corresponds to the maximum number of available processors and
the other corresponds to the small number of processors.

Since all the windows are independent, every window can, certainly, be parsed in parallel. In order to fully utilize all the
available processors, several processors can be made into a group
and each group of processors used for parsing a window. Figure 10
shows speed improvements for different combinations of processors
for parsing 100 window patterns. The numbers used in this figure
indicate the number of processors in a group. Although the curves
of Figures 9 and 10 are collected from the averaged data from grammar G_1, they give some idea about how to design an optimal system.
For a specific problem, the simulation program can be used to design
an optimal system.

3.   ROAD RECOGNITION BY USING AN ERROR-CORRECTING TREE AUTOMATON

The tree systems have been used to recognize roads in LANDSAT
images as described by Lu and Fu (1978), Li and Fu (1976), and Keng
and Fu (1976). According to the chosen primitives and tree structures, different grammars and approaches are used. A new and improved tree system approach is presented in this chapter. This
approach has the following advantages over the previous work:

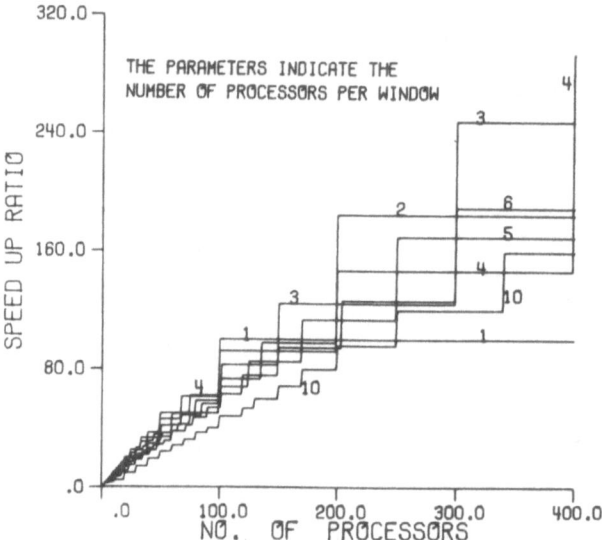

Fig. 10.   Graph of speed-up factor for 100 windows.

1.   Time-consuming preprocessing methods such as the pointwise
     classification used by Lu and Fu (1978) and Li and Fu (1976)
     and the line-smoothing method of Keng and Fu are no longer
     needed.

2.   Grammar rules are constructed in a systematic way such that
     the descriptive capabilities of the grammar rules are fully
     utilized. All the potential road patterns can be described
     by just a few grammar rules.

3.   Less computation time is needed to process the non-overlapped
     windows used in this approach than the overlapped windows of
     a picture. Since all the non-overlapped windows are indepen-
     dent, this approach is more suitable for processing by a multi-
     processing machine.

3.1 Primitive Selection

     Each LANDSAT image usually consists of four color channels.
Two channels are visible bands and the other two channels are infrared bands. LANDSAT images are given in a digitized form by NASA
with spatial resolution of one pixel corresponding to (79m)x(56m)

on the surface of the earth. Due to this resolution, the spectral
signals of small objects are usually composed of the combined re-
flectance information for several different kinds of ground cover.
This uncertainty as well as other forms of noise cause some difficulty in distinguishing roads from other objects. Although both
visible bands are sensitive to the spectrum of concrete, only data
from Channel 2 are used in this approach. Since the original raw
data are used as input, this approach is ready to be implemented in
a real-time processing system.

A digitized picture is divided into non-overlapped 7x7 windows.
The tree structure shown in Figure 11 is used to connect pixels
within a window as a tree representation. For simplicity, only
binary primitives are used. If the gray level of a pixel lies between 26 and 46 (out of 128), it is treated as primitive 1; otherwise, 0. Since an "error-correcting" tree automaton is used in
parsing, the selection of threshold values is not critical. Experimental results show that similar results can be obtained for different threshold values inferred from various training samples.

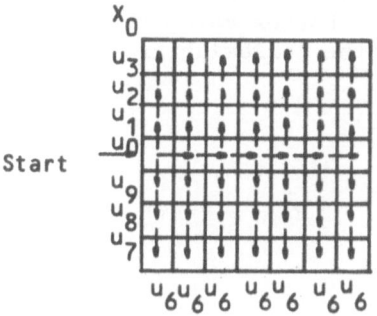

Fig. 11. Tree structure used for
connecting pixels within a 7x7
window.

3.2 Grammar Construction

Since a road is a track for travel between one place and
another, a road must satisfy certain physical and geometric requirements as follows:

1.   The width of a road has a range. For example, the width of a
     four-lane highway is approximately 150 ft.

2.   The local curvature of a road has an upper bound.

3.   Several roads nearby are connected and form a network.

Based on the above requirements and the resolution of LANDSAT images, a model of a typical road pattern within a 7x7 window is proposed as follows:

1. A road is a connected strip both of whose end points are on the edges of a window. The width of the strip is one pixel.

2. The slope of a road within a 7x7 window is either non-positive or non-negative.

3. The junction of two roads is formed from the "inclusive or" of a non-positive slope strip and a non-negative slope strip.

A systematic way to construct tree grammar rules for describing all the strips with non-negative slope is described as follows. A strip with non-negative slope can be considered as a directional strip which enters a window from its left or bottom edge and leaves the window from its right or top edge. The movement of the directional strips are rightward and/or upward. For example, the pattern shown in Figure 12 (left) has the movements shown in Figure 12 (right). The tree grammar which generates all the non-negative slope strips can be easily described by a tree automaton if the tree structure shown in Figure 11 is used. For example, the pattern shown in Figure 13(a) can be generated by the following grammar rules:

$$S \rightarrow \begin{array}{c} \$ \\ | \\ U_0 \end{array}$$

$$U_0 \rightarrow \begin{array}{c} 1 \\ \diagup | \diagdown \\ X_0 \; U_0 \; X_0 \end{array} \quad ; \quad \begin{array}{c} 1 \\ \diagup \diagdown \\ X_0 \; X_0 \end{array}$$

$$X_0 \rightarrow \begin{array}{c} 0 \\ | \\ X_0 \end{array} \; ; \; 0$$

The pattern shown in Figure 13(b) can be generated by the following grammar rules:

$$S \rightarrow \begin{array}{c} \$ \\ | \\ U_1 \end{array}$$

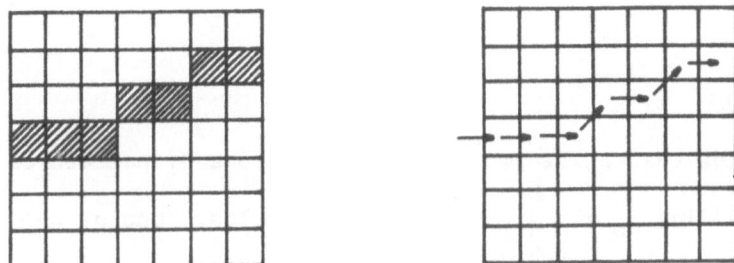

Fig. 12.   A pattern (left) and the corresponding movements (right).

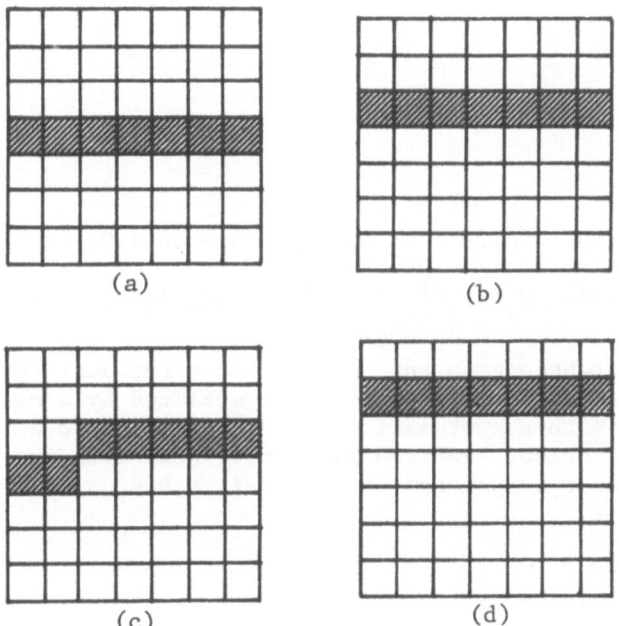

Fig. 13.   Patterns generated using different grammars.

$U_1 \rightarrow$;

$X_1 \rightarrow$

All the strips that shift from a U_0 pattern to a U_1 pattern can be defined by adding $U_0 \rightarrow$ to the above rules. One of the six possible patterns is shown in Figure 13(c). Similarly, the U_2 pattern shown in Figure 13(d) can be generated by the following rules:

$S \rightarrow$

$U_2 \rightarrow$

$X_2 \rightarrow$

All the strips that shift from a U_0 pattern to a U_2 pattern can be described by adding $U_1 \rightarrow$. Figure 12 (left) shows one of the 14 patterns that belong to this group. Thus the descriptive capabilities of the grammar rules are fully utilized by constructing grammar rules in this systematic way. In addition to all the non-negative slope strips, the tree grammar G_1 generates the all patterns of 0's (no road pattern) and the all patterns of 1's (large concrete areas).

3.3 Road Recognition

Since a structure-preserved error-correcting tree automaton is used in parsing, a minimum distance pattern χ_1 described by G_1 can be found for a windowed input pattern χ. Another minimum distance pattern χ_2 described by G_1 can also be found for the 90° rotation of pattern χ. The 270° pattern of χ_2, χ_3, is the minimum distance

strip with non-positive slope for the original input pattern χ. A
third pattern χ_4 can be obtained simply from the logic "inclusive
or" of the results of χ_1 and χ_3. The third pattern χ_4 can be considered as the potential junction of two roads. Therefore, the resultant pattern for the input pattern χ is chosen as the one among
χ_1, χ_3, and χ_4 having minimum error.

Every window of the digitized LANDSAT image shown in Figure 5
(left) was processed using this procedure. A potential road pattern
was recognized from each noisy and distorted window. Although the
recognized roads are connected within each window, there are several
disconnections along edges of consecutive windows. These disconnections can be easily connected by using a simple procedure along the
edges of the windows. For example, a connection procedure can be
applied to every two rows of consecutive edges as follows: For
every road point which is not connected, connect it to the nearest
road point along the opposite edge if their distance is within a
specified limit, say, 3 pixels. This procedure can also be applied
to every two columns of consecutive edges. The connected highway
network is shown in Figure 6 (right). Although different window
assignments may cause different results, the recognized networks
still have the same basic structure.

4.   CONCLUSIONS

By using the road recognition problem as an illustrative example, the operations of applying an error-correcting tree automaton (ECTA) are described in detail in this chapter. A very simple
set of tree grammar rules can be used by an ECTA to recognize roads
in the distorted and noisy LANDSAT images. This approach can be
used to solve other recognition problems as long as the associated
grammar rules can be inferred.

The parsing speed of an ECTA is improved greatly by using the
proposed lexical analyzers for tree grammars during the preprocessing phase and using the corresponding efficient search algorithms
during the parsing phase. A special-purpose multiprocessing system
can be easily implemented for the proposed parallel parsing tree
automata.

5.   ACKNOWLEDGEMENTS

The research reported in this chapter was supported by the
United States Defense Research Projects Agency under grant MDA 903-
77-G-1.

6.   REFERENCES

Bear, J. L., "A Survey of Some Theoretical Aspects of Multipro-
    cessing," Computing Surveys $\underline{5}$(1):31-80 (1973).

Chang, N. S., Mulrooney, T., and Weiderman, N., "An Interactive
    Simulator for Microprogramming Development," Proc. 11th An.
    Simulation Symposium (March 15-17, 1978), pp. 271-282.

Data General Corporation, Microprogramming the ECLIPSE Computer with
    the WCS Feature, Southboro (Aug. 1975).

Fishman, G. S., Concepts and Methods in Discrete Event Digital
    Simulation, New York, Wiley (1973).

Fu, K. S., Syntactic Methods in Pattern Recognition, New York, Aca-
    demic Press (1974).

Fu, K. S., "Tree Languages and Syntactic Pattern Recognition," in
    Pattern Recognition and Artificial Intelligence (Chen, C. H.,
    ed.), New York, Academic Press (1977), pp. 257-291.

Fu, K. S., and Bhargava, B. K., "Tree Systems for Syntactic Pattern
    Recognition," IEEE Trans. Comput. $\underline{C-22}$:1087-1099 (1973).

Keng, J., and Fu, K. S., "A Syntax-Directed Method for Land-Use
    Classification of LANDSAT Images," Proc. Symp. Current Math.
    Prob. Image Science, Monterey (Nov. 10-12, 1976).

Kuck, D. J., "A Survey of Parallel Machine Organization Programming,"
    Computing Surveys $\underline{9}$(1):29-59 (1977).

Li, R. Y., and Fu, K. S., "Tree System Approach for LANDSAT Data
    Interpretation," Proc. Symp. Machine Proc. Remot. Sensed Data,
    Purdue University (1976).

Lu, S. Y., and Fu, K. S., "Structure-Preserved Error-Correcting
    Tree Automata for Syntactic Pattern Recognition," Proc. IEEE
    Conf. Decision and Control (Dec. 1-3, 1976).

Lu, S. Y., and Fu, K. S., "A Syntactic Approach to Texture Analy-
    sis," Comput. Graph. Image Proc. $\underline{7}$:303-330 (1978).

Ritchie, E., and Thompson, K., "The UNIX Time-Sharing System,"
    Comm. ACM $\underline{17}$(7):365-375 (1974).

Rosenfeld, J. L., "A Case Study in Programming for Parallel-
    Processes," Comm. ACM $\underline{12}$(12):645-658 (1969).

Sastry, K. V., and Kain, R. Y., "On the Performance of Certain Multiprocessor Computer Organizations," IEEE Trans. Comput. C-24(11):1066-1074 (1975).

Thurber, K. J., and Wald, L. D., "Associative and Parallel Processors," Computing Surveys 7(4):215-255 (1975).

PIXAL: A HIGH LEVEL LANGUAGE FOR IMAGE PROCESSING

S. Levialdi, A. Maggiolo-Schettini, M. Napoli* and
G. Uccella*

Laboratorio di Cibernetica del C.N.R.

80072 Arco Felice, ITALY

1. INTRODUCTION

In the field of image processing and pattern recognition many
attempts towards the creation of general purpose software have
been made during these last fifteen years according to the type of
computing resources and classes of problems that were confronted
by each working group.

According to the application requirements, images may either
be transformed into other images, to highlight specific features,
or into numerical quantities which codify specific properties
extracted for subsequent classifications. It is therefore highly
desirable to have a language for expressing specific operations on
arrays (representing the digital version of images) and for performing arithmetical computations on integers and reals. Such a
language should enable the programmer to express his algorithms in
a natural way and, at the same time, should be machine independent
in order to achieve maximum transparency and portability.

Since the original work on ILLIAC by McCormick (1963), many
parallel processors have been suggested and built by such workers
as Duff (1973), Kruse (1973), Preston (1977), etc. These machines

---

*Gruppo Nazionale di Informatica Matematica del C.N.R., c/o Isti-
tuto di Scienze dell'Informazione dell'Universita, 84100 Salerno,
ITALY

are especially suited for image processing and, with the advent of
LSI and VLSI, they will become both practically and economically
feasible in the very near future. In many cases it is still an
open question whether a fully parallel process is better than a sequential one and it has been suggested that a compromise between
these two approaches might improve the performance for a certain
class of tasks. For this reason it is also important to design the
language so that it can express both parallel and sequential operations in a very simple and direct way.

At present the existing software systems which either emulate
a parallel machine, e.g., PAX as documented by Johnston (1970) and
Stein (1963), or help in producing an interactive facility for image
transactions, e.g., VICAR developed at JPL, Pasadena, do not satisfy
the above requirements in our opinion since they are not truly languages but packages of an extensible directory of routines. On the
other hand in connection with the use of very recent parallel processors especially designed for image processing, the main software
effort has gone towards the development of languages tailored to a
specific hardware architecture, e.g., PICAP, as described by Kruse
(1976), and CAP4 documented at University College, London (1977).

Another interesting effort is the PICASSO language of Kulpa
and Novicki (1976) which is based on the use of an extensible library of routines written in assembly language. A set of powerful
basic operations is included so that most typical image processing
algorithms may be easily coded. The main advantage of this interactive programming system is the small amount of minicomputer memory which is used.

After considering the different approaches we have oriented
our research effort towards the design of a high-level language for
image processing with both sequential and parallel capability which
is machine independent and which can also run on a medium sized
minicomputer, in particular a HP21MX. It has also been our intention to exploit the experience gained in the formal definition of
algorithmic languages for numerical computation. We have therefore decided not to define a language from the beginning but to
embed within an ALGOL 60 environment some typical constructs involving parallel manipulation of arrays. Since this language will
be mainly used for *pix*el manipulation and is *al*gorithmically oriented we have called it PIXAL. The choice of ALGOL 60 (see Naur,
1963) instead of the more promising Pascal (see Wirth, 1971) was
due mainly to the availability of an ALGOL 60 compiler running on
the HP21MX having 32K words of core memory.

2.   BASIC FEATURES OF THE PIXAL LANGUAGE

Besides the practical motivation mentioned in Section 1, ALGOL 60 has been chosen because of (1) the power of some constructs that were initially introduced in this language (like the *while* statement), (2) the organization of the program in blocks which allows program transparency and better use of memory (particularly useful when large quantities of data are manipulated), and (3) the large number of theoretical studies on this language undertaken over almost twenty years which one could extend to cover parallel constructs. Notice that Pascal does not contain more powerful constructs than ALGOL. The most interesting feature of Pascal is the facility to define structure types. This, however, would not be sufficient for our purposes (see the type *frame* introduced below).

In the following we give rules [in the usual Naur (1963) form] which are intended to be added to the rules of the grammar of ALGOL 60.

2.1 Declaration

Syntax

declaration :: = mask declaration | frame declaration | edge declaration

mask declaration :: = type *mask* mask identifier   index list |
type *mask* mask identifier   index list
of value list

frame declaration :: = *frame* frame identifier   index list

edge declaration :: = *edge-of* array identifier *is* arithmetic expression

type :: = *binary* | *grey*

mask identifier :: = identifier

frame identifier :: = identifier

index list :: = [bound pair list] | [bound pair list]
on [arithmetic expression {, arithmetic expression }]

value list :: = (extended arithmetic expression {, extended arithmetic expression }) extended arithmetic expression :: =   arithmetic expression |

don't care symbol

don't care symbol :: =   ?

procedure declaration :: =   specifier *procedure* procedure

heading   procedure body

specifier :: =   type *mask* | *frame*

Semantics

Mask and frame declarations serve to declare certain identifiers to represent multidimensional structures. The list of indices gives the bounds in the different dimensions and optionally the coordinates of a special element which allows masks and frames to be positioned on an array element whenever the special element is not the geometrical center. The list of values (among which a special "don't care symbol" is included) defines the particular pattern of the mask to be compared with the environment of the element on which the mask is positioned.

The frame declaration allows a submatrix to be defined around the array element on which the frame is positioned (this enables the definition of a neighborhood upon which a parallel operation can be performed). An edge declaration can be used for constraining the image data either to be embedded on a background (0-elements) or in any specific grey level whose value is provided. The default options automatically defaults to the first instance. Note that this feature is generally implemented in the hardware of array processors as pointed out in Duff (1973).

Besides the usual types of ALGOL 60 (real, integer and Boolean) grey and binary types are also included.

Examples

(1) *binary array* A [1:128,1:128]

(2) *edge-of* A *is* 1

(3) *frame* F [1:3,1:3]

By default the center of the frame is intended to be " on [2,2] ."

(4) *binary mask* M [1:3,1:3] *on* [2,2] *of* (0,1,0,0,1,0,0,1,0)

2.2 Parallel Statements

Syntax

parallel statement :: = *par* special statement *parend*

special statement :: = special unconditional statement |
                        special conditional statement

special unconditional statement :: = special assignment statement |
                                       special compound statement |
                                       special for statement |
                                       procedure statement |
                                       empty statement

special assignment statement :: = array identifier := special
                                    arithmetic expression

special compound statement :: = *begin* special statement {; special
                                statement }*end*

special for statement :: = *for* array identifier := special
                           arithmetic expression *until* special
                           arithmetic expression *step* special
                           arithmetic expression *do* special
                           statement

special conditional statement :: = special if statement |
                                     special while statement |
                                     special if statement *else*
                                     special statement

special if statement :: = *if* special Boolean expression *then*
                          unconditional special statement |
                          if special Boolean expression
                          then special while statement

special while statement :: = *while* special Boolean expression
                             do special statement

special arithmetic expression :: = special unconditional arithmetic
expression | *if* special Boolean
expression *then* special unconditional arithmetic expression
else special arithmetic expression

special unconditional arithmetic expression :: = special term |
special unconditional arithmetic expression
adding operator   special term

special term :: = special factor | special term   multiplying
operator   special factor

special factor :: = special primary | special factor | special
primary

special primary :: = identifier | (special arithmetic expres-
sion)

special Boolean expression :: = special Boolean | *if* special
Boolean expression *then* special
Boolean *else* special Boolean
expression

special Boolean :: = special implication | special Boolean ≡
special implication

special implication :: = special Boolean term | special implica-
tion ⊃ special Boolean term

special Boolean term :: = special Boolean factor | special
Boolean term ∨ special Boolean factor

special Boolean factor :: = special Boolean secondary | special
Boolean factor ∧ special Boolean
secondary

special Boolean secondary :: = special Boolean primary | − special
Boolean primary

special Boolean primary :: = special relation | array identi-
fier | variable identifier |
(special Boolean expression)

special relation :: = special unconditional arithmetic expression

relational operator   special unconditional

arithmetic expression

Semantics

The special statements inside *par* ... *parend* are intended to be performed simultaneously for every element of each array appearing within the special statements. These statements are of the form of the statements defined in ALGOL 60 with the following exceptions: (1) labels and goto's are not allowed (in order to forbid jumps in or out of the parallel statement); (2) assignments can be made only to array identifiers and only expressions not containing subscripted variables can be assigned; (3) in the conditional statements tests cannot be performed on subscripted variables.

Example

par

   if A=1 *then* A:=0 *else* A:=1

parend

The elements of the array A (assumed to be declared of type binary) are tested simultaneously and set either to zero or to one depending on the result of the test, i.e., the complement of the content of the array A is performed in parallel.

2.3 Global Assignment

Syntax

assignment statement :: = special assignment statement | array

identifier := arithmetic expression |

procedure identifier := special

arithmetic expression | mask identi-

fier := value list

Semantics

Assignments to arrays can also be performed under sequential control. The value computed by the arithmetic expression (in which subscripted variables can also appear) is assigned to every element of the array. The assignment of a list of values to a mask identifier implies the assignment of the values of the list to the elements of the mask by row order.

Examples

(1)   A:=A+1

All the elements of A, supposed to be declared as an array, are incremented by 1.

(2)   M:=(0,0,0,0,1,1,0,1,?)

The list of values on the right hand side is assigned to M, supposed to be declared as a mask.

2.4   Composition of Sequential and Parallel Operations

Syntax

 unlabelled basic statement :: = parallel statement

 Boolean expression :: = special Boolean expression

Semantics

    Defining parallel statements as unlabeled basic statements allows the composition of sequential and parallel statements.

    Having the special Boolean expression as a Boolean expression allows testing on global properties of arrays in the usual conditional statements.

Example

    if A=0 *then goto* stop *else par*

                    if A≠0 *then* A:=A+1

                parend

The parallel statement is performed only if the array A is not empty. All the values of the array which are different from zero are simultaneously incremented.

2.5   Built-In Functions

    Certain identifiers should be reserved for some useful functions in pattern recognition and image processing. Such functions will be expresed as procedures. The reserved list contains:

first(A)   for the function which gives the first column of the
                array A

last(A)   for the function which gives the last column of the
                array A

top(A)   for the function which gives the first row of the
                array A

bottom(A)   for the function which gives the last row of the
                array A

sum(A)   for the function which gives the sum of all the ele-
                ments of the array A

Inside a parallel statement the following functions can be used:

compare(M,A)   for the Boolean function which, given a mask M
                    and an array A, is true if the submatrix of A,
                    with the same dimensions as M and centered over
                    the current element of A (selected by the parallel
                    control) is equal to M; otherwise, false

overlap(F,A)   for the function which, given a frame F and an
                    array A, provides a new array with the same
                    dimensions as F and whose elements are the same
                    as those in the environment of the current ele-
                    ment of A

overweigh(M,A)   for the function which, given a mask M and an
                    array A, provides a new array with the same di-
                    mensions as M whose elements are the products
                    between M and the corresponding elements around
                    the current element of A

We give now in the following some examples in order to clarify
the use of the new constructs.

Examples

(1) *binary array* A[1:128,1:128];

     frame F[1:3,1:3]

     while first(A) = 0 ∧ last(A) = 0 ∧ bottom(A) ∧ top(A) = 0

     do par

            if A=0   sum(overlap(F,A)) > 0 *then* A:=1

        parend;

     comment: A binary image is expanded until the edge elements

     are touched;

(2) *grey array* A[1:128,1:128];

    integer mask M[1:3,1:3] *on* [1,3] *of* (1,1,0,-1,-1);

    par

        A:=sum(overweigh(M,A))

    parend;

    comment: This program implements a vertical boundary detector corresponding to the linear difference filter of Suenaga et al. (1964).

(3) *grey array* G[1:128,1:128];

    binary array BIN[1:128,1:128];

    integer array HIST [0:63];

    for I:=0 *step* 1 *until* 63 *do*

        begin BIN:=0; *par if* G=I *then* BIN:=1 *period*;

            HIST[I]:= sum(BIN)

      end;

    comment: To each element of the array G having a given grey value I a 1-element is correspondingly assigned in the array BIN. Finally the sum of these elements in BIN is computed and stored in the I-th element of HIST. This is done for all the 64 grey values. In order to detect the valley of a bi-modal histogram the central minimum must be located. This is done by finding the second sign inversion when comparing the contents of adjacent I-locations in HIST. The index I found in this manner corresponds to the location of the central minimum. The value of I found is then used to threshold the values of the elements of the array G. The entire program is as follows:

    I:=0;

    while HIST[I] $\leq$ HIST[I+1] *do* I:=I+1;

    while HIST[I] $\geq$ HIST[I+1] *do* I:=I+1;

    par if G $\geq$ I *then* G:=1 *else* G:=0 *parend*;

3. IMPLEMENTATION CONSIDERATIONS

As mentioned in Section 1 our purpose is to implement a high level language on a minicomputer, in particular on a HP21MX with 32K words of core and a disk operating system. ALGOL 60 is available on this computer with only one major shortcoming, namely, recursive calls to procedures are forbidden. Calls to procedures written in the HP Assembler or in FORTRAN are allowed so that an already existing library of routines for picture processing can be used. The occupation of memory of the HP ALGOL compiler is 6K words and 4K more are available for possible extensions.

A general plan for the implementation of the new constructs is under development. The routines for the analysis of such constructs will be fitted into the existing ALGOL 60 analyzer.

As regards the internal representations which will be generated for the new structures we have chosen templates. The templates for masks and frames will be similar to the ones used for arrays, but for the fact that in a frame template no pointers to the contents of the elements will exist. The array template also contains the information regarding the edge. The edge declaration may be changed on entering a block without changing the declaration of the array (only the information about the edge in the template of the array will be changed). Note that, when neighborhoods partially fall outside the array, the built-in functions use the edge declaration to insure correct operation.

Variables declared as binary will require only one bit of memory; for the ones declared as grey one byte will be needed (the maximum grey level value is $2^8-1=255$).

As regards the implementation of the parallel statement, it will be obviously transformed into a sequential control, independent of the number of statements inside the parallel control. At most one temporary copy will be needed for each different array.

4.   FINAL COMMENTS

We consider this project an experiment on the definition of a high-level language for picture processing. Many improvements of the constructs already defined and many new constructs can be easily envisaged. Besides the HP21MX code we would also like to generate a CAP4 code in order to run our programs on a CLIP4 machine which is able to process a 96x96 array in parallel.

The actual time required by our programs in PIXAL to run on the minicomputer may or may not encourage further extensions of this language, but a real estimate is difficult to establish until implementation is completed and a number of programs have been written and tested using our facilities.

5.   REFERENCES

Duff, M. J. B., "A Cellular Logic Array for Image Processing," Pat.
    Recog. 5:229-247 (1973).

Hewlett Packard ALGOL Programmer's Reference Manual, HP02116-9072,
    Cupertino, California (1971).

Johnston, E. G., "The PAX II Picture Processing System," in Picture
    Processing and Psychopictorics (Lipkin, B., and Rosenfeld, A.,
    eds.), New York, Academic Press (1970), pp. 427-512

Kruse, B., "A Parallel Picture Processing Machine," IEEE Trans.
    Comp. C-22(12):1075-1087 (1973).

Kruse, B., "The PICAP Picture Processing Laboratory," Dept. of
    Electrical Engineering, Linkoping University, 04-23 Intern.
    Skrift, LITH-ISY-I-0096 (1976).

Kulpa, Z., and Novicki, H. T., "Simple Interactive Processing Sys-
    tem, 'PICASSO Show,,'" Proc. 3rd Intern'l Joint Conf. Pattern
    Recog. IEEE 76CH 1140-3C, Coronado (1976), pp. 218-222.

McCormick, B. H., "The Illinois Pattern Recognition Computer-ILLIAC
    III," IEEE Trans. Elect. Comput. EC-12(6):791-813 (1963).

Naur, P. (ed.), "Revised Report on the Algorithmic Language ALGOL
    60," Comm. ACM 6:1-17 (1963).

Preston, K., Jr., "Applications of the Golay Transform to Image
    Analysis in Cytology and Cytogenetics," in Digital Image Pro-
    cessing and Analysis (Simon, J. C., and Rosenfeld, A., eds.),
    Nato Advanced Study Institutes Series-Applied Science-No. 20,
    Leyden, Noorhoof (1977), pp. 401-412.

Rosenfeld, A., and Kak, A. C., Digital Picture Processing, New York,
    Academic Press (1976).

Stein, J. H., "Program Description of PAX, an IBM 7090 Program to
    Simulate the Pattern Articulation Unit of ILLIAC III," Rpt.
    151, Digital Computer Lab., Univ. of Illinois, Urbana (Sept.
    1963).

Suenaga, Y., Toriwaki, J., and Fukumura, T., "Fundamental Study of
    Difference Linear Filters for Processing of Continuous Tone
    Pictures," Trans. IECE, JAPAN, 57D(3):119-126 (1964) (in
    Japanese).

University College London, Image Processing Group, CAP4 Programmer's
    Manual (June 1977).

Wirth, N., "The Programming Language Pascal," Acta Inform. $\underline{1}$:35-63
    (1971).

LANGUAGES FOR PARALLEL PROCESSING OF IMAGES

K. Preston, Jr.

Carnegie-Mellon University

Pittsburgh, PA 15213 USA

1. INTRODUCTION

Parallel processing of large data arrays is characteristic of image analysis. In order to be compatible with typical television systems, the usual image size is 512x512. If the image is a 3-color image, this means that approximately 1 million bytes of data are present in the single-frame output of an ordinary television scanner. If the television scanner operates at 15 frames per second, then the full data transfer rate is more than 10 megabytes per second, i.e., one picture element (pixel) is generated every 100 nanoseconds. In order that a computing system process images at this rate, several picture points operations (pixops) must be performed in this time interval.

Since ordinary general-purpose computers have a single-instruction time which is greater than 100 nanoseconds, special-purpose computers have been designed, which are capable of carrying out instructions in parallel, in order to achieve the data processing rate required for real-time video signal processing. These machines are loosely called "parallel processors" although they usually exhibit different degrees of parallelism as well as different degrees of reconfigurability. The most straightforward of these machines performs the identical operation on each pixel and handles a multiplicity of pixels in parallel. In this case the overall machine is called a SIMDS (Single Instruction Multiple Data Stream) machine, whereas, if the system can perform different operations simultaneously it is a MIMDS (Multiple Instruction Multiple Data Stream) machine. This contrasts with the usual general purpose computer which is SISDS (Single Instruction Single Data Stream).

Programming the SIMDS and the MIMDS machines can, of course, be done in FORTRAN since FORTRAN is a language which readily handles indexing through arrays. However, due to the specialized nature of parallel processing and of image analysis, there has long been a need for a simplified, high-level language which will permit the manipulation of arrays as single entities rather than by using specific indexing methods. One of the earliest of such languages was GLOL (GLOPR Operating Language) which was designed by Preston (1971) in 1968 for the purpose of driving the Golay Logic Processor (GLOPR) designed by the Research Division of the Perkin-Elmer Corporation. This machine was designed specifically for performing logical transforms in the 6-neighbor hexagonal Golay array in parallel. The direct pixop time was 2 microseconds which, after adding input/output overhead, averaged to 3 microseconds. Thus GLOPR performed a Golay transform in 50 milliseconds for a 128x128 image.

In GLOL image arrays were dimensioned simultaneously by the SYSTEM SIZE command and, thereafter, were labeled by the DEFINE command. In this language two previously defined and dimensioned arrays could be equated by simply typing A = B. Similar shorthand notations were available for all logical functions with a somewhat more complicated expression being needed for the Golay transform in that arguments referring to surrounds, subfields, and subfield order were required. GLOL is now used exclusively as the operating and image processing language for the Coulter Biomedical Research Corp. instrument "diff3" used for real-time image analysis in hematology.

It is the purpose of this chapter to review recent developments in designing languages which, like GLOL, have been specifically produced for the purpose of performing image processing functions in some of the newer parallel processors that have been constructed during the late 1970's. A number of these machines are listed in Table 1 along with the names of the languages associated with them (where available) as well as the institution which has developed the machine.

2.   LANGUAGES FOR IMAGE ANALYSIS

Before the advent of special-purpose image-analyzing computers, many command languages were constructed for image analysis using general purpose computers. These languages have formed the basis upon which some of the more modern languages associated with specific parallel processors have been structured. Twenty-seven of these languages are given in Table 2. They have been reviewed in detail by Preston (1980A) as part of the activities of recent workshops on high-level languages and on pattern recognition. The commands in these languages fall into certain broad categories which are given on the facing page.

Table 1 – Languages Designed for Parallel Processing Machines

LANGUAGE	MACHINE	INSTITUTION
MORPHAL	AT4	Center for Mathematical Morphology (France)
CAP 4	CLIP4	University College London (England)
C3PL	CYTOCOMPUTER	Environmental Research Institute of Michigan (USA)
DAP FORTRAN	DAP	ICL and Logica Ltd. (England)
GLOL	DIFF 3	Coulter Biomedical Research Corp. (USA)
–	DIP	Delft University of Technology (Netherlands)
–	FLIP	Institute for Information Processing (Germany)
INTRAC	GOP	Linkoeping University (Sweden)
–	IP	Hitachi Central Research Laboratory (Japan)
PPL	PICAP II	Linkoeping University (Sweden)
–	PPP	Toshiba Research & Development Center (Japan)
–	SYMPATI	CERFIA – UPS (France)

Utilities
   Identifiers
   Executives
   Formaters
   I/O commands
   Test pattern generators
   Help files

Image Display
   CRT
   Hard copy
   Interactive graphics

Arithmetic Operators
   Point
   Line (Vector)
   Matrix
   Complex number
   Boolean

Geometric Manipulation
   Scaling/rotation
   Rectification
   Mosaicing/registration
   Map projection
   Gridding/masking

Image Transforms
   Noise removal
   Fourier analysis and
      other spectral transforms
   Power spectrum
   Filtering
   Cellular logic

Image Measurement
   Histogramming
   Statistical
   Principal components

Decision Theoretic
   Feature select (training)
   Classify (unsupervised)
   Classify (supervised)
   Evaluate results

Table 2 – Parallel Processing Command Languages
Designed for General Purpose Computers

LANGUAGE	REFERENCE
ASTEP (Algorithm Simulation Test and Evaluation Program)	Johnson Space Flight Center
CAPCU	Rutovitz
CELLO	Eriksson et al.
DIMES (Digital Image Manipulation and Enhancement System)	Computer Sciences Corp.
ENUS	Dunham et al.
ERIPS (Earth Resources Interactive Processing System)	IBM Federal Systems Division
FIDIPS	Paton
IDAMS (Image Display and Manipulation System)	Pape and Truitt
IFL	Ralston
IPL	Joyce Loebl Inc.
KANDIDATS (Kansas Digital Image Data System)	Univ. of Kansas
LARSYS (Laboratory for Applications of Remote Sensing)	Purdue Univ.
MAGIC	Taylor
MINBASIC	Alexander
MSFC	Marshall Space Flight Center
Pascal PL	Uhr
PECOS (Picture Enhancement Computer Operating System)	Electro-Magnetic Systems Laboratory
PICASSO/PICASSO-SHOW	Kulpa
PICPAC	Akin and Reddy
PIXAL	Levialdi
PL/S	Krevy et al.
SLIP	Toriwaki et al.
SMIPS	Goddard Space Flight Center
SUPRPIC	Preston
SUSIE	Batchelor
TAL	Vrolijk
VICAR	Castleman

Many of these image processing languages were generated as part of programs conducted by the United States National Aeronautics and Space Administration (USNASA) laboratories for the purpose of analyzing images produced in such space exploration projects as Ranger, Mariner, etc. These languages emphasize the geometric manipulation of images and the correction of images for video malfunctions and long-distance transmission errors. Others of these languages emphasize the segmentation of images according to the pixel multi-spectral signature. Of these languages the most elaborate is MSFC which has large numbers of commands in all of the above image processing categories.

In addition to these USNASA developments, the United States Department of Defense sponsored the development of additional image analyzing languages which, rather than concentrating on geometric manipulation and multi-spectral signature analysis, were designed for the purpose of generating pixel statistics and using both arithmetic and geometric operators and for simple image matching by correlation techniques. Their use was primarily for target detection, recognition, and identification.

3.   LANGUAGES ASSOCIATED WITH PARALLEL PROCESSORS

With the sudden drop in electronic memory and logic circuit costs during the late 1970's accompanying the introduction of LSI, many institutions found it economically feasible to begin the construction of machines which were in some sense parallel processors. These machines include the CLIP-series of machines being constructed by University College London (Duff, 1977), the PICAP-series of machines designed by Kruse (1973) of the University of Linkoeping, and the series of GLOPR-machines constructed by the Perkin-Elmer Corporation and Coulter Biomedical Research Corp. (See Table 3.) Of these machines only the GLOPR incorporated in the Coulter diff3 is in wide commercial production. The DAP (Digital Array Processor) is in limited production whereas the other machines are in use primarily for image processing research.

In this section we review the characteristics (as well is now known at present) of these machines and the languages which have been developed to command them. In doing this we are indebted to the National Research Council (NRC) of Great Britain and, in particular, to the NRC Workshop on High-Level Languages (held at Windsor in June 1979). At this workshop authors of approximately 20 image processing languages were present and furnished the instruction set necessary to perform the exclusive OR (EXOR of image A with image B with placement of the results in C) using their parallel processing language.

Table 3 – Some Specifications of Parallel Processors

MACHINE	IMAGE MEMORY	PROCESSING WINDOW	PIXOP TIME (APPROX.)
AT4	?×256×256×1	3×3	1 μs
CLIP4	32×96×96×1	96×96	3 ns
CYTOCOMPUTER	88×512×4	88×3×3	7 ns
DAP	64×64×4096	64×64	50 ps
DIFF	4×64×64×1	3×3×3	25 ns
DIP	2×256×256×8	up to 16×16	150 ns
FLIP	?×1024×1024	4×4	2 μs
GOP	512×64	up to 64×64	30 ns
IP	3×256×256×8	4×4	170 ns
PICAP II	16×512×512×1	up to 4×4	20 ns
PPP	8×512×512×8	up to 12×12	1 μs
SYMPATI	8×512×512×1	?	2 μs

3.1 MORPHAL

MORPHAL is the command language for the AT4 recently completed by the Center for Mathematical Morphology at Fontainebleu. The system is used primarily for logical image transforms over the hexagonal neighborhood. The basic machine cycle (over a 256x256 array) is 20 milliseconds. The AT4 not only operates upon many pixels simultaneously but upon several images in parallel. For example, the AT4 can construct the union of several images and intersect the result with the union of several other images in a single machine cycle. Complicated operations, such as skeletonization, are performed in subfields of 6 and require, therefore, 120 milliseconds. Operations such as gradient detection require 40 milliseconds when using two thresholds.

The MORPHAL code for the exclusive OR is given below:

```
SFLD (MA,MB)(128,128)(1,1)      Dimension A and B
NØL ((MA,MG,JN)-(MA,MB,IN)(MC)) XOR into C
```

3.2 CAP4

CAP4 is an assembly-like language. A programmers manual for CAP4 is published by the University College London. CAP4 has been used to program the CLIP-series of machines which are basically software compatible. The most recent of these machines is the CLIP4 which is an array of 9162 individual processors (96x96) operating with a basic cycle time of 25 microseconds. As with the AT4 this machine is used exclusively for logical transforms. In one cycle it is capable of performing the logical AND, OR, etc. More than one cycle is required for a single step of skeletonization. The most efficient skeletonizing program requires 8 full cycles as it operates in subfields.

In order to compare CAP4 with other languages given in this section the code for the exclusive OR is given below:

```
C$IN                              Input first array
.WORD 1
SET A
PST 0                             Save Data
C$IN                              Input second array
.WORD 2
SET A
PST 1                             Save Data
;
SET P@A                           Set up EXOR function
LDA 0
LDA 1
PST 2                             Execute and save results
.END
```

3.3 C3PL

C3PL is the language developed at the Environmental Research Institute of Michigan for the purpose of commanding the Cytocomputer (Sternberg, 1980). The Cytocomputer is a pipelined machine which is comprised of a total of 113 stages of which 88 are logical and 25 are integer. All stages operate over a 3x3 neighborhood. A single stage of the Cytocomputer functions in 650 nanoseconds. Therefore, in order to obtain the full rate of operation from the pipeline configuration there must be at least 113 operations in the command string. Otherwise, the pipeline is only partially filled and the effective cycle time over an image is relatively slower.

The C3PL code for the exclusive OR is given as follows:

```
TAPEIN                            Reads 1st image from tape.  Auto-
                                  dimensioned.
SAVE 1                            Stores active image file into disc
                                  file 1.
TAPEIN                            Read 2nd image.
COMBINE 1,2                       Combines both images in active
                                  file.
BITXOR 2,1                        Performs Exclusive OR.
```

C3PL supports 52 commands and is strong in almost all image processing categories but is lacking in decision theoretic capabilities.

3.4   DAP FORTRAN

The DAP (Digital Array Processor) is a commercial machine consisting of a stack of 4096 bit-planes each configured in a 64x64
array. In order to program in DAP FORTRAN one must write both a
host program and an image processor program. An example (for the
exclusive OR) is given below:

HOST PROGRAM
COMMON/IMAGE/A(64,64,4),B(64,64,4)   Dimension arrays
LOGICAL A(64,64,4),B(64,64,4)   Declare type
READ (5,1)A,B   Read images
1 FORMAT (L1)   Specify format
CALL DAPXOR   Call DAP routine
WRITE(6,100)A   Write results

IMAGE PROCESSOR (DAP)
SUBROUTINE DAPXOR   Use 4 64x64 subarrays to obtain
                                            128x128 result

COMMON/IMAGE/A(,,4),B(,,4)
DO 10 I=1,4
CALL CONVMFL(A(,,I))   Convert "horizontal" to "ver-
                                            tical" format

CALL CONVFML(B(,,I))
A(,,I)=A(,,I) .LNEQ. B(,,I)   Do XOR
CALL CONVMFL(A(,,I))   Convert format
10 CONTINUE
RETURN
END

DAP FORTRAN supports 52 commands (as well as the standard
FORTRAN commands). All but 7 of these 52 commands are arithmetic
operators. The DAP does not support commands for image display,
geometric manipulation, image transforms, measurements, or decision theory.

3.5   GLOL

GLOL is the command language for the Coulter Biomedical Research Corp. diff3 (GLOPR). This machine has been in commercial
production for about 4 years and several dozen diff3's are now de-
ployed in the United States and overseas. They are found in hos-
pital hematology laboratories where they are used for the automatic
analysis of images of human white blood cells at the rate of approximately 4000 images per hour. The GLOL commands for the exclusive
OR are as follows:

```
SET SYS 128                 Dimension 128x128
DEF A,B,C                   Name images A,B,C
SEARCH A,N1                 Put image labeled N1 in A
SEARCH B,N2                 Put image labeled N2 in B
C=A-B                       Exclusive OR of A,B stored in C
```

GLOL supports 38 commands which fill all image processing categories. The structure of GLOL has been detailed by Preston (1971)
in a paper illustrating the use of this language in performing cellular logic transforms.

3.6 PPL

PPL (Picture Processing Language) has two forms, namely, PICAP
version I and PICAP version II. In this chapter we discuss version
II only as it is the language of choice and is being used to command the PICAP II computer system recently completed by the University of Linkoeping. The PICAP II machine has 16 512x512 one-bit-
per-pixel image memories. It has a basic cycle time of 10 milliseconds. This machine can perform both logical transforms and integer convolutions (using a 4x4 window) and comprises a modern and
flexible system. The exclusive or in PPL-PICAP version II is given
as follows:

```
PROGRAM EXOR
BPICT P1,P2(128,128)        Dimension
BEGIN
P1=XOR(P1,P2)               Do EXOR
DISP(P1,4)                  Display Result (on TV monitor 4)
END
```

This programming language supports 32 commands which are concentrated primarily in the utility area. The language is particularly strong in branching and conditional manipulations. Commands
for geometric manipulation and image transforms are written as
chains of commands for arithmetic operations. There are 9 commands
in the image measurement area and none in the decision theoretic
area.

3.7 Other Languages

Languages for DIP (Delft Image Processor), FLIP (Flexible
Image Processor), GOP (General Operator Processor), and SYMPATI
(Systeme Multi-Processeur Apaté Traitement Image) are not presently
documented. The language INTRAC, which commands the GOP (General
Operator Processor) of the University of Linkoeping is mentioned
in a publication by Granlund (1980). The languages which command

the IP of the Hitachi Central Research Laboratory and the PPP (Parallel Pattern Processor) of the Toshiba Research and Development Center are documented in companion chapters in this book. Specific examples of IP code are given for thresholding and filtering operations (but not for the exclusive OR).

This chapter does not treat the many parallel processing languages that run on general purpose computers which are listed in Table 2 as these have been covered to a large extent in a previous paper by Preston (1980A).

4.   EXAMPLES OF USE

Color Plates 9 through 11* illustrate the use of a parallel processing language (SUPRPIC) for image analysis. The original three-color image is of fatty human liver tissue digitized at 512x 512 pixels using the Automatic Light Microscope Scanner-Model 2- of the Jet Propulsion Laboratory.

In visually processing this image the pathologist spends a few seconds on the field of view presented (0.5x0.5mm). In order to locate all cellular components (cell nuclei) in the image and determine their architectural arrangement by exoskeletonizing (Color Plate 10) approximately 2 billion instructions must be executed using a general purpose computer. Another 8 billion instructions are required to extract fatty cells, vessels, sinusoids, and to construct size histograms of the cellular components and the tiles of the exoskeleton (Color Plate 11). These 10 billion instructions (1600 lines of SUPRPIC code) require several hours of CPU time. This makes machine image analysis totally impractical on even one of the most modern general purpose machines.

Using CLIP4 or the Cytocomputer, for example, which are the highest speed cellular logic machines in existence, only a few seconds, approximately, would be needed to complete the same analysis. This dramatic speed increase indicates how modern parallel processors will make image processing feasible on a large and economic scale.

5.   CONCLUSIONS

As is obvious from the above discussion there is a great diversity in parallel processing languages; as much diversity as in the machines which they command. A careful review of about 50 of

*The color plates will be found following page 46.

these languages discloses certain common command categories which
are given in Section 2. A complete language requires commands in
all categories in order to permit the user to utilize the command
structure directly in short, powerful programs without coding detail subroutines to perform command functions which are not directly
available.

Some languages are extremely powerful and are of an APL-like
nature. (An example is GLOL.) Other languages, such as CAP4 are
close to assembly language in their structure and high-level commands may evolve by writing assembly language "macros" using chains
of assembler commands. Some languages have deliberate roots in
FORTRAN (an example is DAP FORTRAN) while others indicate entirely
new directions. (See Fu, 1978.)

To a large extent each language is strongly machine-dependent
in order that the full capabilities of the hardware are utilized.
It is evident that no effort is presently being made to coordinate
command mnemonics. This is unfortunate in that it makes the transfer of image processing software impossible. If there is a general
recommendation that results from the discussion presented in this
chapter it is that more attention be given to standardization of
mnemonics so that software transfer will be possible to some extent.
Otherwise the many man-years of coding which go into applications
command code for array processing machines will be lost.

6.   REFERENCES

Akin, O., and Reddy, R., "Knowledge Acquisition for Image Under-
    standing Research," Comput. Graph. Image Proc. 6:307-334 (1977).

Alexander, I., Brunel University (personal communication).

Asada, H., Tabata, M., Kidode, M., and Watanabe, S., "New Image
    Processing Hardwares and Their Applications to Industrial
    Automation," in Imaging Applications for Automated Industrial
    Inspection and Assembly, Vol. 182, Soc. Photo-Opt. Instr.
    Engrs. (1979), pp. 14-21.

Balston, J., Plessey Electronic Systems Research (personal communi-
    cation).

Basille, J. L. (verbal presentation), CNR Workshop on New Computer
    Architectures for Image Processing, Ischia (June 1980).

Batchelor, B. G., University of Southampton (personal communication).

Castleman, K. R., Digital Image Processing, Englewood Cliffs,
    Prentice Hall (1979), pp. 401-411.

Center for Mathematical Morphology, Lantuejoul, C. (personal com-
    munication).

Computer Sciences Corp., DIMES Users Handbook (1973).

Duff, M. J. B., "Geometrical Analysis of Image Parts," in Digital
    Image Processing and Analysis (Simon, J. C. and Rosenfeld, A.,
    eds.), Leyden, Noordhoff (1977).

Dunham, R. G., Line, B. R., and Johnston, G. S., "A Comprehensive
    System for Producing Functional Maps," Proc. 7th Symp. Comput.
    Prog. Tech. (1978).

Electro-Magnetic Systems Laboratory, The PECOS System (1973).

Eriksson, O., Holmquist, J., Bengstsson, E., and Mordin, B. "CELLO--
    An Interactive Image Analysis System," Proc. DEC Users Society,
    Copenhagen (1978).

Fu, K. S., "Special Computer Architectures for Pattern Recognition
    and Image Processing--An Overview," Proc. Nat'l Comput. Conf.
    (1978), pp. 1003-1013.

Gemmar, P., "FLIP: A Multiprocessor System with Flexible Structure
    for Image Processing," in Computer Architectures for Image
    Processing (Levialdi, S., ed.), in preparation.

Gerritson, F. A., and Monhemius, R. D., "Evaluation of the Delft
    Image Processor DIP-1," in Computer Architectures for Image
    Processing (Levialdi, S., ed.), in preparation.

Goddard Space Flight Center, "Small Interactive Image Processing
    System: Users Manual," (1973).

Granlund, G. H., "An Architecture of a Picture Processor Using a
    Parallel General Operator," Proc. 4th Internat. Joint Conf.
    Pattern Recog., Kyoto (Nov. 1978).

IBM Federal Systems Division, Users Guide, Earth Resources Inter-
    active Processing System (1972).

Johnson Space Flight Center, Users Guide and Software Documentation
    for the Algorithm Simulation Test and Evaluation Program (1973).

Johnston, E. G., "The PAX II Picture Processing System," in Picture
    Processing and Psychopictorics (Lipkin, B. S. and Rosenfeld,
    A., eds.), New York, Academic Press (1970).

Joyce Loebl, Inc., Image Processing Library, Programmers Manual
    (Jan. 1979).

Krevy, R. H., Deveau, L. A., Alpert, N. M., and Brownell, G. L.,
    "PL/S: A Higher Level Language for Image Processing," Phys.
    Res. Lab., Mass. Gen. Hosp. (1977).

Kruse, B., "A Parallel Picture Processing Machine," IEEE Trans.
    Comput. C-22:1075 (1973).

Kulpa, Z., Institute for Biocybernetics (personal communication).

Levialdi, S. (see chapter, this book).

Logica Ltd., Redstone, P. (personal communication).

Marshall Space Flight Center, Numerical Analysis and Digital Com-
    puter Processing of Pictorial Imagery (1973).

Pape, A. E., and Truitt, D. L., "The Earth Resources Interactive
    Processing System (ERIPS) Image Data Access Method (IDAM),"
    Symp. Mach. Proc. Remotely Sensed Data, Purdue Univ. (1976).

Paton, K., Medical Research Council, England (personal communica-
    tion).

Preston, K., Jr., "Feature Extraction by Golay Hexagonal Pattern
    Transforms," IEEE Trans. Comput. C-20:1007-1014 (1971).

Preston, K., Jr., "Image Manipulative Languages: A Preliminary
    Survey," in Pattern Recognition in Practice (Gelsema, E. S.,
    ed.), Amsterdam, North-Holland (1980A).

Preston, K., Jr., "Interactive System for Medical Image Processing,"
    in Real-Time Medical Image Processing (Onoe, M., Preston, K.,
    Jr., and Rosenfeld, A., eds.), New York, Plenum Press (1980B).

Purdue University, LARSYS Users Manual (1973).

Rutovitz, D., Medical Research Council, Scotland (personal communi-
    cation).

Sternberg, S. (see chapter, this book).

Taylor, C. J., Manchester University (personal communication).

Toriwaki, J-i., Shiomi, Y., and Fukumura, T., "On the Subroutine
    Library for Image Processing SLIP," Tech. Comm. Pattern Recog.
    Learning (PRL78-69), Inst. Elect. Comm. Engrs. Japan (Jan.
    1979)(in Japanese).

Uhr, L., "A Language for Parallel Processing of Arrays Embedded in
    Pascal," Comput. Sci. Tech. Rpt. #365, Univ. of Wisconsin
    (Sept. 1979).

University College London, Image Processing Group, CAP4 Programmers
    Manual (1977).

University of Kansas, "KANDIDATS: Kansas Digital Image Data System"
    (1971).

Vrolijk, H., University of Leyden (personal communication).

REAL TIME REGION ANALYSIS FOR IMAGE DATA

Y. Fukada

Central Research Laboratory, Mitsubishi Electric Corp.

Amagasaki, JAPAN

1. INTRODUCTION

An important and difficult problem in image processing is to
segment an image into several different kinds of regions. Unless
we isolate regions, we cannot perform further processing for
pattern recognition, scene analysis, image interpretation, etc.
Besides, if the segmentation phase can be done at high speed, this
reduces the total processing time greatly.

Usually we may assume that measurements on pixels (picture
elements) in one region are similar and those in different regions
are dissimilar to each other. In such a situation, we can adopt
a clustering technique to solve the segmentation problem, because
it forms clusters by grouping similar species. Many clustering
algorithms have been developed to date (see, for example, Duda and
Hart, 1973), but almost all of them are inapplicable to those
objects which have a large number of species to be partitioned or
to those whose cluster sizes differ greatly as is the case with
images. Ingenious spatial clustering methods have been applied to
images by Haralick and Kelly (1969) taking into account the coordinate information of the pixels. However, they do not guarantee
the appropriateness of the merging distance in measurement space
with which they merge pixels. Moreover, it is time-consuming to
search for pixels that connect some region spatially and must be
selected for merger one by one.

We have developed an algorithm which searches for kernel candidate vectors of regions and then finds kernels by clustering those
candidates with some proper merging distance for a given criterion.

We then classify all pixels in an image by comparing their distances from these kernels. Two simple characteristic parameters, mean and variance, are used in finding kernel candidates. We find that candidates to be clustered are few. Since, after finding the kernels, we classify pixels using only these kernels the number of comparisons can be greatly reduced.

The above algorithm has been implemented in FORTRAN, assembly language, and micro-program. FORTRAN is used only to control the flow of procedures, i.e., what routine to call next, how many times to call routines, etc. In order to achieve high-speed execution, every single-purpose routine has been implemented in assembly language or micro-program, especially those routines that are time-consuming or are called many times. Thus we have demonstrated the efficiency of firmwarization for picture processing.

2. REGION SEGMENTATION

This section provides basic assumptions and definitions. It then describes the spatial clustering algorithm in detail.

2.1 Properties of Images and Regions

In order to clarify our basic viewpoint, we define some key words which appear frequently in this chapter and enumerate the hypotheses about the images and the regions with which our algorithm deals.

Definitions:

Pixel   : A picture element which is denoted by a n-tuple vector $X = (x_1, x_2, \ldots, x_n)^T$.

Feature space   : An n-dimensional space spanned by n components of X's.

Subimage   : A spatially connected group of pixels of an image.

Region   : A subimage which satisfies hypotheses H-2 and H-3 (below).

Subregion   : A subimage which is contained inside some region spatially.

Boundary region: A subimage which consists of at least two different regions.

Dispersion   : Trace (summation of the diagonal com-
                    ponents) of a covariance matrix of
                    the X's.

Hypotheses:

H-1   : Feature space is isotropic.

H-2   : The dispersion of any subregion is less
                    than the dispersion of any boundary
                    region.

H-3   : Every subregion in the same region has
                    the same probability density function
                    of X in the feature space.

A simple example showing boundary regions, subregions, subimages, and regions is shown diagrammatically in Figure 1.

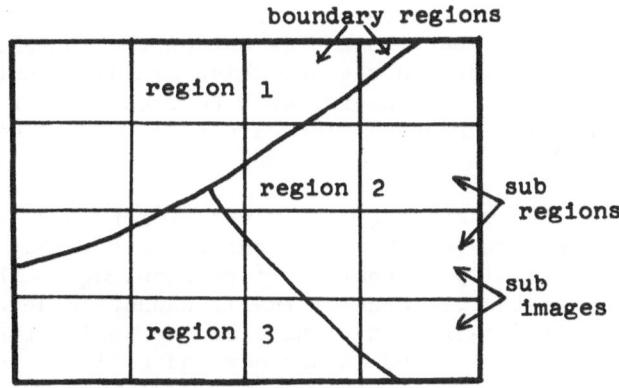

Fig. 1. Example of image analysis format.

2.2 Spatial Clustering Algorithm

Since we know neither the positions nor the extents of the
regions in the image, we initially divide the given image into subimages of appropriate size (Figure 1). We now have both subregions
and boundary regions. If we can distinguish only the subregions
among the subimages, we can find the kernels of the regions by
averaging the pixel measurements over these subregions. Unfortunately we cannot always do this. Therefore, we judge each subimage
as to whether or not it is a subregion as follows: If the dispersion of the pixel measurements is less than some threshold θ it is
a subregion; otherwise, it is not.

As a result of this judgement those subimages that have dispersions less than θ are misjudged to be subregions, even if in reality they are boundary regions. The mean vectors of these misjudged subimages differ from the mean vectors of primary regions. We can derive the largest (worst) deviation δ which may be caused by θ using the method of Fukada (1978). The quantity δ is a monotone increasing function of θ because the mean vector of a subimage deviates more and more from the correct mean vector of the region if it contains many pixels belonging to different regions; in such a case the dispersion becomes larger.

Next, we consider only those subimages whose dispersions are less than θ. Among the M different regions in the image, we denote the two regions that have a nearest distance a_0 in the feature space as Region 1 and Region 2. Since the mean vectors are distributed within δ from the correct mean vectors of regions, then the distances between any two mean vectors within some region are less than 2δ and the distances between any two mean vectors in region i and region j ($i{\neq}j$) are greater than $a_0-2\delta$. (See Figure 2.) Therefore, if $2\delta < (a_0-2\delta)$ we can merge them without risk.

The above considerations suggest the following. When θ is in a suitable range, the mean vectors are merged correctly with the proper merging distance, but, once θ is out of the suitable range, the mean vectors belonging to regions which are different are merged with one another. Therefore, as θ becomes greater, the number of clusters becomes greater. Thereafter the number of clusters becomes smaller.

Although the mean vectors of the clusters represent the regions reasonably well, we cannot distinguish whether they differ from each other in reality without *a priori* knowledge. The degree of inter-regional distance, large or small, cannot be determined without any real-world criterion. Therefore we assign the number of regions that we want to find in advance and call this number C.

The algorithm which we use consists of the following steps:

Step 1:   Divide the given image into square, non-overlapping subimages with the proper minimum size for the system considered. Calculate the dispersion and the mean vector of every subimage.

Step 2:   Adopt some dispersion value θ and merge the mean vectors of the subimages whose dispersions are less than θ with the proper merging distance d_0. Exercise this operation from the smallest to the largest dispersion of the subimages one after another. When the largest number of

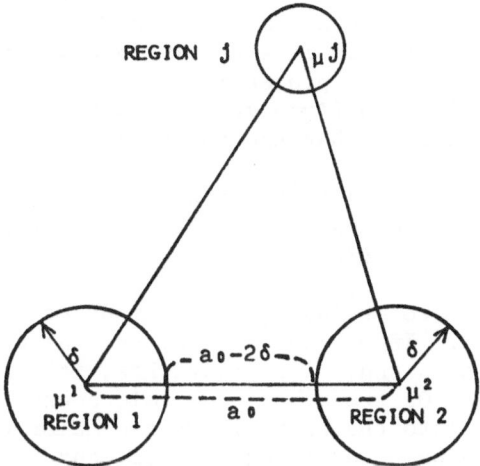

Fig. 2. The range of mean vectors for three regions.

clusters is obtained at θ_0, we call the mean
vectors of these clusters the kernel candidate vectors.

Step 3:   Merge the kernel candidate vectors so as to
be less than or equal to C according to the
distances between them. Calculate the mean
vectors of the merged vectors. Call these mean
vectors the kernels.

Step 4:   Classify every pixel based on the distances
between this pixel and the kernels. In this
operation, we also allow an unknown region.
If the minimum distance between some pixel and
the kernels is greater than the shortest distance between kernels, we classify this pixel
as belonging to the unknown region.

3.   PICTURE PROCESSING FIRMWARE

This section describes and compares methods for algorithm
implementation using a medium scale computer, the MELCOM-COSMO 500.

3.1   Characteristics of Firmware

Algorithms can be implemented in the following ways: (1) software (higher level programming languages), (2) firmware (microprogramming technique), (3) hardware (special purpose processors).

Implementation with software and maintenance of software are very easy. Unfortunately software is severely inferior to the other methods of implementation in processing speed. Concerning the other two methods, although firmware is less efficient in processing speed than hardware, it has the following advantages:

(1) Firmware is a program written with micro-instructions. It is comparatively easy to develop and maintain; it is flexible.

(2) In picture processing, various kinds of algorithms are needed. If we have as many special hardware processors as there are algorithms, it will be too expensive. On the other hand, micro-programs can be loaded into the Writable Control Storage (WCS) by overlaying. We can execute as many micro-programs as we want using only one or, at most, a few WCS's.

3.2   Support Software

In developing and using Picture Processing Firmware (PPF) the following problems should be addressed as discussed by Hirayama (1976):

(1) The interface between PPF and user programs written in higher-level languages should be simple; users should not have to use complicated procedures when they want to link PPF to their programs.

(2) Users should be free from controlling the WCS; in order to achieve high speed processing, each PPF has to share or overlay the WCS. It is troublesome and inefficient for users to control the WCS by themselves.

3.2.1   *Relocatable Object Module Generator*

The relocatable object module generator takes the micro-program object code and several input parameters and generates a standard relocatable object module. In consequence, we can call and link any PPF subroutines like other FORTRAN or assembler subroutines and create executable modules easily. Once they are registered in the program library, FORTRAN users can call PPF subroutines as follows:

(1)   Subroutine subprogram

        CALL PPF1(X1,X2,...,Xn)

(2)   Function subprogram

        X = PPF2(X1,X2,...,Xm)

3.2.2 *Micro Monitor*

Since the WCS is one of the resources of a computer system, many computer users can utilize the WCS simultaneously in multi-job mode. In this situation, one cannot expect to load the PPF into the WCS only once, because the PPF may be destroyed at any time. Although loading the PPF every time it is called is simple, the overhead time is not negligible.

In order to control the WCS efficiently, a Micro Monitor has been developed. Under the control of Micro Monitor, we need not check to see whether the PPF required to execute our program exists in the WCS. Also it is not necessary to reload micro-programs into the WCS when they are already loaded.

3.3 Evaluation of Picture Processing Firmware

The principal reasons for the efficiency of micro-programs compared to FORTRAN or assembler programs are diagrammed in Table 1. We have developed many micro-programs and have applied them to several examples. Taking those which are used in region analysis Table 2 shows a comparison of the execution times of PPF subroutines with FORTRAN or assembler using the same computer facilities. A two times to twenty times speed increase is typical.

4.   APPLICATION TO IMAGE DATA

We have applied our algorithm to real image data such as land-use maps, aerial photographs, fabric designs, LSI masks, etc. This section demonstrates results of the land-use map analyses.

Color Plate 12* illustrates part of the land-use map of Tsuchiura in the Kanto district in Japan published by Geographical Survey Institute. Each region in a land use map is distinguished by color and/or hatching. Usually information is also included in the form

*The color plates will be found following page 46.

Table 1 - The Reasons for PPF Efficiency (MELCOM-COSMO 500)

	FORTRAN or Assembler	Micro-Program
Instruction Operand fetch	0.8 µsec (Main Memory request)	0.3 µsec (WCS request)
Register	Few (especially index registers)	Many
Execution	Sequential INSTRUCTION-FETCH (MM) DECODE OPERAND-FETCH (MM) ALU-operation	Parallel MICRO- INSTRUCTION ALU- FETCH operation

of characters, numbers, special symbols, contours, etc. This presents an obstacle to our algorithm, because we assume that pixels in the same region have similar vectors.

Color Plates 13 and 14 show the results of clustering. The former is the result in the case C=4; the latter, for C=5. The regions 1, 2, 3 in Color Plate 12 are recognized as one region for C=4 (Color Plate 13) because they have similar colors. For C=5 the farthest region among them (region 2) is distinguished (Color Plate 14).

Color Plate 15 shows similar results for an aerial photograph for C=6. The mean vectors of the regions are presented in Table 3. The distinguishing colors for each region are given in Table 4. Colors assigned in this case have no specific meaning, since we assigned them at random for the purpose of presenting the results of image segmentation.

Since we use only simple parameters, our algorithm is very efficient. For the aerial photograph, it takes about 50 seconds to process the picture totally. The image has a 256x256 pixel format and is stored in the external disc memory. The size of the subimages employed is 8x8.

Table 2 - Comparisons of Execution Times (MELCOM-COSMO 500)

Routine	Description (image size or number of pixels)	SOFTWARE Assembler: A FORTRAN: F	FIRMWARE	SPEED-UP RATIO
HISTFW	1024 point histogram	16.1 msec (A)	5.1 msec	3.2
		90.0 msec (F)		17.6
PAVSIG	Histogram range 0 ~ 255	15.2 msec (A)	4.3 msec	3.5
SORTFF	Sort random variables (512 point)	5943 msec (A)	2255 msec	2.6
ISLND4	Find 4-connected regions 128x128 (8 islands, 6656 point)	1620 msec (F)	131 msec	12.4
CLSFWB	Classify pixels given kernels 64x64 (3 dimensions, 5 classes)	1673 msec (A)	365 msec	4.6
PACK8	1024 words (word → byte packing)	11.8 msec (A)	1.8 msec	6.6
UNPAC8	1024 bytes (byte → word unpacking)	12.0 msec (A)	2.1 msec	5.7
NOISEB	128x128 (1638 point) noise removal	206 msec (A)	110 msec	1.9
EXPNDB	128x128 (4096 point) enlarge image	503 msec (A)	250 msec	2.0

Table 3 – The Mean Vectors of Regions

Region	1	2	3	4	5	6
blue	92	96	99	75	167	176
green	85	101	95	90	111	120
red	78	95	81	80	109	129

Table 4 – Distinguished Regions in Aerial Photograph

white	blue, white and brown roofs and roads
yellow	red roofs
purple	grass lands and fields
cyan	bare fields
green	forest
red	shadows of houses and forest
black	unknown regions

5.   CONCLUSIONS

We have developed a spatial clustering algorithm for image data using the Euclidian distance in feature space as the measure of the pairwise similarity of pixels. The distance threshold used depends upon a pre-assigned criterion, i.e., the number of regions. The algorithm is useful and efficient, because it uses only simple statistical parameters, means and dispersions, as features. Moreover, since our algorithm has been implemented in micro-programs and assembly language, we have been able to realize real-time processing for region analysis.

6.   ACKNOWLEDGEMENTS

The author thanks Messrs. A. Oouchi, Mitsubishi Research Institute, M. Hirayama, S. Ikebata and H. Nakajima, Mitsubishi Electric Corporation, for their kind support, fruitful suggestions, and encouragement.

This research has been supported by the Pattern Information Processing System Project of the Agency of Industrial Science and Technology of the Ministry of International Trade and Industry of Japan.

7.   REFERENCES

Duda, R. O. and Hart, P. E., Pattern Classification and Scene Analysis, New York, Wiley (1973).

Fukada, Y., "Spatial Clustering Procedures for Region Analysis," Proc. 4th Int. Joint Conf. on Pattern Recog. (1978).

Fukunaga, K., Introduction to Statistical Pattern Recognition, New York, Academic Press (1972).

Haralick, R. M. and Kelly, G. L., "Pattern Recognition with Measurement Space and Spatial Space for Multiple Images," Proc. IEEE 57(4):654-665 (1969).

Hirayama, M., "Microprogram Support Softwares on a Medium Scale Computer," Inform. Proc. Soc. Japan CA 23-2 (1976) (in Japanese).

Knuth, D. E., The Art of Computer Programming, Vol. 3, Boston, Addison-Wesley (1973).

HYBRID IMAGE PROCESSING USING A SIMPLE OPTICAL TECHNIQUE

Y. Ichioka and S. Kawata

Faculty of Engineering, Osaka University

Suita, Osaka, JAPAN

1. INTRODUCTION

A digital image processor can carry out arithmetic operations and nonlinear operations with noise-free processing but takes much computer time to calculate the two-dimensional (2-D) convolution often used for image processing. On the other hand, an optical processor can do the 2-D convolution at very high speeds but not with great arithmetic precision. Division is especially difficult by means of optical methods. The fundamentals of hybrid processing presented in this chapter make good use of the capability for high-speed 2-D convolution (2-D low pass filtering) by an incoherent optical system in the course of digital processing.

Hybrid processing causes 2-D convolution to be carried out by a simple optical technique and arithmetic operations and nonlinear processing by digital methods. The features of such hybrid image processing are: (1) the total processing time can be greatly shortened as compared with pure digital methods and highly precise processing is attainable and (2) interactive image processing, which requires the iterative calculations of many 2-D convolutions, is realizable efficiently because the 2-D convolution for different processing parameters can be performed by an optical method in a short time.

To perform hybrid processing, no special optics are needed except for the imaging system attached to the TV camera, which is the primary image input system of the interactive image processing system. With the proper arrangement of the vidicon camera and the CRT monitor for real-time display as a feedback loop and use of a simple optical technique one can attain such hybrid processing efficiently. For 2-D convolution by an optical system one obtains the defocused version of the input image to be processed or the intermediate processed result displayed on the CRT monitor. Such a defocused image serves as the reference or standard signal of an arithmetic operation in subsequent digital processing.

In this chapter, we describe briefly the interactive image processing system which we have developed (Kawata et al., 1978) and then show the applicability of such hybrid image processing to experiments of (1) adaptive binarization as described by Tokumitsu et al. (1978), (2) contrast improvement and feature extraction, and (3) constant variance enhancement of dislocation lines in the electron micrograph images (Kawata et al., 1979).

2.   INTERACTIVE IMAGE PROCESSING SYSTEM

A small, low-cost interactive image processing system using a minicomputer has been developed by Kawata et al. (1978). Figure 1 shows the block diagram of the system developed. The system consists of the minicomputer (CEC Model 555H, with 20K-word memory) and a refreshed CRT monitor for real-time display, a vidicon camera, self-scanning solid-state image sensors, and interfaces designed and constructed by the authors. This system does not contain an external buffer memory to store one display frame but the main computer memory is parallelly connected to the CPU, the input system, and the CRT monitor by direct memory access (DMA).

The system has two inputs. One is the high-speed system which includes the self-scanning solid-state image sensors and interfaces. Image data acquired are directly transferred to the main memory through the DMA channel. The other input consists of a vidicon camera and its interface with data acquisition made under program control during the timed memory access of the CPU. The output system is a CRT monitor for real-time display to which the content of the main memory is directly transferred by using the DMA data transfer facility which permits display of the visible image without flicker.

The main operational feature of the system developed is the time-shared operation of three peripheral systems, i.e., the image input system, the CPU, and the CRT monitor. Figure 2 provides a timing diagram showing the relative timing between the operation of the main memory and the peripheral systems. Use of this time-shared operation enables the CRT monitor to display either the image that is being detected or the image being processed.

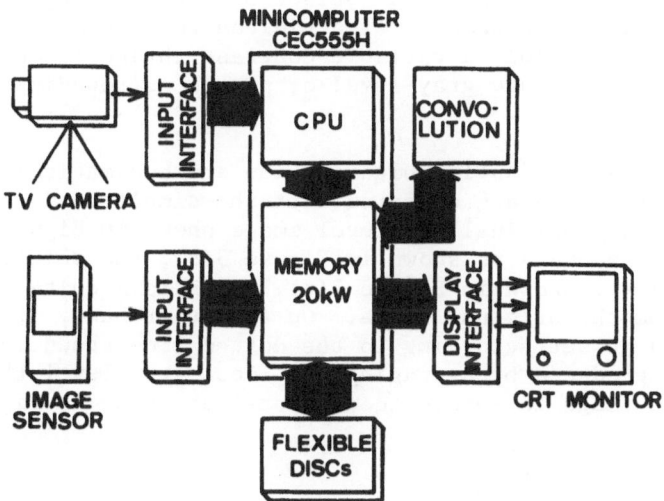

Fig. 1. Block diagram of the interactive image processing system developed.

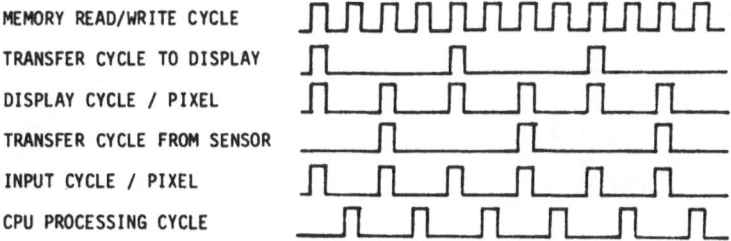

Fig. 2. Timing diagram showing the relative timing between the operation of the main memory and those of the three peripheral systems.

3.   ADAPTIVE BINARIZATION USING HYBRID PROCESSING

Adaptive binarization is used, as described by Andrews (1976), to binarize a multi-level image of an original two-level which is regionally contaminated by dust, nonisotropic processes, illumination conditions, or the shading of a vidicon tube in a TV camera. In adaptive binarization, a variable-contrast threshold must be determined as the average gray-level of pixels surrounding the pixel examined.

The features of adaptive binarization are schematically illustrated in Figure 3 and are compared with the simple hard clipping technique. If the original two-level image shown in Figure 3(a) is detected or recorded as shown in Figure 3(b), the result is the hard clipped image shown in Figure 3(c). This is a quite different appearance from the original image. On the other hand, if an adaptive threshold is set according to the dotted line shown in Figure 3(b) which is produced by averaging the local gray-level the original two-level image is recovered correctly as shown in Figure 3(d).

3.1 Mathematical Analysis

The equation describing adaptive binarization is straightforward. Let $f(x)$ and $\bar{f}(x)$ be the input two-level image to be processed and the spatially weighted local average of $f(x)$, respectively. For the 1-D case the image processed by adaptive binarization is given by

$$g(x) = \varepsilon/2 \ \{ \ 1 + \text{sgn} \ [\ f(x) - \bar{f}(x) \ ] \ \} \ , \tag{1}$$

where

$$\bar{f}(x) = f(x)*h(x),$$

$$\text{sgn}(x) = \begin{cases} 1 & \text{for } x>0 \\ 0 & \text{for } x=0 \\ -1 & \text{for } x<0, \end{cases}$$

and $\varepsilon=1$ for x=0 and $\varepsilon=2$ for x=0, $h(x)$ is the spatially weighting function, and * denotes convolution. The input image to be processed, $f(x)$, is generally represented by the original two-level image, $s(x)$, superposed by contamination, $c(x)$, that is,

$$f(x) = s(x) + c(x). \tag{2}$$

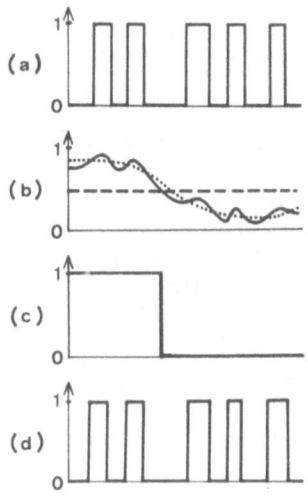

Fig. 3. Steps illustrating adaptive binarization: (a) the original two-level image; (b) the input image degraded by contamination (solid line), constant threshold (dashed line), and variable contrast threshold (dotted line); (c) the hard clipped image; (d) the image processed by adaptive binarization.

Then

$$f(x) = s(x) + c(x),\qquad\qquad(3)$$

where Equation (3) is comprised of the spatial local averages of $s(x)$ and $c(x)$. If $c(x)$ contains lower spatial frequencies than those contained in $s(x)$, $c(x)$ and $c(x)$ can be replaced by the constant value c_k inside a small area centered on the kth pixel. Using this replacement and equations (2) and (3), equation (1) becomes

$$g(x) = \varepsilon/2 \{ 1 + \text{sgn} [s(x) - s(x)] \} .\qquad\qquad(4)$$

As shown by Tokumitsu et al. (1978) equation (4) also holds in the case that contamination shows multiplicative characteristics.

3.2 Optical Implementation

When adaptive binarization is carried out by hybrid processing, an optical processor is applied to obtain the 2-D variable-contrast threshold, which is a defocused version of the input image. Data acquisition of the defocused image is made by defocusing the imaging system on a vidicon camera. This simple optical processing permits $c(x)$ and $c(x)$ in equations (2) and (3) to be replaced automatically by c_k with the help of convolution in the space domain, because, under the geometrical approximation, the PSF for the defocused system may be uniform irradiance within a disk-like area. Binarization is carried out by subtracting this variable-contrast threshold from the corresponding pixels in the input image stored previously by means of digital techniques. The main advantage of adaptive binarization using hybrid processing is the capability of

rapid interactive image processing by simply readjusting the variable-contrast threshold by changing the defocus value of the
imaging system.

Figure 4 shows the experimental verification of the usefulness of hybrid processing for adaptive binarization. Figure 4(a)
is a low contrast Siemens star contaminated by a nonisotropic
process due both to development and to shading of the vidicon tube
during the data detecting process. Figure 4(b) is the hard clipped
image of Figure 4(a). The regional contrast irregularity affects
the processed result. Figure 4(c) is the sampled version of the
defocused image of Figure 4(a). Figure 4(d) is the result processed by adaptive binarization. Details over the entire image
are now clearly recovered and enhanced. Figure 4(e) is the processed image with pepper and salt noise removed by digital processing.

4.   CONTRAST IMPROVEMENT AND FEATURE EXTRACTION BY HYBRID PRO-
     CESSING

Feature extraction is one important technique in image processing. From the point of view of practical use, it is significant to extract fine structures in the input image in which there
is a large regional dynamic range and a large local contrast variation. The technique of hybrid processing described in Section 3
can be also applied to feature extraction of fine detail. Two
examples of the extraction of dislocation lines in electron micrograph images are presented here.

4.1 High Voltage Electron Microscope Images

The high voltage electron microscope (HVEM) whose beam accele-
rating voltages are over 1MV has the capability of focusing upon
fine structures inside thick specimens. Observation of the three-
dimensional distribution of the dislocation lines existing in
thick specimens of a metal is an important subject in metallurgy.
However, if the thick specimen is a crystal which has been bent,
the contrast of the observed image is locally variable and dislocation lines in the brighter background regions are imaged with
very low contrast. This is because the Bragg diffraction condition
varied over the specimen due to bending of the crystal lattice.
The variation of the contrast of the dislocation lines in such a
situation is schematically depicted in Figure 5.

To extract the dislocation lines from such an electron micrograph image in which there is a large regional variation in gray
level, a simple processing technique is utilized. Consider the

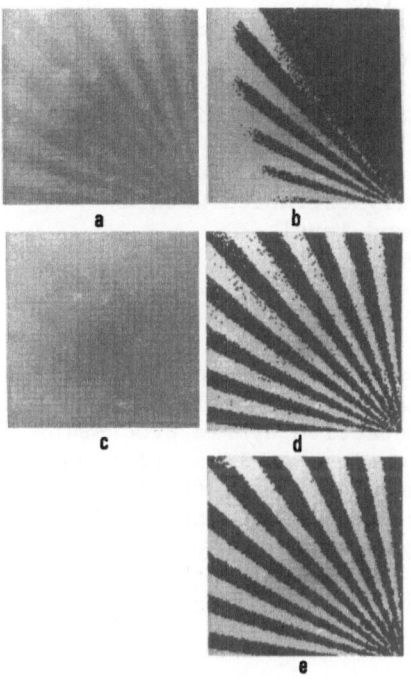

Fig. 4. Experimental results of adaptive binarization: (a) the low-contrast input image to be processed; (b) the hard clipped image; (c) the defocused version of (a); (d) the image processed by adaptive binarization; and (e) the image with pepper and salt noise removed.

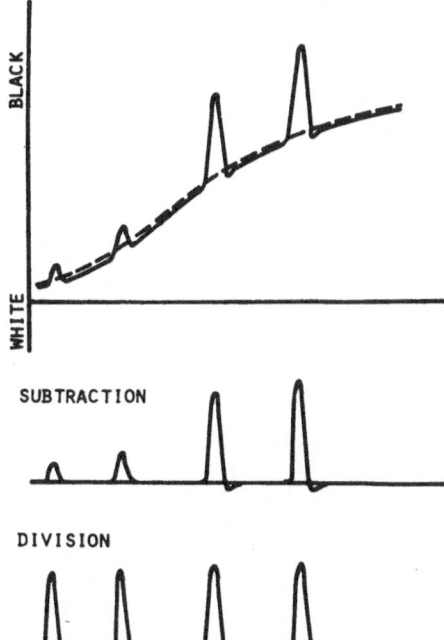

Fig. 5. Steps illustrating feature extraction of fine structures in a variable gray level image; (upper) the image to be processed (solid line) and the local average (dashed line); (middle) feature extraction obtained by subtraction of the local average from the input image; (bottom) fine structures obtained by dividing the input image by the local average.

local average of the input image shown by the broken line in Figure 5 which is easily obtained by low pass filtering. Subtraction
of this local average from the input image can extract fine details
over an entire image. This situation is shown in Figure 5(b).
Even in Figure 5(b) the contrast of extracted features is different.
To equalize the contrast simple nonlinear processing can be applied
in which the input image is divided by the local average. By this
processing, fine details can be successfully extracted with a
moderate contrast in the constant background level.

4.2   Extraction of Dislocation Lines

     Processing can be done, of course, by digital processing alone
but it is a time consuming task to calculate the variable-contrast ,
threshold of the input image. Use of interactive hybrid processing
can reduce the total processing time extremely. As in Section 3,
the local average is given by the defocused version of the input
image using the optical method. As before the input image and the
defocused version of it are separately acquired into the computer
main memory or external auxiliary memory. Then, using this stored
data, the simple arithmetic operations presented above are attained
by the digital method.

     Figure 6 shows experimental results in extracting the dislo-
cation lines in the electron micrograph image of a bent crystal of
aluminum by hybrid processing. Figure 6(a) is the sampled, input
image (192x128 pixels) in which there are two different contrast
areas. Dislocation lines in bright background are hardly recognizable. Figure 6(b) is the local average of the input image which
is obtained by defocusing the imaging system. Figure 6(c) shows
the processed image obtained by subtracting the defocused version
of Figure 6(b) from the input image shown in Figure 6(a). Figure
6(d) is the processed image which is division of the input image
by the defocused version. Either arithmetic operation gives a
reasonable result in which the dislocation lines distributed over
an entire image are extracted and enhanced.

5.   CONSTANT VARIANCE ENHANCEMENT BY HYBRID PROCESSING

     Constant variance enhancement (CVE) as described by Harris
(1977) is another useful processing technique for image enhancement which increases the information extractable by the human
visual system. The main feature of the technique is that it can
efficiently extract information having high spatial frequency components in the image in which there is a large regional dynamic
range coupled with a large local contrast variation. The formulation describing CVE is straightforward and for 1-D case it is given
by

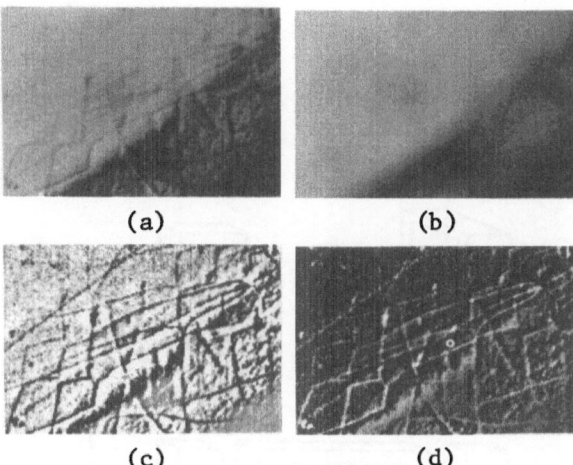

(a)   (b)

(c)   (d)

Fig. 6. Experimental results of feature extraction by hybrid processing: (a) the input image to be processed; (b) the defocused
image of (a); (c) the image processed by subtracting (b) from (a);
and (d) the image processed by dividing (a) by (b).

$$CVE = \frac{f(x) - \bar{f}(x)}{\{[\ f(x) - \bar{f}(x)\ ]^2 *h(x)\ \}^{1/2}} \tag{5}$$

The numerator of equation (5) shows the image processed by high-
pass filtering in the spatial frequency domain. The denominator
represents the local standard deviation of the signals to be enhanced.

Hybrid processing using the interactive image processing system together with simple optical techniques facilitate the execution of CVE. It becomes an especially useful processing method
when a vidicon camera and a CRT monitor are arranged in a feedback
loop as shown in Figure 7.

The steps carried out for this implementation of CVE by hybrid
processing are as follows:

Step 1:   The input image is read by a primary image
            detecting system (a vidicon camera and inter-
            face) into the main memory of the computer and
            stored on a flexible disc. The vidicon camera
            is rearranged to be able to detect the CRT
            monitor display.

Step 2:   The input image is immediately displayed on
            the CRT monitor as a visible image.

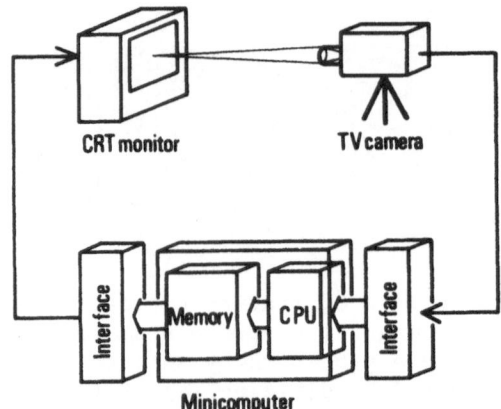

Fig. 7. Arrangement of a vidicon camera and a CRT monitor in a feedback loop.

Step 3:   The displayed input image is detected by the
            vidicon camera whose image system is properly
            defocused. Thus the defocused version $f(x)$ is
            acquired into the computer memory.

Step 4:   Using the computer $f(x)-f(x)$ is calculated
            and the answer is also stored on a flexible
            disc.

Step 5:   The power of $[f(x)-f(x)]$ is calculated by
            the computer and the result is immediately dis-
            played on the CRT monitor as a visible image.

Step 6:   The displayed image $[f(x)-f(x)]^2$ is detected
            again by the vidicon camera which keeps the
            same defocus condition as in Step 3. The signal
            acquired into the computer main memory is now
            $\{f(x)-f(x)\}^2 *h(x)$. The square root of this
            quantity is calculated, which is a local stan-
            dard deviation of signals to be enhanced.

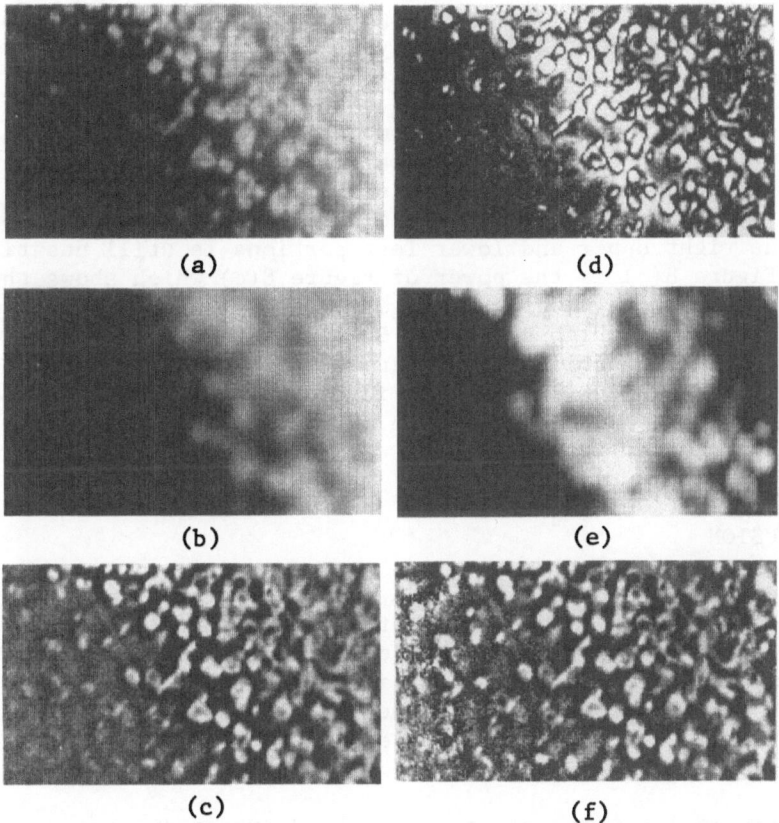

(a)   (d)

(b)   (e)

(c)   (f)

Fig. 8. Experimental results of CVE by hybrid processing: (a) the image to be processed; (b) the defocused image of (a); (c) the intermediate processed result, which is the input image minus the defocused image; (d) the power of (c); (e) the defocused version of (d); and (e) the final result of CVE processing.

Step 7:   Dividing the signal $f(x)-\bar{f}(x)$ stored on the flexible disc by $[f(x)-\bar{f}(x)]^2*h(x)$, CVE is calculated. This result is immediately displayed on the CRT monitor as a visible image.

Step 8:   The image displayed is observed by an operator. If he judges that the processed result is a reasonable one, the operation is completed. If not, the operation can be started again at Step 2. This is repeated until a satisfactory result is obtained.

Figure 8 shows experimental results describing the effectiveness of hybrid processing for CVE. Figure 8(a) shows the sampled

version of an electron micrograph image of a crystal of aluminum.
The contrast of the image is locally variable with fine structure
distributed over a large regional dynamic range, which includes
three background areas with different contrast. Figure 8(b) is
the defocused version of Figure 8(a) which shows the local average
of the input image. Figure 8(c) is the intermediate processed result which is the original image minus the local average. Fine
structures over the entire image are clearly enhanced but the contrast in the right upper and lower left portions is still unsatis-
factory. Figure 8(d) is the power of Figure 8(c) which shows the
variance from the local average. Figure 8(e) is the defocused version of Figure 8(d), which shows the local variance of signals to
be enhanced in the input image. Figure 8(f) is the final processed
result by CVE, in which fine details in the original image are successfully recovered and enhanced with a moderate contrast over the
entire results image.

6.   CONCLUSION

Hybrid image processing combines the speed of optical data
processing with the precision of digital processing. The applicability of hybrid processing has been confirmed through experiments
demonstrating adaptive binarization, image enhancement, and feature
extraction. Simple procedures are used. A specific example is the
constant variance enhancement of the dislocation lines in electron
micrographs.

The principal advantages of such hybrid processing are economy
of the total processing time, ease of data handling, the low cost.
Complicated optical systems and delicate optical techniques are not
necessary. The hybrid processing methods presented here are especially useful for interactive image processing in which iterative
two-dimensional convolution calculations are required.

7.   REFERENCES

Andrews, H. C., "Monochrome Digital Image Enhancement," Appl. Opt.
    15:495-503 (1976).

Harris, J. L., Sr., "Constant Variance Enhancement: A Digital
    Processing Technique," Appl. Opt. 16:1268-1271 (1977).

Kawata, S., Ichioka, Y., and Suzuki, T., "Hybrid Image Processing
    for Constant Variance Enhancement," Opt. Acta 26:1549-1556
    (1979).

Kawata, S., Ichioka, Y., and Suzuki, T., "Man-Machine Conversational Image-Processing System," J. Phys. E: Sci. Instrum.
11:1191-1194 (1978).

Tokumitsu, J., Kawata, S., Ichioka, Y., and Suzuki, T., "Adaptive
Binarization Using a Hybrid Image Processing System," Appl.
Opt. 17:2655-2657 (1978).

FOCUS OF ATTENTION IN THE ANALYSIS OF COMPLEX PICTURES SUCH AS

AERIAL PHOTOGRAPHS

M. Nagao

Department of Electrical Engineering

Kyoto University, Kyoto, JAPAN

1.   INTRODUCTION

The analysis of aerial photographs has intrinsic problems which are not encountered in many other areas of photograph analysis, such as medical and industrial applications. The problems are that the size of the photographs are very large and that we can do nothing to achieve better conditions when photographing the surface of the earth. Photography can be controlled in medical and industrial applications and some others, but it is impossible in the case of aerial photographs. Another problem in aerial photograph analysis is that, as there are so many different situations on the surface of the earth, it seems hardly possible to establish models which describe the state of the earth's surface.

The techniques adopted in the analysis of aerial photographs in the past did not attempt to recognize the existing structures of the earth's surface. They handled each sample point of the aerial photograph by its spectral intensity only and classified the point into categories whose reference spectral intensity values are obtained from the same photograph. The classification was usually done by the maximum likelihood method assuming a normal distribution for the samples in each category. By this method each point is classified by its spectral intensity alone and no consideration is given to its spatial relations with neighboring points.

However, there have been in the past few years some efforts to utilize spatial relations for aerial photograph analysis, among which texture appears to be the most hopeful and successful property, at least up to the present moment. Another direction of these

185

efforts is to consider spatial structures in aerial photographs.
Lines are extracted and roads, rivers, and coastlines are detected
by the help of spatial relations with other components, as well as
their spectral intensity information. City blocks, houses, and
many smaller objects, such as cars and ships, are also detected.
Sometimes map information is utilized as a powerful knowledge
source to assist the analysis. These approaches are very interesting and seem to be very hopeful for the analysis of complex
aerial photographs. But the problem remains that the methods are
much too complicated, take too much time to execute, and we still
do not know whether we can include all possible structural situations of the earth's surface in a computer program.

2.   SELECTIVE AREA ANALYSIS

      Because of the above-mentioned complexities and difficulties
of describing and analyzing aerial photographs, we have to change
our point of view. That is, we do not always need to analyze all
the areas in a photograph, but need to analyze and detect only some
specific areas which interest us. We are very often interested in
specific fields of crops, big roads, residential areas, and so on,
but not in other areas. In fact, in aerial photographs there are
many parts which are very difficult to classify into particular
known categories. For example, there may be an area consisting
partly of bare soil, having some wild grasses, somewhat swampy,
and, to some extent used as a dump area for waste materials. It is
very difficult to classify areas of this kind into definite categories. If we are not particularly interested in these specific
types of areas, we need not analyze them.

      The areas we are interested in do not comprise the whole area
of a photograph, but some small portions of it. If we can easily
and rapidly limit the areas of interest to some small portions of a
photograph, we can concentrate on the detailed structural analysis
of these areas. We may then be able to apply very sophisticated,
time-consuming artificial intelligence techniques to the analysis
and detection of targets and objects in these regions.

      Target detection cannot be done well if we only utilize
knowledge about the target itself. We have to utilize information
about the environment in which the target is embedded. For example,
it is very difficult to detect cars by the simple knowledge that
they are rectangular without knowing that they are on the roads or
in the parking lots. Many other objects exist which look rectangular when seen from above. Therefore the detection of cars will
entail the recognition of roads and parking lots. Environmental
knowledge of this kind makes it easy and reliable to recognize the
detected rectangles as cars. There are many other examples of this

kind. Ships will be found on the water surface, rivers may have crossing bridges, houses may have roads and gardens adjacent to them, and so on.

2.1 Use of Context in Target Detection

When we examine the existing conditions of the targets in detail, we can very often find some contextual information which is helpful for the detection of these targets. Syntactic and semantic constraints of this sort can be described to some extent for each object in a picture. The problem is how to confine the analysis to very small restricted areas wherein we are able to apply complicated syntactic and semantic analysis processes to detect the objects which we want to obtain. This is in a sense the problem of focusing attention which we as human beings ordinarily do in the course of object recognition.

Human perception of a scene is very complex. It has not been made clear how perception functions, what one sees in a picture, and how one understands the whole picture. It is almost certain that one carries out a very quick trial and error process, starting from the detection of gross prominent features and then analyzing details, using one's knowledge of the world. The process of noticing gross prominent portions of a photograph is a kind of simulation of the human behavior of "focus of attention" on interesting regions. This mechanism is especially useful for the analysis of complex pictures such as aerial photographs, where we do not need to analyze all the areas of a photograph, but to get the details of the interesting areas only.

In the process of focusing attention we usually do not use any particular syntactic and semantic information about the targets at all, because at first glance we do not know what is in the picture. When the analysis stage comes down to detail, knowledge about the objects in the picture will be utilized and the interpretation of the picture components will be performed according to this knowledge. In using this top-down approach, our interest centers on the question of what detailed stages we can go into without particular knowledge of the targets present. Another problem concerns the interactions between the stages of analysis with and without knowledge, and to what extent feedback analysis is required in the system. It will be true that feedback paths are necessary, at least from the systems viewpoint. But it is not good to rely too much on feedback paths because the feedback process wastes time and it is therefore uneconomical. Small local feedback loops will be much better than a large feedback loop which includes many stages. Using small feedback process small errors are corrected at each stage, the reliability of each stage becomes higher, and almost no feedback processes using large loops are necessary.

2.2 Planning Methods for Focusing Attention

The attempt to "plan" for the detailed analysis may be regarded as one of the first trials of the principle focusing of attention. The original picture is reduced N times horizontally and vertically. Line extraction is performed on this reduced picture, and the extracted areas defined by significant lines are regarded as reference areas where detailed algorithms are applied to find detailed and exact lines. In a companion chapter, Rosenfeld describes the pyramid method which may be regarded as an extension of the idea of planning. The reduced pictures obtained by shrinking N times with N = 2, 3, 4, ... are a sequence of averaged pictures of the original. The objects in the original are blurred in the reduced pictures. The strong edge features in the original are blurred and, sometimes, the edge detection fails.

To overcome this problem we have developed a method called "edge preserving smoothing" which has the effect of smoothing flat areas and sharpening slopes. The principle is to apply a rectangular mask to a point and to compute the variance of the intensity within that mask area. This is done for all directions from the point as the center. We find the rectangle which has the smallest variance of intensity change and the average intensity value of this rectangular portion is given to the point as its new value. The operation is applied to all the points of the picture and repeated several times. The result is that the picture quality is largely improved, with the edges becoming sharper and noise being removed from the flat areas.

2.3 Application of Edge Preservation Smoothing

Aerial photographs of complex landscapes, such as Japan, where landuse units are small and complex, need detailed structural and semantic analysis. Even artificial intelligence techniques need to be introduced to get any satisfactory analysis results. The first stage of our approach to aerial photograph analysis is based on the following principles, which are related to focusing attention. They are simple, do not require long analytical processes compared to the planning and pyramid methods and they have the possibility of being easily implemented in parallel hardware.

    Step 1:   Strong and typical features are first
               extracted from the aerial photographs. These
               are primary features of the targets to be
               recognized. These features must be very
               prominent and easy to extract. This stage
               corresponds to the stage of focusing atten-
               tion.

Step 2:   The extracted portions are further analyzed
            in detail using knowledge about the objects.

Step 3:   In Step 2 we have minor feedback loops such
            as the following. Typical objects which are
            prominent are first extracted using default
            parameter values. Next the parameter values
            are actually calculated from the extracted
            objects and the initial default parameter
            values are replaced by the extracted para-
            meter values. Object detection is repeated
            for new objects of the same kind using these
            new parameter values. This process is re-
            peated by modifying the parameter values each
            time until no new objects of the same kind are
            detected. This interaction process works very
            nicely.

We are interested in using this method to find forest areas,
crop and vegetable fields, houses and roads. For the detection of
those typical areas, we adopted the following features to limit
the target areas during the first stage of "focus of attention."

1.   Large areas of uniform smoothness (vegetable
       field, water surface, etc.)

2.   Long thin regions (roads, rivers, railroads,
       etc.)

3.   Some stable color features (green for forests,
       blue for water, etc.)

4.   Heavily textured areas (forest, vegetable
       field, urban area, residential area, etc.)

All of these features, of course, depend on the resolution of
the photographs, but we can generally utilize information about the
resolution and apply it into the analysis algorithms. A simple
threshold operation on the intensity axis can be applied to obtain
large uniform areas. The threshold value can be obtained by checking the intensity histogram of each area. However, if the areas
have texture, this simple-minded operation does not work well. We
then have to apply more sophisticated texture operations. Using
the Fourier transformation and low pass filtering to eliminate texture, we can extract large uniformly textured areas easily. Long
straight lines can also be found by the Fourier transform operation.
In the Fourier domain we may find energy sharply concentrated in
some directions. If so, we keep these portions filtering out other
components and perform the inverse Fourier transform. We then

obtain the approximate areas of straight lines. If we wish to detect curved line areas, we may apply a differentiation operation and
find line segments. These line segments can be connected easily to
form long continuous lines.

To obtain specific spectral intensity areas we can apply thresholding techniques to the red, green, blue and infrared spectral
bands of each picture and perform a logical AND operation to the results. We may also be able to get better region segmentation by
using hue information. Various texture parameters can also be applied to extract heavily textured areas. Fourier transform and
bandpass filtering operations can be applied to get specific frequency textures. A simpler way to compute this is to differentiate
the picture, to extract lines, and then to apply the augment-reduce
operations. We then get areas where short line components exist
very densely. The same operation is also useful for the detection
of a long line from many short line components. It has the effect
of eliminating noise.

3.   CONCLUSIONS

The operations described above are very simple and focus attention very effectively to limited areas in a large aerial photograph
(Nagao et al., 1978). Parallel operations are generally applicable
to many of these operations, especially for preprocessing and extracting gross features which restrict the analysis areas where no
special knowledge concerning targets is required. Examining these
operations, we can see that the parallel operations should include
not only logical and arithmetic operations among neighboring points,
but also some comparison and decision functions among the candidate
values which are obtained from spatial combinations of the values of
the neighboring points. For example, if we want to have the steep-
est slope at a point, we have to calculate the slope values for all
directions and choose the maximum. Another example is edge preserving smoothing as described by Nagao and Matsuyama (1978), where
variances have to be calculated for eight directions at each point.
To perform functions of this kind, each parallel logical element
must be provided with some local memory, and must perform sophisticated operations. That is to say, many powerful processing units
as well as memory capability must be connected in a two-dimensional
parallel manner.

To extend the ability of parallel operations from spatially
uniform operations to locally specific ones, it is very effective
to introduce the concept of a production system. A production
system is a set of condition and action pairs where any pair of
rules will be activated if the situation satisfies the condition.
The picture plane acts as the "blackboard." The realization of

this system in parallel hardware architecture is very difficult
because many powerful central processing units must be connected
in a two-dimensional way. Also the local range for a production
rule may not be restricted to the eight-adjacency. It will be
better to develop a different hardware architecture to realize
such complicated functions as a production system, for example a
high-speed, two-dimensional shift register could be provided and
all the production rules attached to this shift register in logic.
The picture would enter this high-speed shift register many times
repeatedly. In today's high-speed shift register a few tens of
repetitions of the scan of a large picture can be cycled in one
second. The problem here is that we have to investigate whether
a particular analysis process can be handled by a production system.
Even if final stages of some sequential algorithms remain which are
very difficult to write in a production system, the major parts of
the image processing program can be successfully realized in a
production system. However, hardware realization of a production
system which is applicable to image processing remains to be
demonstrated.

A more flexible structure for performing the function which
we call "focus of attention" in parallel is to provide a few
microprocessors which can access a large picture memory independently. Each microprocessor is loaded with a particular feature
extraction function and overall control is handled by a central
control system. By sending different feature extraction algorithms
to different microprocessors as the picture processing proceeds,
central control can manage the whole system most efficiently. The
problem here is to extract tasks which can be performed in parallel,
and to do the dynamic scheduling of assigning these tasks to all
the micro computers.

4.   REFERENCES

Nagao, M. and Matsuyama, T., "Edge Preserving Smoothing," Proc.
    4th Internat'l Joint Conf. Pattern Recog. (1978).

Nagao, M., Matsuyama, T. and Ikeda, Y., "Region Extraction and
    Shape Analysis of Aerial Photographs," Proc. 4th Internat'l
    Joint Conf. Pattern Recog. (1978).

Nagao, M. and Matsuyama, T., A Structural Analysis of Complex
    Aerial Photographs, New York, Plenum Press (1980).

THE VIRTUAL PLANE CONCEPT IN IMAGE PROCESSING

T. Soma, T. Ida, N. Inada and M. Idesawa

The Institute of Physical and Chemical Research

Wako-shi, Saitama 351, JAPAN

1.   INTRODUCTION

Although the storage and retrieval system for image data and the image data processing system itself should be considered as a whole in image processing systems, it is important to be able to write efficient image processing algorithms without having to worry about detailed data handling instructions, especially in handling a large volume of image data which cannot be accommodated in the main storage or even in the high speed auxiliary storage. Providing a method for accessing image data, no matter how advanced in its capability, not only relieves the programmer of the burden of handling image data in writing an efficient program but also provides a method for describing pure algorithms; allowing to standardize the description of image processing algorithms (although there may be some algorithms which depend upon data structures) or allowing improved transportability of the image processing program package. Regarding transportability, the need for interchange of image data will also be the natural consequence. Also the separation of processing from the handling of image data appears to encourage the standardization of image data. Standardization of image processing algorithms and image data format is one of the recent topics in the image processing community as discussed by Onoe and Shirai (1976) and in several workshop memoranda and proceedings (1975-1978).

Table 1 summarizes image data access methods adopted in several image processing systems described by Johnston (1970), Lerman (1972), Shirai et al. (1975), Pape and Truitt (1976), Selzer, Sakurai et al. (1977), and Tamura and Mori (1978). In this chapter, an image data handling algorithm is developed based upon the

Table 1 - Image Data Access Methods

No.	System/Method	Sub-Picture	Buffer Management	Operation	Accessing	Reference
1	PAX		least recently used first out	sequential	virtual array	Johnston (1970)
2	AI-Lab (MIT, ETL)		least recently used first out	tracking	function reference	Lerman (1972)
3	ERIPS-IDAM		look-ahead loading	sequential/tracking	local array	Pape and Truitt (1976)
4	mini-VICAR		double buffer	sequential/tracking	local array	Selzer
5	IPCR		farthest first-out	tracking	function reference	Sakurai et al. (1977)
6	LEXEC		on demand	sequential	local array	Tamuri and Mori (1977)

virtual plane (V-plane) concept; buffer management using virtual
memory with look-ahead swapping capability, providing a means for
the user to control some functions in buffer management to enhance
the performance of some picture processing operations. In Section
2, after introducing the concept of the virtual tape (V-tape), some
capabilities of the V-tape, relevant to succeeding discussions are
explained. In Section 3, the concept of the virtual plane (V-plane)
is developed as a two-dimensional extension of a V-tape. Finally,
in Section 4, the V-plane features of a machine called "FLATS" now
under development at the Institute of Physical and Chemical Research
is described. FLATS is efficiently implemented in FORTRAN to run
both numerical and algebraic programs as originally presented by
Goto et al. (1979).

2.   VIRTUAL TAPE

The V-tape concept of Itano and Goto expresses a much wider
concept than physical tape and is an abstraction of the sequentially
accessed linear array of data, such as stacks, queues, and tapes.
The concept also includes buffering strategies for enhancing performance in a hierarchical memory environment, handling data too
large to reside in the high speed main memory. V-tape consists of
consecutive data cells (tape) and an access mechanism (head) which
moves along the tape via which a datum can be read from or written
into the cell. The V-tape can be modeled as the buffered I/O operations between the main and auxiliary memory, such as magnetic disks
or tapes.

The physical entity of the V-tape is divided into pages each
having a fixed size. The page containing the cell to which the head
is pointing and its neighboring pages are opened in the buffer area
in the main memory. Since the head moves on a step-by-step basis,
the missing page problem can be reduced or eliminated by applying
look-ahead swapping strategy in buffer management. The look-ahead
swapping for the V-tape can be effected by assigning to each page
the distance d from the page on which the head resides. Figure 1
shows the state of consecutive pages just after the head passed the
page boundary B with a swap-in operation requested for the page
assigned d = 1 and swap-out for the page with d = 3. For read-only
pages the swapping-out operation can be skipped. By providing a
buffer accommodating 4 pages as shown in Figure 1, unnecessary swapping operations can be avoided, even when the head moves back and
forth across the boundary B. The V-tape can be expanded or shortened in either direction by assigning a new page from the page pool
or returning the used page to the page pool.

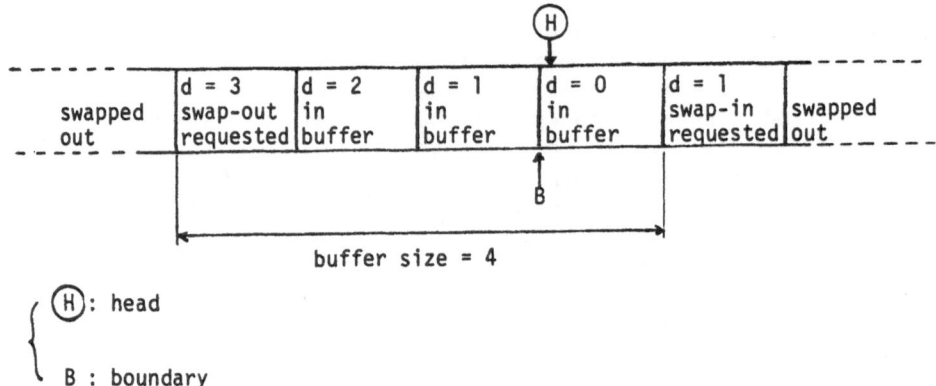

buffer size = 4

$\left\{\begin{array}{l} \text{(H): head} \\ \\ \text{B : boundary} \end{array}\right.$

Fig. 1.   Look-ahead swapping for virtual tape.

Some elaborations to the V-tape as defined above, such as the inclusion of multi-head and marking capability or random access capability within a page, are treated in Itano and Goto (1979) and in Sassa and Goto (1977).

3.   VIRTUAL PLANE

As a two-dimensional extension of the V-tape concept, one can consider the virtual plane (V-plane) to model the picture data handling system. By dividing the picture into sub-pictures of appropriate size (corresponding to the V-tape pages), the distance d between sub-pictures can be defined as, for example, the maximum of d_x and d_y as

$$d = \max(d_x, d_y)$$

where d_x and d_y are the absolute values of the x and y coordinate differences of the representative point in each sub-picture. Figure 2 shows the state of the pages with a buffer of size 16 pages. The state of each page can be represented by a diad (d, r) comprised of distance d from the head-pointed page, and the request number r representing the request conditions as

$$r = \left\{\begin{array}{ll} -1 & \text{swap-out requested} \\ 0 & \text{no request} \\ 1 & \text{swap-in requested.} \end{array}\right.$$

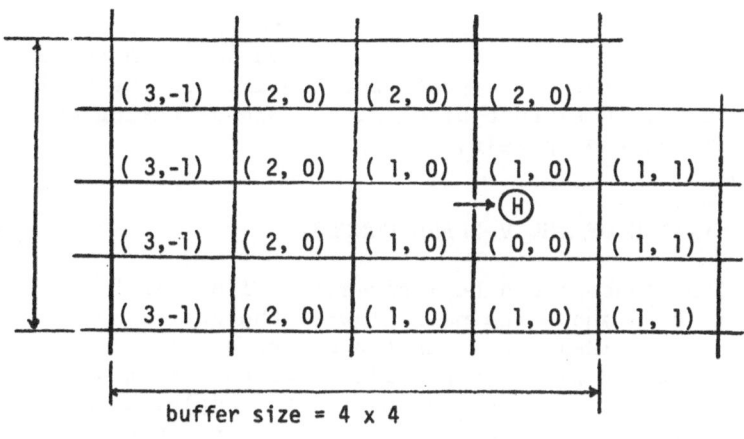

buffer size = 4 x 4

(H): head

Fig. 2. Look-ahead swapping in virtual plane.

The swap-in request is issued for those pages whose distance changes
from 2 to 1, and the swap-out request for those pages whose distance
changes from 2 to 3.

    Unlike the V-tape, several pages fall into the swapping request
state at the same time, allowing a choice in issuing the swapping
request. It should be noted that the order in which the request is
issued can be decisive in determining the efficiency of image processing. Some means must be provided to predict the next head movement from its past movement, and to select the most probable page to
be visited next. The user should be able to specify the order of
swapping requests; for sequential operations unnecessary swapping
can be avoided by restricting the swapping operation to those pages
which are in the direction of sequence. Another situation not found
in the V-tape is the possibility of cutting out the sub-picture
whose axis is not parallel to the axis of the original picture.
Such a non-parallel sub-picture is known to be useful in some tracking operations as mentioned in Shirai et al. (1975).

    According to the V-plane concept, a description of the picture
processing algorithm can be considered to consist of two parts:
those describing the processing for pixels in the processing window;
those specifying the movement of the processing window. The head of
the V-plane corresponds to the processing window and the pixels in
that window can be accessed referring to the representative pixel in
that window. Further, the movement of the window can be defined by
specifying the translation and orientation. It is convenient if the
orientation of the sub-picture is coincident with that of the processing window. It is further convenient if there is some means to
change the orientation of the sub-picture according to the changes

in orientation of the processing window as the process proceeds.
There is a capability in the V-tape to retain pages in the buffer
for future access. The corresponding capability in the V-plane
seems to be useful in picture processing, such as taking correla-
tions with some distant pixels.

4.   IMPLEMENTATION OF THE V-PLANE CONCEPT

The V-plane concept can be implemented either in hardware or
software. Figure 3 shows the main flow of the buffer management of
the V-plane. In FORTRAN, in which the execution of multi-processes
is not allowed, the V-plane cannot always be efficiently implemented,
even though it is logically equivalent; for example, a statement must
be included for each reference of a pixel to check whether the page
which contains the requested pixel is open. Without some hardware
support, it is difficult to realize pixel access times comparable to
that of ordinary memory access.

In what follows we consider the V-plane concept in FLATS which
is under development at our Institute. Table 2 shows the V-plane
features in FLATS Fortran. The declaration specifies the picture
size, the size of sub-pictures and buffers, and a look-ahead swapping

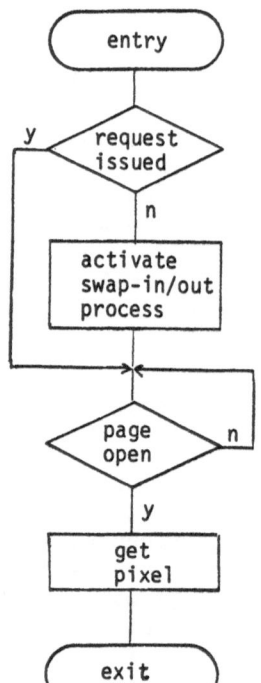

Fig. 3. Buffer management flow
chart for virtual plane.

Table 2 - V-Plane Feature in FLATS Fortran

Declaration

VPLANE VP*n (p_x, p_y, s_x, s_y, b, ℓ)

n : pixel size in bytes

p_x, p_y : plane size

s_x, s_y : sub-picture size

b : buffer size in sub-picture number

ℓ : look-ahead pattern specification

Pixel reference

VP(I, J) = VP(I, J) + ...

Freezing and releasing of sub picture in buffer

CALL VPFIX (VP(I, J))

freeze sub-picture containing pixel VP (I, J)

CALL VPRLS (VP(I, J))

release sub-picture containing pixel VP(I, J)

---

pattern. As stated before, the problem is to realize efficient checks on the page residence. As a minimum requirement for hardware implementation, we imposed the condition that no additional time is allowed for accessing pixels when they are in a buffer.

In FLATS a high-speed bit table is provided for specifying the memory partitions and use is made of this bit table for buffer management of the V-plane, assigning each bit the status of the page (Figure 4). Since the bit table search occurs in parallel to the memory access, the status of the page can be read by sending the page address to the bit table. The page address is obtained by dividing the logical address a_L of the pixel by the factor 2^m

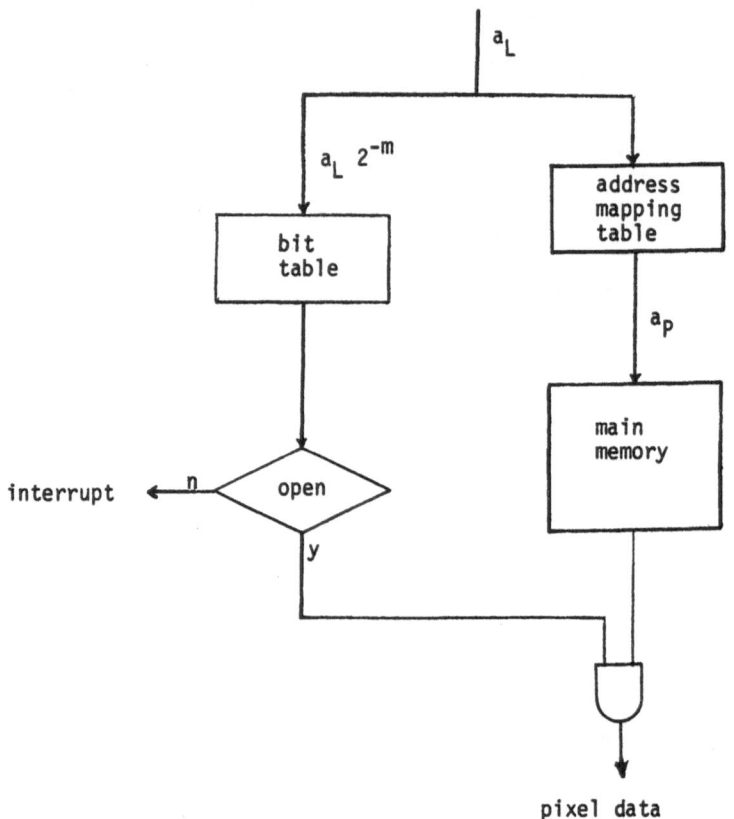

Fig. 4. Use of bit table in virtual tape management.

(assuming the page size has been taken as a multiple of 2^m). If
the page containing the pixel to be accessed is open, the ordinary
memory access proceeds, but if the page is not open, a control signal is generated, which causes an interrupt and an appropriate
action to be taken, by transferring control to the V-plane management routine.

5.   CONCLUSIONS

     The concept of a virtual plane, the two-dimensional extension
of virtual tape, is formulated and shown to be useful in implementing image data access methods for handling large image data bases,
which allows look-ahead swapping by introducing a distance between
sub-pictures. The concept of the V-plane is formulated by introducing a distance between sub-pictures on the V-plane, and is
applied to image data handling. By extracting memory management

information from the V-tape specifications and from associated V-tape handling algorithms, look-ahead swapping between the memory hierarchies is made possible.

A proposal is made for implementing the V-plane concept in our FLATS machine, using a high-speed bit table for buffer management. It is shown that the V-plane concept encourages standardization of both the description of image processing algorithms and image data format.

With the increase in the use of LSI technology, the handling of images of 1M pixels using main memory will become popular. The V-plane concept will still be useful for handling images containing much larger numbers of pixels, in which case the processing speed will be crucial in determining throughput.

6. ACKNOWLEDGEMENTS

The authors would like to acknowledge Dr. E. Goto, Chief Scientist, the Institute of Physical and Chemical Research, for his encouragement and critical comments; they are also grateful to the members of the Working Group on Image Processing of IPSJ (The Information Processing Society of Japan) headed by Dr. M. Onoe, for their stimulating discussions, without which the present chapter would never have come into existence.

7. REFERENCES

Goto, E., Ida, T., and Hiraki, K., "FLATS, A Machine for Numerical, Symbolic and Associative Computing," Proc. IEEE Internat. Symp. Comput. Arch. (1979), pp. 102-110.

Itano, K., and Goto, E., "Realization of a Processor with Virtual Tapes and Its Evaluation," J. Info. Proc. 2:65-71 (1979).

Johnston, E. G., "The PAX II Picture Processing System," In Picture Processing and Psychopictorics (Lipkin, B. S. and Rosenfeld, A., eds.) (1970).

Lerman, J., "Using the Vidisector and the Stored Picture Facility," Vision Flash 27, AI Lab., Mass. Inst. Tech. (1972).

Onoe, M., and Shirai, Y., "Standard Format for Digital Image Data," Rp. 9-1, Working Group on Image Processing, Info. Proc. Soc. Japan (1976).

Pape, A. E., and Truitt, D. L., "The Earth Resources Interactive Processing System (ERIPS) Image Data Access Method (IDAM)," Symp. on Machine Proc. Remotely Sensed Data, Purdue Univ. (1976).

Sakurai, T., Date, M., Kobayashi, K., Watanabe, Y., Soma, T., and Aoki, K., "Pattern Representation and Manipulation in Crystallography," J. Info. Proc. Soc. Japan (Joho-shori), 18: 149–157 (1977) (in Japanese).

Sassa, M., and Goto, E., "V-Tape, A Virtual Memory Oriented Data Type and Its Resource Requirements," Info. Proc. Ltrs. 6:50–55 (1977).

Selzer, R. H., Jet Propulsion Lab., private communication.

Shirai, Y., Suwa, M., Tsukiyama, T., and Kyura, N., "The Stored Picture Facility for Object Recognition," Bul. Elec. Tech. Lab. (1975), pp. 785–805 (in Japanese).

Tamura, H., and Mori, S., "A Data Management System for Manipulating Large Images," Proc. Workshop on Picture Data Description and Management, IEEE Comput. Soc., Catl. No. 77CH1187-4C (1977), pp. 45–54.

Working Group on Image Processing Memoranda, Info. Proc. Soc. Japan (1975–1978).

Workshop on Picture Data Description and Management, IEEE Comput. Soc. Catl. No. 77CH1187-4C (1977).

Workshop on Standards for Image Pattern Recognition, U.S. Nat'l Bur. Stds. Special Publ. 500-8 (1977).

A MULTI-MICROPROCESSOR ARCHITECTURE FOR ASSOCIATIVE PROCESSING OF IMAGE DATA

H. Aiso, K. Sakamura and T. Ichikawa*

Department of Electrical Engineering

Keio University, Keio, JAPAN

1. INTRODUCTION

For the organization of image data retrieval systems, one of the most important features to be realized is a large scale semantic data base operating in a real-time environment. This calls for the development of a high-speed associative processor incorporating recent LSI technology. Therefore, we are developing a multi-microprocessor system designed to satisfy these requirements. This system, called "ARES," was originally described by Ichikawa et al. (1977, 1978). ARES is essentially capable of associating stored data with a query item. Therefore, it is a problem-oriented computer designed for processing pattern or image data.

The necessity for the development of such a problem-oriented computer stems from the following considerations. Taking a simple example from the field of character recognition, let a character be expressed by a 54-bit word, and let the space of the characters be 8K bytes, i.e., the space contains 1170 characters. To recognize a single character by the method of heuristic search, the average time required for association is 1.05×10^5 μs on the NOVA 01 computer and 2.49×10^4 μs on the UNIVAC 1108 computer. However, on-line applications such as the recognition of handwritten characters or associative processing in semantic data bases may demand much faster computers than these conventional ones. We are trying to develop a very high-speed computer which can process data more than 30 times faster than the UNIVAC 1108.

---

*Research and Development Laboratories, Kokusai Denshin and Denwa Co., Ltd.

For applications based on associative processing in semantic
data bases, the heuristic control of the number of associations is
a very important function which is not incorporated in the conventional computers. ARES provides a heuristic search mechanism using
a hard-wired multiple response resolver and an effective associative memory, both controlled by microprograms.

2. PRINCIPLE OF ASSOCIATION

The operation of ARES is based on a new principle of association invented by Ichikawa (1978).

The functional block diagram of ARES is shown in Figure 1. To
look for associations with the stored information, the input data X,
which is composed of p blocks of a code, is fed into the Error Correcting Array through the Search Key Register. At the Error Correcting Array, an error corrected code vector IX is derived from X
by applying an error correction procedure to each block of X separately. Consequently, IX is composed of p blocks of error corrected
code vectors, and is called the index of X for the association.

Suppose that A's are pieces of information, which are quite
similar to X in terms of the Lee distance, stored in the Contents
Array and the corresponding error correcting code vectors of the
A's, denoted as IA's, are assumed to have been obtained previously
and to have been kept in the Index Array.

The blockwise coincidence of IX and the IA's is checked and
then the number of blocks coinciding with each other is counted for
all IA's stored in the Index Array. This is compared with a predetermined value θ. When the value is equal to or greater than θ,
a flag is set at the output of the index cell which contains the
corresponding IA. The corresponding information in the Contents
Array is selected as a candidate for association with the information IX.

However, for most practical applications, there will be a
limit δ on the number of words of associated information, and θ is
adaptively selected at every step of the association process through
trial and error. This requires ARES to be capable of very high-
speed processing of index identification with probably a larger
number of modules than conventional content addressable memories.
Therefore, an all-bits-in-parallel comparison scheme is adopted
for index identification.

In order to control θ, the number of flags appearing at the
outputs of the index cells is counted and compared with δ at the
Multiple Response Resolver. If the number of flags is greater than

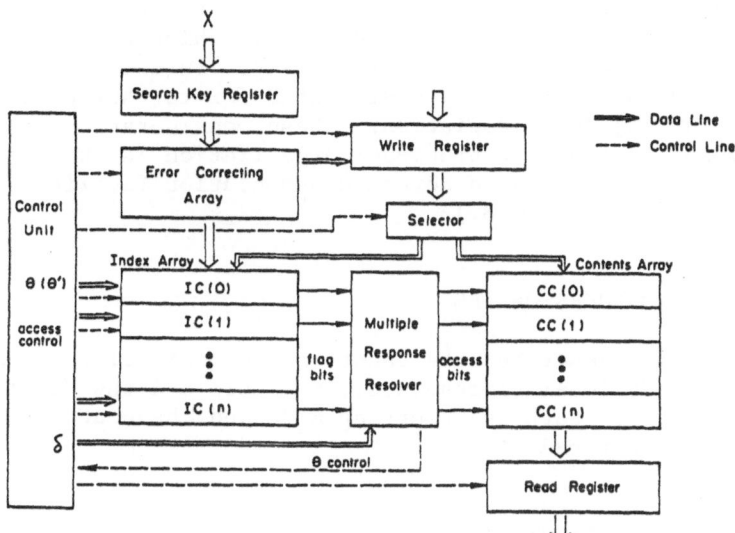

Fig. 1. Functional block diagram of ARES.

δ, i.e., the number of words of information to be associated, θ is increased by one. Otherwise, it is decreased by one. Thus, θ is heuristically selected so that the greatest possible number of flags not exceeding δ is obtained and then the Contents Array is accessed by the flag bits.

The input of new information is controlled by the Control Unit and the information is transferred to both the Index Array and the Contents Array through the Write Register.

3. ASSOCIATIVE PROCESSING IN A SEMANTIC DATA BASE

In order to be able to carry out an associative search on a semantic data base, the following functions should be provided at the computer architecture level:

(1) Hierarchical blurring capability

(2) Fundamental content-addressing capability

(3) Heuristic control mechanism to obtain a reasonable number of associated data items

In principle, these functions can be accomplished by the mechanism which is incorporated in ARES which was originally designed for the recognition of hand-written characters. However, in order

to facilitate repeated blurring and to create blurred data spaces,
it is extremely desirable to provide a logic which works over many
sets of blocks of data words simultaneously. Furthermore, a blurred
data space should be spatially distributed for further parallel execution of blurring and association. These considerations lead to
the conclusion that a multiprocessor organization is the best approach to attain the desired advanced association facility. (See
Ichikawa et al., 1977, 1978.)

4.   MULTI-MICROPROCESSOR ARES

This section describes the functional organization of ARES and
of its major functional blocks, i.e., the Cell and the Multiple
Response Resolver. Both the blurring mode and the content-addressable mode are described.

4.1 Functional Configuration

The functional block diagram of the multiprocessor organization
of ARES is shown in Figure 2.

The host computer gives macro commands to ARES which handles a
semantic data base. ARES is composed of a Master Control, eight
Cluster Controls and sixty-four hierarchically structured Cells.
The Master Control synchronizes the operation of all the logic units
and issues operational instructions to each of the Cluster Controls.
Therefore, up to eight different programs may be processed in paral-
lel. Each Cluster Control has eight logic Cells and enables the
Cells to carry out parallel execution of blurring and association.

Table 1 shows typical instructions provided by ARES. It should
be noticed that the microprograms of the Cluster Control control the
operation of the Cells. The microprograms can be modified by applying a LOAD micro instruction guaranteeing functional flexibility.

4.2 Cell Structure

The internal structure of a Cell is shown in Figure 3, which
is the nucleus of the association logic. The original data item
and its blurred representations are kept in a cell which can be considered as a conventional associative memory.

The registers called Block Size and Word Size indicate the
length of a blurring block and of a data item to be performed, respectively. Since processing is carried out on 16-bit data in the
Cell, the Local Sequencer which is implemented in wired-logic splits

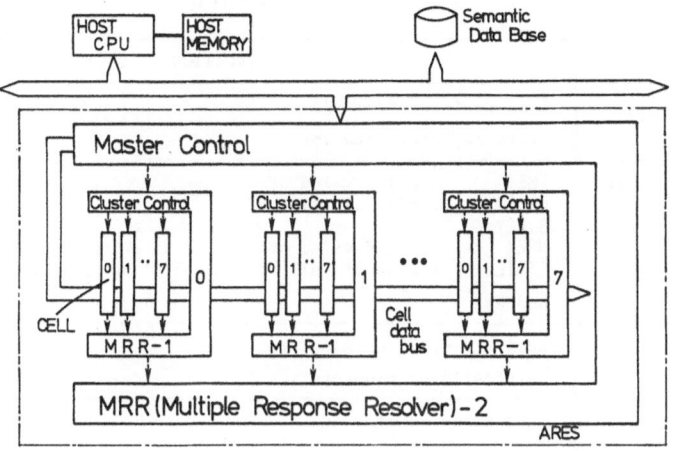

Fig. 2.   Functional configuration of ARES.

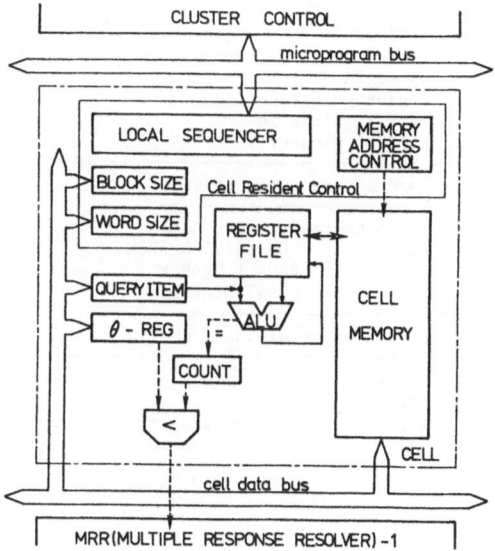

Fig. 3.   Cell structure.

Table 1 - Typical Instructions

MNEMONIC	INSTRUCTIONS
L reg,adr	load register from adr
ST reg,adr	store register into adr
where reg :	THETA initial value of θ DELTA threshold value of δ S.KEY search key MRRi counted value of MRRi $(i = 0,1,...,7)$ STATEi status of cell or cluster i $(i = 0,1,...,7)$
adr :	address
ACT $c_1,...,c_n$	activate cells or clusters
HLT $c_1,...,c_n$	halt cells or clusters
where c_i :	cell id or cluster id
READ HIT reg	read a hitmarked-item
READ FAIL reg	read an unhitmarked-item
where reg :	work register id
DEFINE SPACE s,r	define associative space region
CREATE SPACE s	create associative space
APPEND SPACE s,reg	append contents of work register to associative space
DELETE SPACE s	delete associative space
CLOSE SPACE s	close associative space
LOAD SPACE s,c	load cell memory
ACTIVATE SPACE s	activate object associative subspace for operation
where s : r : reg : c :	space id region specification work register id cell or cluster id
LOAD micro	load microprogram routine
where micro :	microprogram routine name e.g. HEURISTIC SEARCH RANDOM SEARCH COUNT SEARCH
BRANCH adr,cond	branch on condition
where adr : cond :	address condition e.g. COMPi operation at cell or cluster i is completed COUNTi,n MRRi reaches n

the data to be processed into blocks of 16-bit data referring to the Block Size and Word Size registers.

When the Cluster Control initiates the operation of association in the corresponding Cells, the Cells continue to carry out associative processing under the control of the Cell Resident Control until it accepts an interrupt.

4.3 Multiple Response Resolver

The Multiple Response Resolver (MRR) is a parallel counter with multiple inputs and is constructed from the MRR-1's and the MRR-2 as shown in Figure 4. Each MRR-1 receives the hit marks from the Cells attached to the corresponding Cluster Control, and the contents of the MRR-1's are collected by the MRR-2. The output of the MRR-2 is compared with the predetermined value of δ. Based on the result of this comparison, the Master Control decides what operation should take place next.

The output value of the MRR-2 represents the number of data words on which associative processing is being executed, and it is kept in the comparator to be compared with the value of δ until the content-addressing on all the blurred spaces is completed.

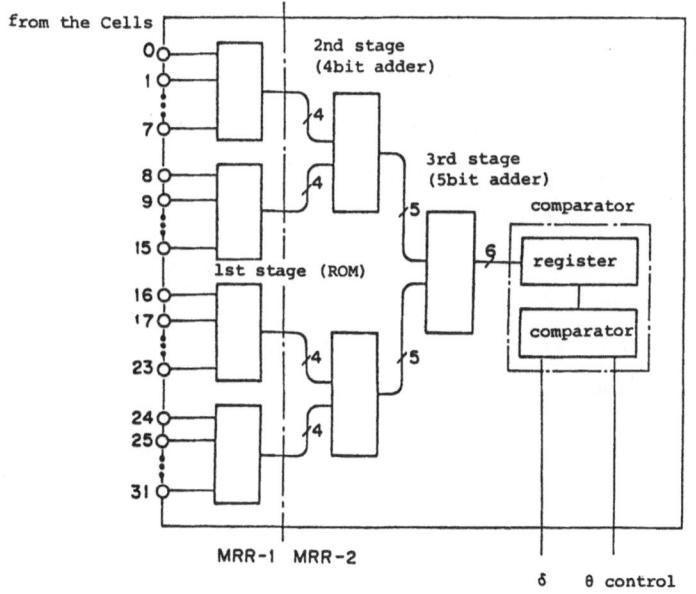

Fig. 4. Multiple Response Resolver.

If the contents of the register exceed the value of δ, then the value of θ is increased immediately by one in order to proceed with further associative processing. When the content-addressing on all the blurred spaces comes to an end, the final number of data with a hit mark is compared with the value of δ. If the number of hit marks is less than the value of δ, the value of θ is lowered by one to perform the associative processing again. The value of θ is controlled in a heuristic way. The following is an ALGOL-like description of the control algorithm in the MRR:

```
begin
    procedure INITIAL (θ);
    begin parallel
        pointer0 := 0;   θreg0 := θ;
                    ⋮
        pointern := 0;   θregn := θ;
    coend INITIAL;

    procedure SEARCH(cell ID,hiti,completei);
    if all data space in the cell have been processed
        then completei := 1
        else begin
            completei := 0; N := 0;
            for j := 1 until Number of blocks in a word
                do begin if Key(pointer) = Query Item
                    then N := N + 1;
                pointer := pointer + 1;
                end;
            hiti := if N>θ then 1 else 0;
        end SEARCH;
```

```
         θ := θ₀;
INIT:     sw := 0;
          hit number := 0;
          if θ>p then goto UNRECOGNIZED;
          INITIAL(θ);
CELL:     begin parallel
               SEARCH(0,hit0,complete0);
                         ⋮
               SEARCH(n-1,hitn-1,completen-1);
          coend;
MRR:      begin parallel
               hit number := hit number + ∑ hiti;
          coend;
          if hit number > δ then
               begin
                    sw := 1;
                    θ := θ + 1;
                    goto INIT;
               end;
          if all cells have been completed then
               begin
                    if hit number = δ then goto END;
                    if hit number < δ then
                         if sw = 1 then goto END
                         else begin
                              sw := 0;
                              θ := θ - 1;
                              goto INIT;
                         end;
               end else
                    goto CELL;
UNRECOGNIZED: unrecognized;
END: end of algorithm;
end;
```

The clustering of cells guarantees high-speed capability, high reliability and functional flexibility; the loss of a single cell function or MRR-1 results in a slight decrease of processing power but not in a complete system shut-down.

4.4 Distribution of Data in a Semantic Data Base

The relationship between the blurred data spaces and the corresponding Cell in the blurring mode is shown in Figure 5. Each blurred data space is divided into subspaces and data items contained in a space are distributed over the cell memories so that they can be processed in parallel. The error correction for blurring is accomplished through the ALU shown in Figure 3 and is controlled by the microprograms in the Cluster Control.

Figure 6 shows the same relationship between the blurred data spaces and the corresponding Cells in the content-addressing mode. A query item and the value of θ are sent in parallel to each Cell by the Master Control. Content-addressing is then performed in the blurred data space Bs* which is distributed over the cell memories.

5. EVALUATION OF THE OPTIMUM NUMBER OF CELLS

As shown in Figures 2 and 4, ARES has 32 Cells. We determined the optimum number of Cells from the results of a simulation of the trade-off between the functional requirements imposed and the cost of implementation. In this simulation, for the sake of simplicity, we excluded the blurring process and limited ourselves to the content-addressing in the blurred data space Bs* and the heuristic determination of θ. Hereinafter, we refer to Bs* as data space V.

We assumed that data space V consisted of 10^3, 10^5, and 10^7 data words, and that each of the data words was constructed from either 23 bits x 9 blocks or 7 bits x 30 blocks. The number of bits in a block, i.e., 23 bits or 7 bits, corresponds to the length of the (23,12,3) Gray code or to the length of the (7,4,1) Hamming code, respectively. The parameters and the corresponding values used for the simulation are listed in Table 2. Furthermore, we assumed that the following conditions are satisfied:

(1)   the bit pattern in the data space has the characteristics of the uniform distribution,

(2)   the number of blocks which exceeds the value of θ is determined by the binomial distribution, and

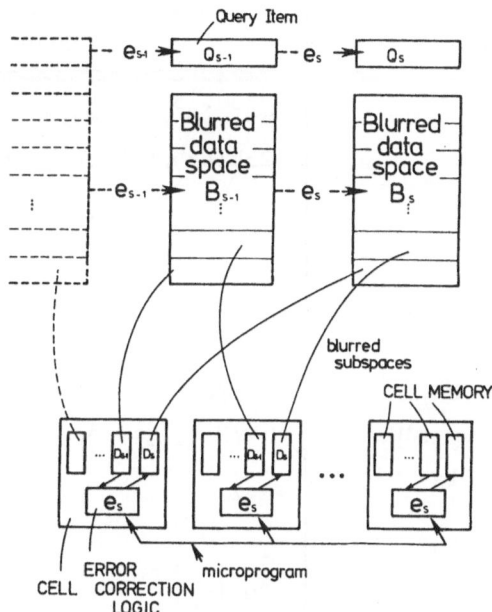

Fig. 5.   ARES in blurring mode.

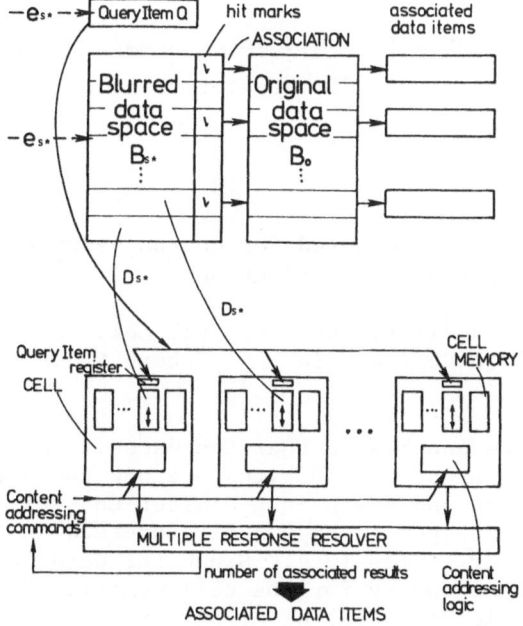

Fig. 6.   ARES in content-addressing mode.

Table 2 – Parameters and Values Used for Simulation

PARAMETERS	VALUE
data space : V	$10^3, 10^5, 10^7$ words
data structure	23bits x 9blocks, 7bits x 30 blocks
upper bounds of number of association output words δ	100
operation unit of cell	16bits
number of cells : N_C	1 - 128
number of cluster : N_{CL}	$\lfloor N_C/8 \rfloor$ *
amount of cell memory : W_C	$\lfloor V/N \rfloor_{256}$ **
memory/processor cycle ratio : R	1.2
initial value of θ : θ_0	1
adapting time for θ : T_{CL1}	20 $[p_c]$ ***
processing time for hit data : T_{CL2}	80 $[p_c]$

* the greatest integer less than or equal to $N_C/8$

** the least multiple of 256 greater than or equal to V/N_C

*** processor cycle

(3) the processing time required for content-addressing is determined by the Pascal distribution.

Since data space V is distributed over the Cell memories, the amount
of data space in each Cell equals V/N_C, where N_C is the total number
of Cells.

In practice, we can assume that the capacity of the Cell memory
can be increased in steps of 256 bytes. Since the bit length of the
ALU in the Cell is 16 and since each Cluster Control manages 8 Cells,
a data word of 23 bits per block can be processed in 2 units x 9
blocks = 18 processor cycles (p_c). Since the word length is 207
bits, 26 bytes are necessary for the cell memory. Therefore, if we
assume that the cycle ratio (R) of memory speed to processing speed
is 1.2, the time required to access a data word in the cell memory,

i.e., the search cycle time (T_C), becomes $18 + 13R = 33.6$ (p_c).
Similarly, for a data word of 7 bits per block, T_C becomes 46.8
(p_c).

For a data space of 10^5 words, we assume that the time required
to obtain the optimal value of θ in the Cluster Control is 20 (p_c)
and that the processing time for a data word with a hit mark is 80
(p_c). Suppose that the initial value of θ is 1, and that the value
changes from 1 to 128, and also suppose that the association is
carried out 100 times for each value of θ.

The total cost (C) of the ARES hardware can be expressed by

$$C = N_C \times (C_C + C_M \times W_C) + N_{CL} \times C_{CL} ,$$

where C_C, C_M and C_{CL} are the hardware cost of the Cells, the Cell
memory and the Cluster Control respectively.

Under these assumptions, we obtained a graph of the products
of cost and average execution time versus number of Cells as shown
in Figure 7. Figure 7 suggests that from a viewpoint of cost performance, the optimum number of Cells is 32. Figure 8 shows how
the optimum number of Cells varies, when the data space changes.
Starting with $V = 10^3$, the broken line represents the case that the
optimum number of Cells is proportional to the number of words of
the data space. For a block consisting of 7 bits, we obtained
almost the same results. As shown in Figure 8, even if the data
space becomes large, the optimum number of Cells remains a constant
between 20 and 40. Therefore, we decided to provide 32 Cells which
are controlled by 4 cluster controls.

From the cost performance point of view, 32 Cells are enough
for large data spaces with contents on the order of 10^7 words (approximately 2×10^2 bits/word). This proves that the ARES multiprocessor configuration is suitable for associative processing of
large sets of data.

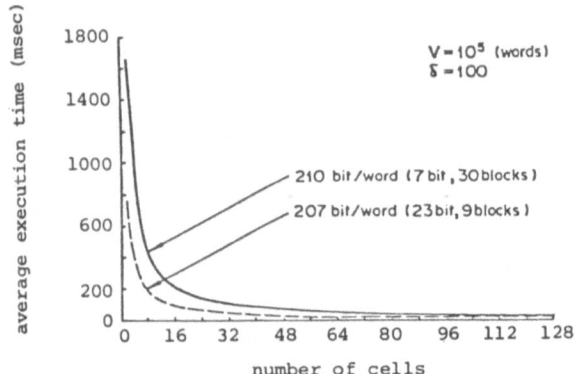

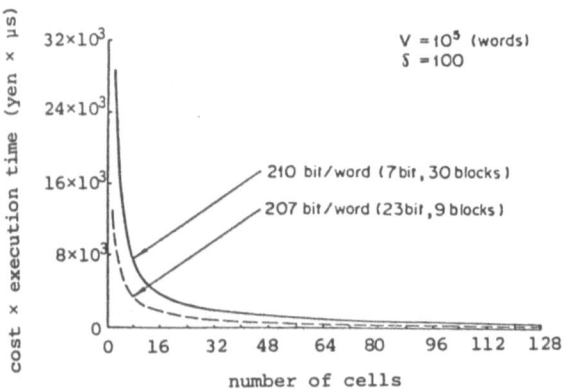

Fig. 7.   Results of simulation study.

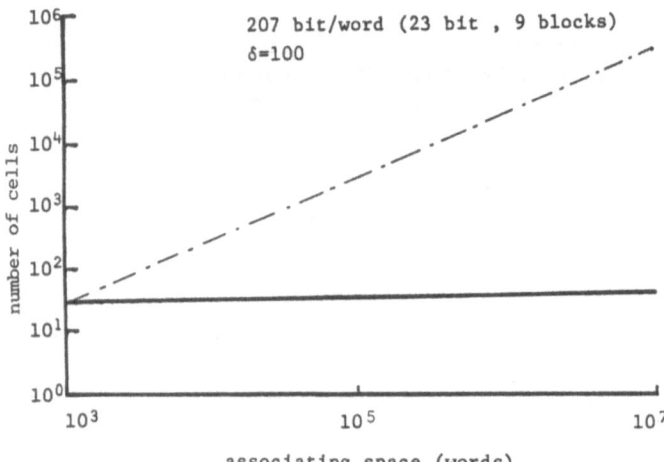

Fig. 8.   Optimum number of cells.

6.   CONCLUSION

In this chapter, we have presented an architectural description of a processor called ARES specially designed for associative processing of pattern or image data. ARES is a multi-microprocessor based on the SIMDS (Single-Instruction-Multiple-Data-Stream) concept. However, it can be used as an MIMDS (Multiple-Instruction-Multiple-Data-Stream) type of processor, because of its capability for parallel processing.

After a brief explanation of the principle of associative processing using ARES, we described the structure of the Cell and the Multiple Response Resolver which is one of the important features incorporated in ARES. We also explained how the data in a semantic data base are processed in parallel in the Cells. Finally, we described the results of a simulation to determine the optimum number of Cells to be implemented from a viewpoint of cost performance. This simulation proved that in most cases encountered in image processing, 32 Cells are sufficient.

It should be noticed that all the complex control mechanisms including the proposed heuristic search, which is carried out through the hardwired Multiple Response Resolver, are to be implemented in firmware. We are now trying to realize the required hardware with relatively inexpensive LSI devices. We believe that ARES is universal enough to be applicable to a wide variety of applications including associative processing of image data.

7.   ACKNOWLEDGEMENTS

We would like to thank Mr. Tatsushi Morokuma and Mr. Hiroyuki Yamaishi for their enthusiastic discussions and assistance.

8.   REFERENCES

Ichikawa, T., Sakamura, K., and Aiso, H., "ARES--A Memory, Capable of Associating Stored Information through Relevancy Estimation," AFIPS Conf. Proc. 46:947-954 (1977).

Ichikawa, T., Sakamura, K., and Aiso, H., "A Multi-microprocessor ARES with Associative Processing Capability on Semantic Data Bases," AFIPS Conf. Proc. 47:1033-1039 (1978).

AN INTERACTIVE IMAGE PROCESSING AND ANALYSIS SYSTEM

S. Hanaki

Central Research Labs, Nippon Electric Co., Ltd.

4-1-1 Miyazaki, Takatsu-ku, Kawasaki-city, JAPAN

1.   INTRODUCTION

Image processing is a computer application which requires tremendous storage capacity and large amounts of processing. Man-machine interactive analysis and processing are preferred to batch processing for more efficient image analysis because of the wide variety of image characteristics. Human judgement of resultant images and human selection of processing algorithms and/or parameters make computer processing very efficient.

Man-machine interactive processing requires image output which is one of the most easily comprehensible data representation forms for human recognition, with a short response time. This class of image processing computer system can be named I^2PAS (Interactive Image Processing and Analysis System). Several commercial or experimental examples of I^2PAS have appeared for LANDSAT image processing and analysis, e.g., the GE/Image 100 and the Bendix/MDAS.

2.   INTERACTIVE IMAGE PROCESSING AND ANALYSIS SYSTEMS

Many efforts have been made in order to improve I^2PAS capabilities. Four processing factors can be specified when discussing I^2PAS improvement: interactive processing, flexibility, speed up, and cost reduction.

2.1 Man-Machine Interaction

Due to the wide variety of image characteristics, human judgement about processing procedures from the interim to the final image
will greatly help to make processing efficient by exchanging and/or
modifying the algorithm and its parameters. Man-machine interactive
processing requires image output with a short response time. Image
output devices require computer-controlled image display.

2.2 Speeding Up Processing

Special purpose hardware for image processing is provided to
support general purpose computers. Some hardware units are hardwired to realize faster processing rates. Pipelining technique,
parallelism, and special addressing schemes have been employed for
processing speed-up.

2.3 Flexibility

Hardwired logic often performs very specialized functions and
is of no use in other types of image processing. Thus, programmable
processors have been built. They usually sacrifice processing speed
in order to obtain processing flexibiliy.

2.4 Cost Reduction

The fundamental logic speed governs everything. The pipeline
technique is subject to logic speed. Another speedup approach,
i.e., parallelism, seems to be less affected by logic speed than
does pipelining. However, serious economy problems appear with
parallelism.

2.5 Summary

Based on these four factors, the following steps are distinguished in the progress of the Nippon Electric Company/Central Research Laboratories (NEC/CRL) image processing facilities.

(1) Application software is developed on a general
     purpose computer to which is attached special
     peripheral devices, such as storage-type or
     simple-refresh-type displays.

(2) Both programmable and hardwired image processors
     are provided. A programmable processor, the

MP-16, was developed and put to use for fingerprint image processing as discussed by Asai (1975). The processor itself has been described by Ishikura (1976).

(3) The T-configuration system is employed, as shown in Figure 1. A major part of the T-configuration consists of a large MOS IC memory to which image I/O devices and special purpose hardwired processors are directly connected. The IC memory is used not only as a refresh memory but also as working bulk memory for image processing.

3. AN I²PAS AT NEC/CRL

This section describes both the color image display and the computer system (hardware and software) developed at NEC for real-time interactive image processing.

3.1 Color Image Display

A block diagram of the color image display is shown in Figure 2. It has two image memory banks, memory-A and memory B. Memory capacities are 512x512x8 bits = 512x256x16 bits = 256K bytes for memory-A and 256x256x8 bits = 64K bytes for memory-B.

Memory-A is used either in direct mode or indirect mode. In the direct mode, memory-A is folded into 512x256x16 bits, which provides two 256x256x16 bit memories. Each 256x256x16 bit memory is large enough to store both an image of 256x256 pixels with red, green and blue (5 bits each) and 1-bit plane for graphic overlay.

In the indirect mode, for 512x512x8 bit memory-A or 256x256x8 bit memory-B, the 8-bit value of each pixel is used for addressing 256 words of an 18-bit-word lookup table, called the color converter. There are two color converters corresponding to memory-A and memory-B, respectively. One word of lookup table is divided into three 6-bit fields, in which 6-bit components, expressing the R,G and B intensity, are stored to drive the CRT monitor through the corresponding DA converters. In order to match the slow memory read cycle time of 1.6μs and the fast pixel data output cycle time of 100ns/pixel for display, parallel data reading is employed with data for 16 pixels being read from memory at one time.

A black-and-white image can also be displayed by concurrently driving the 3 DA converters for R,G and B with single band image data. A masking circuit gates parallel memory output data for each pixel to select a specific parallel bit width.

INPUT DEVICES   OUTPUT DEVICES

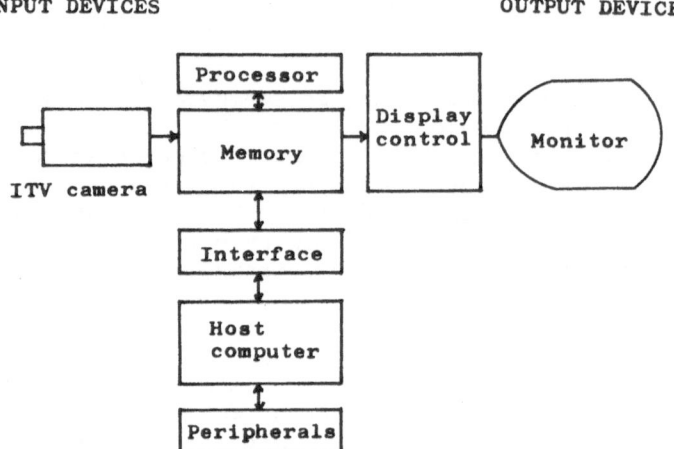

Fig. 1.   Block diagram of the T-configuration system.

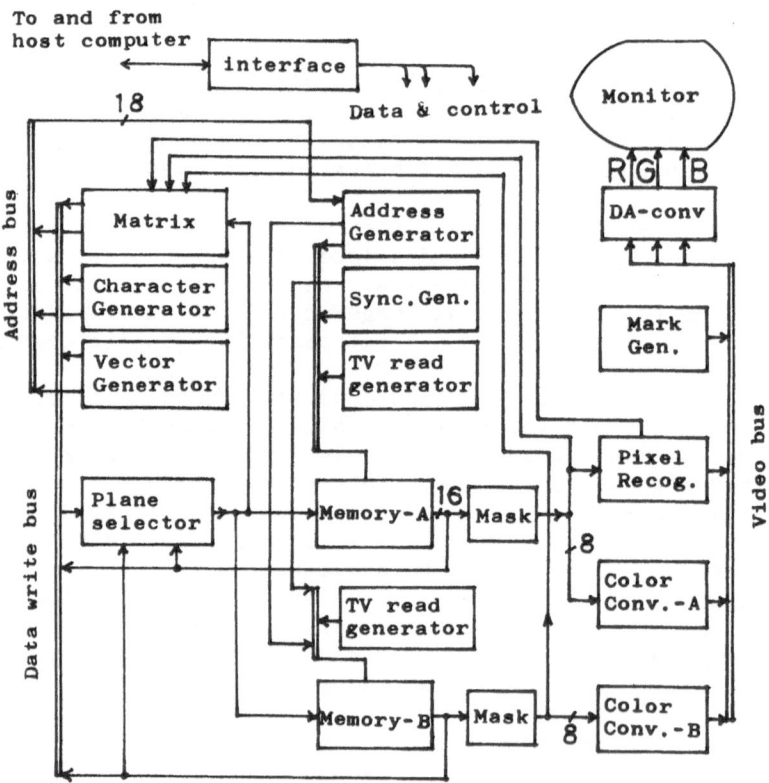

Fig. 2.   Image display system block diagram

When data are written in either memory-A or B, they are supplied via a 16-bit data-write bus through a plane selector which
shifts and/or masks each pixel according to a predetermined control
given by the host computer. Therefore, a memory bank can be sliced
so that several image planes and/or graphic planes are provided.

When image data are read out from either memory-A or B for
display, the data can be modified by hardware logic to obtain a
shifted or zoomed image or to recognize a specific image area and
change its color. Zooming is done by pixel duplication to magnify
the displayed image 2,3,4,..., or 16 times. Recognition logic detects only those pixels whose (R,G,B) components satisfy the following conditions

$$R_1 \leq R \leq R_2, \qquad G_1 \leq G \leq G_2 \qquad \text{and} \qquad B_1 \leq B \leq B_2$$

for predetermined upper and lower limits in each component. These
upper or lower limiting values can be used in an arbitrary combination.

Table lookup operations, shifting, magnification or pixel recognition is carried out in real time (1/30 second for 512x512
image). Both a vector generator and a character generator are provided to write vectors and characters on a graphic plane. Also, a
cursor generator is provided. They support graphic capabilities.

3.2   Computer and Peripherals

The image display is connected to the host computer which is
the NEAC-2200/200. This computer has 64K characters of core memory, 5 magnetic tape units, 2 disk drives, a line printer, a storage display and a CCTV camera. A drum-type color reflection scanner is also available on the other computer.

3.3   Software and Applications

Color analysis software (CIPS-I) was first developed as given
by Tajima (1977). The main CIPS-I programs are listed in Table 1.

Later, the system was expanded to interactively handle LANDSAT
and airborne MSS data by making use of the color image display,
data tablet, and keyboard as described by Naito (1977). A hierarchical arrangement of image data was employed in order to reduce
the effect of disk storage capacity limitations in the interactive
processing.

Table 1 - CIPS-I Basic Programs for Color Analysis

Program name	Description
*IDAVR	Color image input from the TV camera (including averaging process)
*NORC3	Normalization of every primitive color data
*CNVHV	Transformation of data format
*LUMCH	Lightness and chromaticity calculation
*CLDSH	Interactive region analysis on tristimulus values
*CLCHM	Interactive region analysis on lightness and chromaticity
*ATRN2	Classification result printing

In the disk file image sizes of both 803x588 and 256x256 were employed in consideration of airborne MSS data handling. In the disk file, an arbitrarily selected window can be extracted from an 803x588 image to form a 256x256 image. From the disk, a 256x256 image is transferred to the image display memory. Either all of the data or a portion of the data (zoom mode) stored in the display memory can be displayed. An 803x588 image is directly loaded into a disk file from an airborne MSS CCT or is extracted from LANDSAT CCT by arbitrarily specifying the image location in the original LANDSAT image. Besides this, a 256x256 subimage may be extracted directly from a LANDSAT CCT and stored on magnetic tape, which is loaded into a disk file and provides a LANDSAT menu image display to make interaction easier.

Several applications have been investigated using the I^2PAS.

(1) Road extraction from aerial thermal photo-maps (see Naito, 1976).

(2) Color analysis on lightness and chromaticity values (see Tajima, 1977).

(3) LANDSAT image analysis (see Naito, 1978).

(4) Airborne MSS geometric correction, using platform attitude information (see Funo, 1978).

(5) Color image mosiac formation by computer (see Tajima, 1976).

4. FUTURE DEVELOPMENTS

Several additional hardware facilities are under consideration, including (1) enlargement of image at an arbitrary scale, (2) image rotation, (3) histogram display, and (4) profile display.

Among the speedup techniques mentioned in Section 2, pipe-lining is subject to device speed limitations. The other approach to achieving speedup is parallelism, which leads to a serious prob-lem of economy when one tries to provide a large amount of parallel hardware. Thus efforts should be made to reduce unit hardware cost. One promising approach seems to involve developing an LSI processing unit. This may be possible with cost reduction which is occurring in manufacturing customized LSI's.

5. CONCLUSION

The T-configuration, whose major part consists of a large mem-ory to which image I/O devices are directly connected and an image processor controlled by a general purpose computer, works quite efficiently as an I^2PAS.

Further development of processing capabilities is required to achieve flexibility, man-machine interactive processing, speedup, and cost reduction. Speedup and economic problems should be seri-ously considered. LSI processor development seems to be one of the most promising approaches to these problems.

6. ACKNOWLEDGEMENT

The author would like to thank his colleagues at NEC Central Research Laboratories for their support in preparing this chapter.

7. REFERENCES

Arakawa, T., "Color Display for Color Image Processing," J. Insti-tute of Electronics and Communication Engineers of Japan, IE-76-83 (1976).

Asai, K., "Fingerprint Identification System," UJCC (1975), p. 30.

Funo, Y., "Airborne MSS Geometric Correction and Its Evaluation,"
    J. Info. Proc. Soc., Image Processing I9-1 (1978).

Ishikura, A., "An Application of Microprocessor to Pattern Recog-
    nition," J. Television Broad. Soc. Japan, TBS-28-5 (1976).

Tajima, J., "Color Analysis by Color Information Processing System
    (CIPS-I)," Nippon Elec. Co. R&D Rpt. 47 (1977), pp. 13-19.

Naito, K., "An MSS Data Analysis System," J. Institute of Electron-
    ics and Communication Engineers of Japan, Info. Workshop
    (1977), p. 252.

Naito, K., "Road Extraction from Aerial Thermal Photo-maps," The
    US-Japan Seminar on Image Processing in Remote Sensing (Nov.
    1976).

Naito, K., "Description of LANDSAT Image," J. Institute of Electron-
    ics and Communication Engineers of Japan S-53 Convention (1978),
    pp. 5-49.

Tajima, J., "An Experiment on Color Image Mosaicing," J. 17th Con-
    vention, Info. Proc. Soc. Japan (1976), pp. 169-170.

INTERPOLATION TO REDUCE DIFFICULTY IN D/A CONVERSION

M. Hatori and Y. Taki

Department of Electrical Engineering, University of Tokyo

Hongo, Bunkyo-ku, Tokyo, JAPAN

1. INTRODUCTION

It is difficult to design an analogue low-pass filter for D/A conversion if the sampling frequency is very near to Nyquist's frequency; even if the sampling frequency is not near the Nyquist frequency, it is still difficult to avoid the distortion which arises from low-pass filtering in D/A conversion. This difficulty must be overcome in the design of PCM transmission systems for colour television. This problem is especially serious in the systems like television where we cannot avoid the repeated use of A/D and D/A conversions 8 or 9 times over, for example. The difficulty can be overcome by interpolation of the samples. This technique was proposed by the authors (Taki et al., 1978) and independently by Ninomiya (1978). Real-time digital interpolation for D/A conversion of the PCM'd NTSC video signal was also demonstrated in Taki et al. (1978).

2. INTERPOLATION OF SAMPLES

Let the spectrum $X(f)$ of the signal $x(t)$ be band-limited from $-fo$ to fo. The spectrum of the sampled signal appears around the integral multiples $\{mfs\}$ of the sampling frequency fs as shown in Figure 1 (b). It is theoretically possible to reproduce the original signal $x(t)$ from the sampled signal by ideal low-pass filtering, even if the sampling frequency fs is chosen equal to Nyquist's frequency $2fo$. In actuality this is practically impossible and the usual tactic to escape from this difficulty is to choose a sufficiently larger sampling frequency.

227

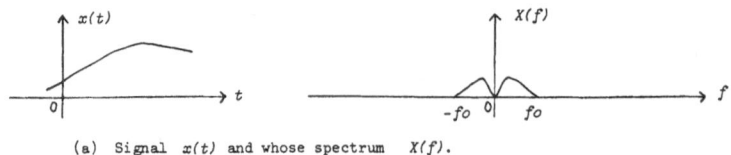

(a) Signal $x(t)$ and whose spectrum $X(f)$.

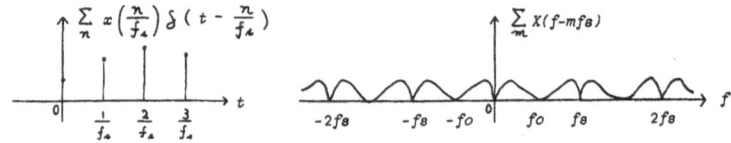

(b) Sampled signal $\sum_{n} x\left(\frac{n}{f_s}\right)\delta\left(t-\frac{n}{f_s}\right)$ and whose spectrum $\sum_{m} X(f-mf_s)$

    sampling frequency $f_s = 2f_0$.

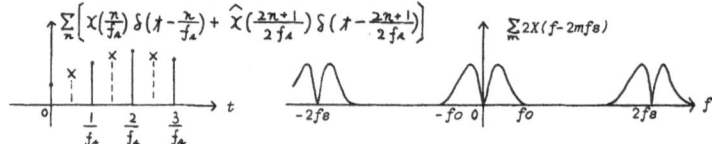

(c) Sampled signal with interpolated samples

$$\sum_{n}\left[x(\tfrac{n}{f_s})\,\delta(t-\tfrac{n}{f_s}) + \hat{x}(\tfrac{2n+1}{2f_s})\,\delta(t-\tfrac{2n+1}{2f_s})\right] \quad \text{and whose}$$

spectrum $\sum_{m} X(f-2mf_s)$ sampling frequency $f_s = 2f_0$

Figure 1.   Demonstration of transmission using Nyquist's sampling rate made possible by the interpolation technique.

Figure 1 (c) explains how the interpolation technique reduces the difficulty of low-pass filtering. If one sample is interpolated in each interval between the original sampling points, the sampled signal with interpolated samples is expressed as

$$\sum_{n}\left[x(\tfrac{n}{fs})\delta(t-\tfrac{n}{fs}) + \hat{x}(\tfrac{2n+1}{2fs})\delta(t-\tfrac{2n+1}{2fs})\right] \tag{1}$$

$$= \sum_{n} x(\tfrac{n}{2fs})\delta(t-\tfrac{n}{2fs}) + \sum_{n}\left[\hat{x}(\tfrac{2n+1}{2fs})-x(\tfrac{2n+1}{2fs})\right]\delta(t-\tfrac{2n+1}{2fs}) \tag{1}'$$

$$= \sum_{n} x(\tfrac{n}{2fs})\delta(t-\tfrac{n}{2fs}) \tag{1}''$$

where $x(\frac{n}{fs})$ is the value of the original sample and $\hat{x}(\frac{2n+1}{2fs})$ is the value of the interpolated sample estimated from neighboring original samples $\{x(\frac{n}{fs})\}$. An error-free estimate of $\hat{x}(\frac{2n+1}{2fs})$ is calculable if the original signal $x(t)$ is band limited from $-f_0$ to

fo, if the sampling frequency fs is larger than Nyquist's frequency $2fo$, and if a sufficiently large number of neighboring original samples are used. Therefore, the second term of the equation (1)'; may be made to vanish and equation (1)" obtained.

The spectrum of the sampled signal with interpolated samples is obtained as

$$\sum_m \left[X(f-mfs) + (-1)^m \hat{X}(f-mfs) \right] \qquad (2)$$

$$\sum_m 2X(f-2mfs) + \sum_m \left[(-1)^m \hat{X}(f-mfs) - X(f-mfs) \right] \qquad (2)'$$

$$\sum_m 2X(f-2mfs) \qquad (2)''$$

where $X(f)$ is the Fourier transform of the original signal $x(t)$ and $\hat{X}(f)$ is the Fourier transform of the absolute value of the expression

$$\sum_m \hat{x}\left(\frac{2n+1}{2fs}\right) \ sin \ \pi fs\left(t-\frac{2n+1}{2fs}\right) \ / \ \pi fs\left(t-\frac{2n+1}{2fs}\right) \qquad (2a)$$

(We do not use the notation $\hat{x}(t)$ because we want to emphasize that only a numerable number of samples $\{\hat{x}(\frac{2n+1}{2fs})\}$ at discrete sampling points are to be estimated and interpolated.) For the same reason mentioned regarding equation (1)', the second term of the equation (2)' may be made to vanish to obtain equation (2)".

Equation (2)" or Figure 1 (c) shows that the spectrum of the sampled signal with the interpolated samples is zero for $-2fs+fo<f<-fo$ and for $fo<f<2fs-fo$. We now have a sufficiently wide guard band, even if we choose the sampling frequency as Nyquist's frequency $2fo$. The guard band produced by the interpolation technique reduces the difficulty of designing an analogue low-pass filter for D/A conversion and makes it possible to transmit PCM signals using Nyquist's sampling rate. If we want a wider guard band, it is possible to erase the spectrum from $\pm fo$ to $\pm(kfs-fo)$ by interpolating k samples in each interval between the original sampling points.

3.   ESTIMATION OF THE VALUE TO BE INTERPOLATED

We are going to interpolate samples by a 2N tapped digital filter of the transversal type (Figures 2 and 3). The estimated value to be interpolated should be of the form

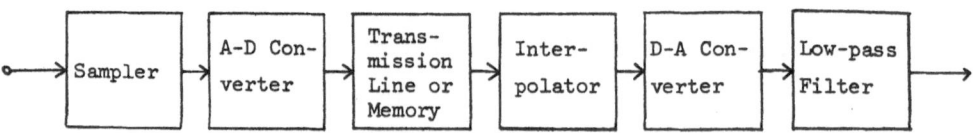

Fig. 2.   Interpolation as used in a PCM system.

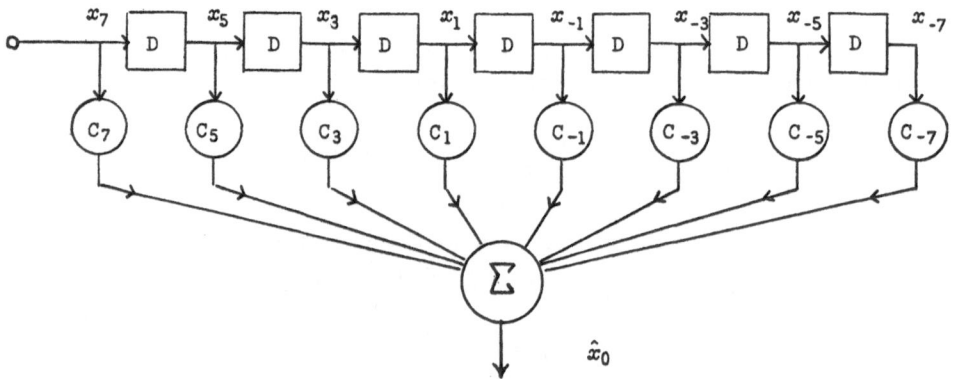

Fig. 3.   Schematic of the interpolator.

$$\hat{x}\left(\frac{2n+1}{2fs}\right) = \sum_{m=-N+1}^{N} C_m x\left(\frac{n+m}{fs}\right) \tag{3}$$

In order to simplify the notation, let us shift the time scale by $\frac{1}{2fs}$ [sec], and consider the interpolation at the time $t = 0$.

$$\hat{x}(0) = \sum_{i=-N}^{-1} C_{2i+1} x\left(\frac{2i+1}{2fs}\right) + \sum_{i=1}^{N} C_{2i-1} x\left(\frac{2i-1}{2fs}\right) \tag{3}'$$

$$\hat{x}_0 = \sum_{i=-N}^{-1} C_{2i+1}\, x_{2i+1} + \sum_{i=1}^{N} C_{2i-1}\, x_{2i-1} \tag{3}''$$

As the signal $x(t)$ is written mathematically as

$$x(t) = \sum_{i=-\infty}^{-1} x(\frac{2i+1}{2fs}) \frac{sin\ \pi fs(t-\frac{2i+1}{2fs})}{\pi fs(t-\frac{2i+1}{2fs})}$$

$$+ \sum_{i=1}^{\infty} x(\frac{2i-1}{2fs}) \frac{sin\ \pi fs(t-\frac{2i-1}{2fs})}{\pi fs(t-\frac{2i-1}{2fs})} \tag{4}$$

the signal value at the time $t = 0$ can be represented as

$$x(0) = \sum_{i=-\infty}^{-1} x(\frac{2i+1}{2fs}) \frac{sin\pi(i+\frac{1}{2})}{\pi(i+\frac{1}{2})}$$

$$+ \sum_{i=1}^{\infty} x(\frac{2i-1}{2fs}) \frac{sin\pi(i-\frac{1}{2})}{\pi(i-\frac{1}{2})} \tag{5}$$

Rewriting this, using the notation used in the equation (3)", we obtain

$$x(0) = \sum_{i=-\infty}^{-1} C_{2i+1}\ x_{2i+1} + \sum_{i=1}^{\infty} C_{2i-1}\ x_{2i-1} \tag{5}'$$

where

$$C_{2i+1} = \frac{(-1)^i}{\pi(i+\frac{1}{2})} = C_{2|i|}\ -1, \qquad i \leq -1$$

$$C_{2i-1} = \frac{(-1)^{i+1}}{\pi(i-\frac{1}{2})}, \qquad\qquad i \geq 1.$$

Substituting these coefficients in equation (3)" using equation (6) and letting $N \to \infty$, we obtain

$$\lim_{N \to \infty} \hat{x}_o = X(0). \tag{7}$$

Equation (7) means that an error-free estimation $\hat{x}_o$ of $x(0)$ is calculable if a sufficiently large number of neighboring original samples are used.

Next, let us consider how to define the coefficients C_{2i+1} and C_{2i-1} in equation (3)" when only a finite number 2N of neighboring original samples can be used. The coefficients which minimize the estimation error are obtained by Wiener's algorithm; the estimation error

$$E[(x(0) - \hat{x}_o)^2] \qquad (8)$$

is minimized, when the coefficients C_{2i+1} and C_{2i-1} are defined by

$$
\begin{bmatrix}
C_{-(2N-1)} \\
\cdot \\
\cdot \\
\cdot \\
C_{-3} \\
C_{-1} \\
C_1 \\
C_3 \\
\cdot \\
\cdot \\
\cdot \\
C_{2N-1}
\end{bmatrix}
=
\begin{bmatrix}
R_{-(2N-1),-(2N-1)} \cdots R_{(2N-1),-2N-1)} \\
\cdot \qquad\qquad\qquad \cdot \\
\cdot \qquad\qquad\qquad \cdot \\
\cdot \qquad\qquad\qquad \cdot \\
\\
R_{-(2N-1),(2N-1)} \cdots R_{(2N-1),(2N-1)}
\end{bmatrix}^{-1}
\begin{bmatrix}
R_{0,-(2N-1)} \\
\cdot \\
\cdot \\
\cdot \\
R_{0,(2N-1)}
\end{bmatrix}
\qquad (9)
$$

where $R_{i,j}$ is the crosscorrelation of x_i and x_j

$$R_{ij} = E[x_i \ x_j]. \qquad (10)$$

The optimum coefficient values calculated from equation (9) are shown in Table 1, where the crosscorrelation of the NTSC signal of the SMPTE slide "Couple" samples by 8.59 [MHz] is used. The interesting property is that the coefficient value converges to that of the equation (6) independent of the crosscorrelation $R_{i,j}$ as the number of original samples 2N increases without limit. The estimation error is shown in Figure 4. The transfer characteristics $H(f)$ of the interpolator are shown in Figure 5, where $H(f)$ is defined by

$$H(f) = \frac{1}{2} \frac{\sum_m [X(f-mfs) + (-1)^m \hat{X}(f-mfs)]}{\sum_m [X(f \ mfs)]} \qquad (11)$$

Table 1 - Coefficient of Estimation

Number of Taps 2N	Coefficient	Value of Co-efficient (NTSC Signal)	Value of Co-efficient (Eq. (6))
2	$C_{-1} = C_1$	0. 5 0 7 5 3 0	0. 6 3 6 6 2 0
4	$C_{-1} = C_1$	0. 7 2 6 6 6 0	0. 6 3 6 6 2 0
	$C_{-3} = C_3$	-0. 2 3 1 2 1 0	-0. 2 1 2 2 1 0
6	$C_{-1} = C_1$	0. 6 5 1 1 9 0	0. 6 3 6 6 2 0
	$C_{-3} = C_3$	-0. 2 6 3 5 1 0	-0. 2 1 2 2 1 0
	$C_{-5} = C_5$	0. 1 1 4 1 4 0	0. 1 2 7 3 2 0
8	$C_{-1} = C_1$	0. 6 3 6 5 3 0	0. 6 3 6 6 2 0
	$C_{-3} = C_3$	-0. 2 1 1 6 8 0	-0. 2 1 2 2 1 0
	$C_{-5} = C_5$	0. 1 3 2 4 5 0	0. 1 2 7 3 2 0
	$C_{-7} = C_7$	-0. 0 5 8 0 9 7	-0. 0 9 0 9 4 6
10	$C_{-1} = C_1$	0. 6 3 0 6 1 0	0. 6 3 6 6 2 0
	$C_{-3} = C_3$	-0. 1 9 7 9 6 0	-0. 2 1 2 2 1 0
	$C_{-5} = C_5$	0. 1 0 8 6 8 0	0. 1 2 7 3 2 0
	$C_{-7} = C_7$	-0. 0 7 8 2 5 0	-0. 0 9 0 9 4 6
	$C_{-9} = C_9$	0. 0 3 7 4 0 9	0. 0 7 0 7 3 6
20	$C_{-1} = C_1$	0. 6 3 5 8 2 0	0. 6 3 6 6 2 0
	$C_{-3} = C_3$	-0. 2 1 0 1 3 0	-0. 2 1 2 2 1 0
	$C_{-5} = C_5$	0. 1 2 4 0 6 0	0. 1 2 7 3 2 0
	$C_{-7} = C_7$	-0. 0 8 3 8 2 7	-0. 0 9 0 9 4 6
	$C_{-9} = C_9$	0. 0 5 9 6 3 1	0. 0 7 0 7 3 6
	$C_{-11} = C_{11}$	-0. 0 4 4 1 1 8	-0. 0 5 7 8 7 5
	$C_{-13} = C_{13}$	0. 0 3 4 1 0 7	0. 0 4 8 9 7 1
	$C_{-15} = C_{15}$	-0. 0 2 7 4 4 6	-0. 0 4 2 4 4 1
	$C_{-17} = C_{17}$	0. 0 2 6 3 0 8	0. 0 3 7 4 4 8
	$C_{-19} = C_{19}$	-0. 0 1 4 5 9 7	-0. 0 3 3 5 0 6

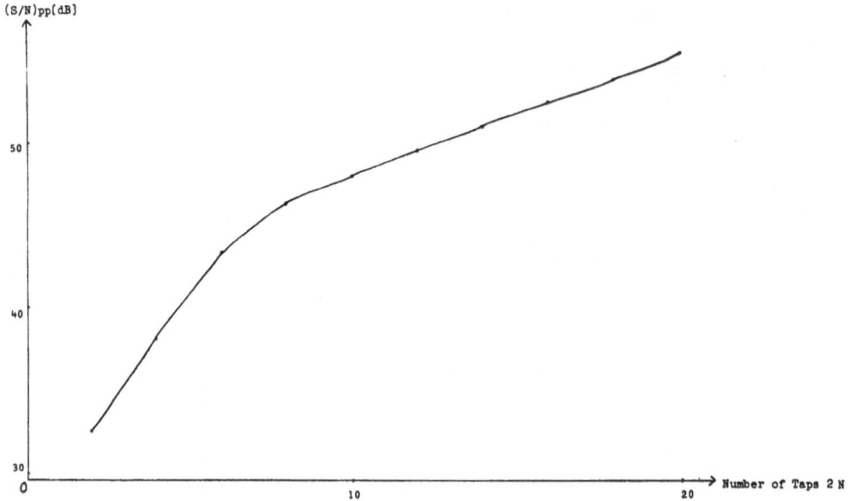

Fig. 4.   Graph of estimation error as a function of the number of taps.

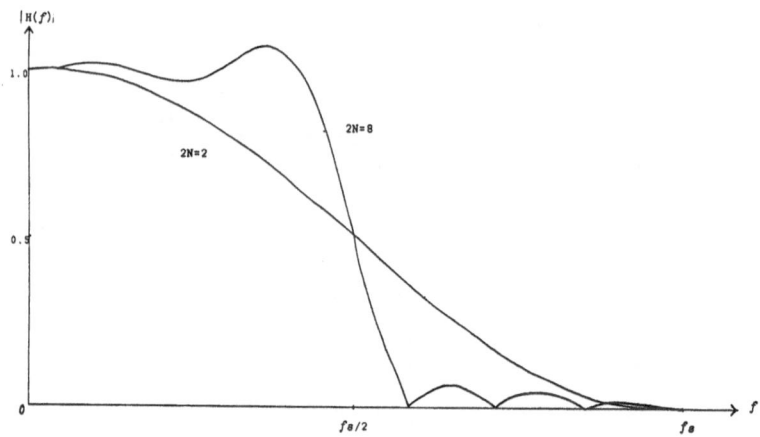

Fig. 5.   Transfer characteristic of the interpolator.

4.   IMPLEMENTATION OF THE INTERPOLATOR

An experimental digital interpolator for use in decoding the PCM'd NTSC video signal was made. Its specifications are listed in Table 2. The sampling frequency fs is 8.59[MHz], which is very close to Nyquist's frequency $2fo$ = 8.4[MHz] and which is locked to the line frequency fh and the color subcarrier frequency fsc, i.e., $fs = \dfrac{12}{5}fsc = \dfrac{12}{5} \times \dfrac{455}{2}fh.$ (A sampling frequency of 10.7[MHz] = $3fsc$ is usually used.)

Table 2 - Specification of the Interpolator

Sampling frequency fs = 8.59[MHz]
A/D conversion : 8[bit/sample], 8.59[M sample/sec]
Interpolation : One sample in each interval
D/A conversion : 8[bit/sample], 17.18[M sample/sec]
Interpolator : 8 tapped of transversal type
Weighting coefficient : as shown in Table 1
Multiplication : 8 bit in 12 bit out
Multiplier : 8 bit in 4 bit out ROM × 3 × 8
Addition : 12 bit in 12 bit out
Adder : 4 bit full adder × 3 × (4 + 2 + 1)

The interpolator is a digital filter of the transversal type. The number of taps is 2N = 8. (This number may not be sufficient; see Figure 4.) Our first interest was to confirm the operation of the circuit. We are now planning to increase the number of taps to 2N = 16. The transversal implementation has the advantage over the recursive implementation that, if one must build a high-speed digital filter, one is not concerned with the accumulation of delay time in the operations which are carried out.

A ROM is used as the weighting coefficient multiplier, because its operational speed is faster than that of the logical multiplier. The addition is performed in parallel and so we use 4 + 2 + 1 = 7 full adders to sum the 8 samples.

Figure 6 shows the implementation of the interpolator. The latches are used to perform serial-parallel conversion to avoid erroneous operations arising from the difference of delay time in the circuits. The limiter is used to remove over flow and under flow. A selector is used to switch between the original samples and the interpolated samples.

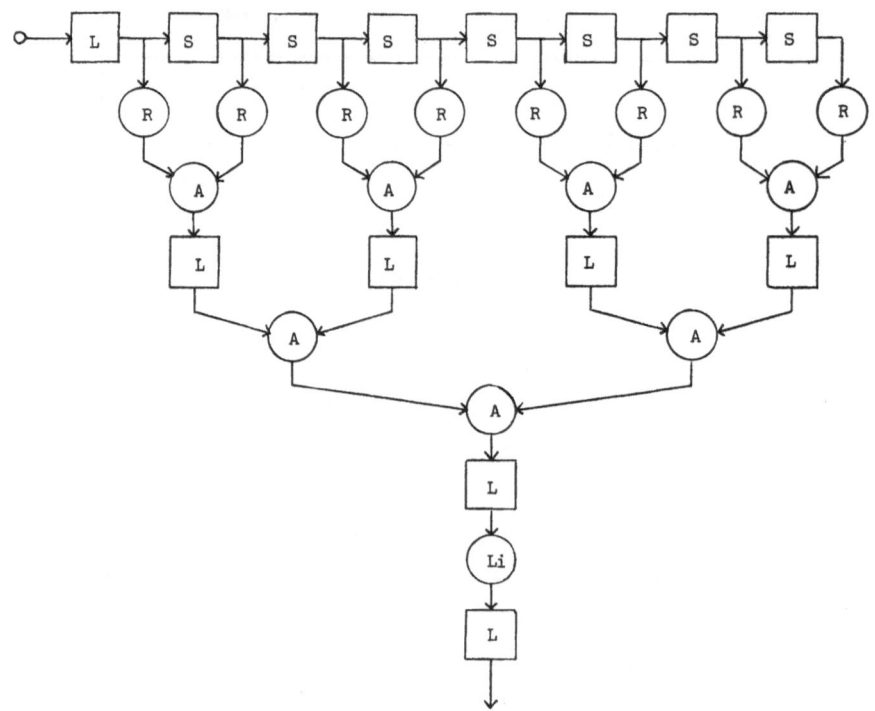

Fig. 6. Implementation of the interpolator: (S) shift register; (R) ROM; (A) adder; (L) latch; (Li) limiter.

5.   EXPERIMENTAL RESULT

Figure 7(a) shows the spectrum of NTSC video signal and Figure 7(b) shows the waveform of the color burst signal. Figure 8(a) shows the spectrum of the low-pass filtered signal (cut-off frequency 4.53[MHz]) of the sampled (sampling frequency is 8.59[MHz]) NTSC video signal where the leaked spectrum is observed. Its color burst waveform is shown in Figure 8(b), where degradation of the waveform is observed. Finally, Figure 9(a) is the spectrum of the low-pass filtered signal (cut-off frequency 4.53[MHz]) of the sampled NTSC video signal (sampling frequency 8.59[Mhz]) with interpolated samples, where we observe only a very small leaked spectrum. Its color burst waveform is shown in Figure 9(b), where degradation of the waveform is small.

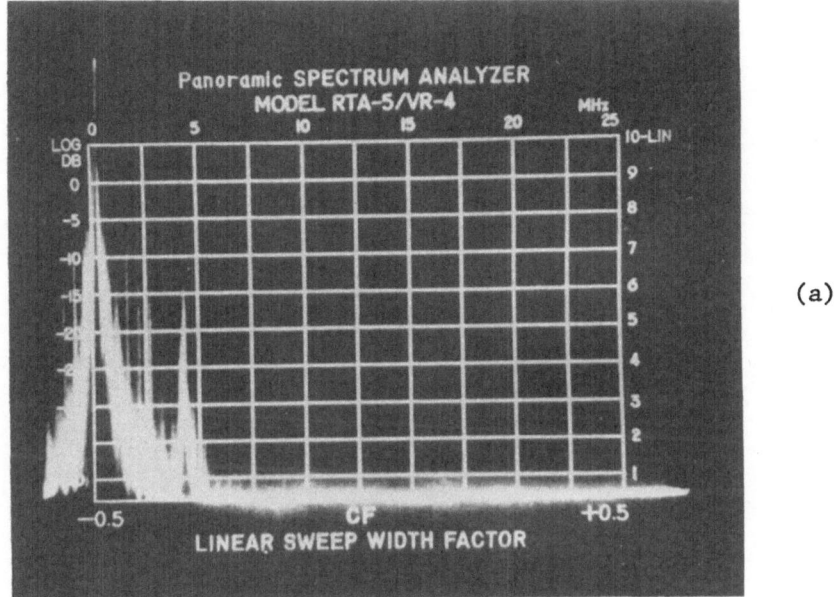

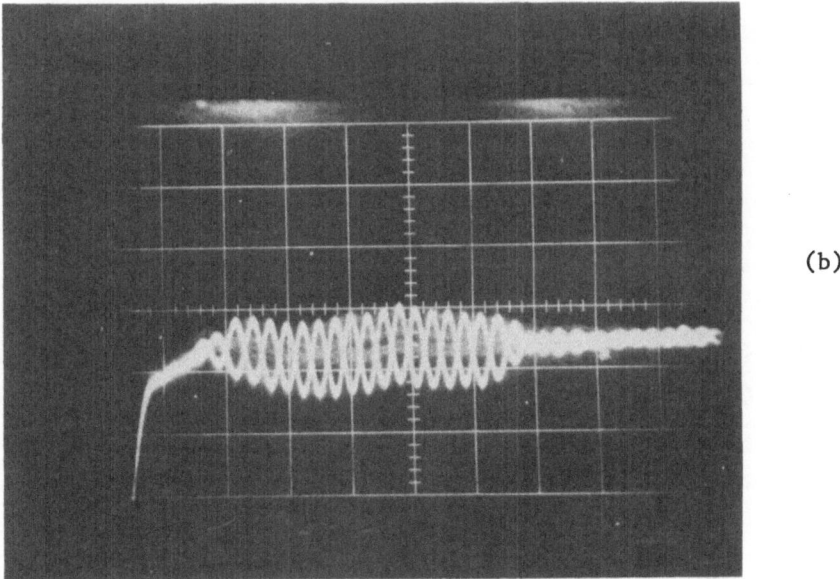

Fig. 7. NTSC video signal: (a) spectrum; (b) NTSC video signal.

(a)

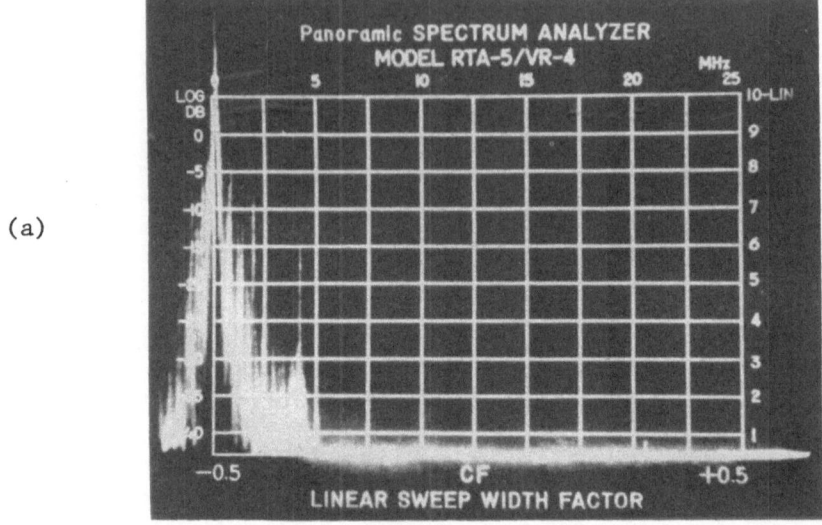

(b)

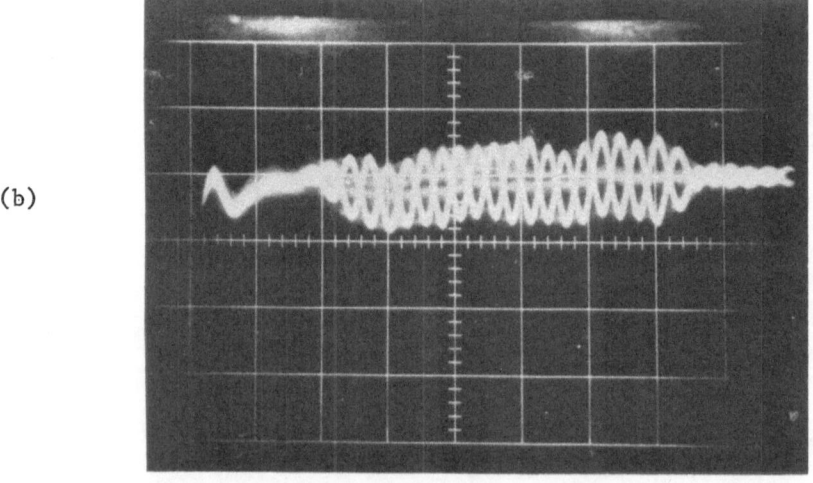

Fig. 8. Sampled (A/D and D/A) NTSC video signal at output of low-pass filter: (a) spectrum; (b) color burst.

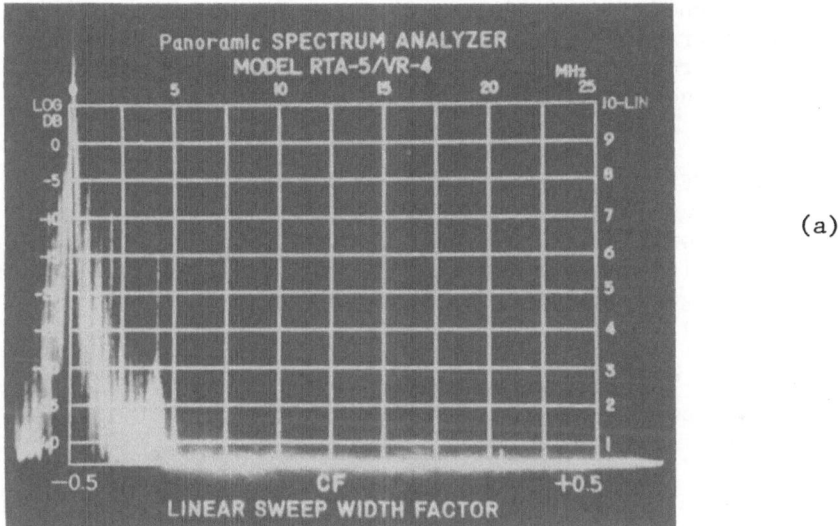

(a)

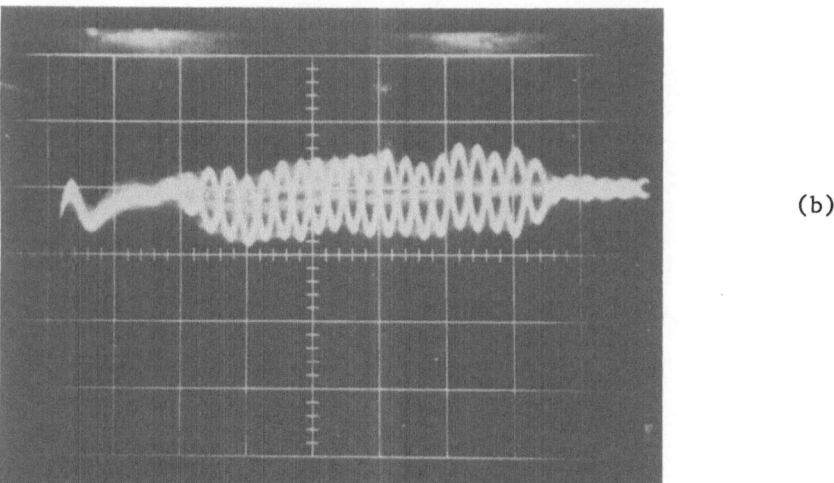

(b)

Fig. 9. Sampled and interpolated NTSC video signal at output of
low–pass filter: (a) spectrum; (b) color burst.

6.   CONCLUSION

The interpolation technique reduces the difficulty of designing an analogue low-pass filter for D/A conversion, which makes it
possible to transmit signals using a sampling frequency very near
the Nyquist frequency. This also reduces signal degradation arising from the non-ideal analogue low-pass filter.

7.   REFERENCES

Taki, Y., Hatori, M., Hatakeyama, Y., and Sano, J., "Demodulation
     of PCM Signal by Interpolation Technique," Inst. Electronics
     Comm. Engrs. Japan, Proc. Ann. Conf. Comm. (1978), p. 526 (in
     Japanese).

Ninomiya, Y., "Digital Interpolation in D/A Conversion," Inst.
     Television Engrs. Japan, Proc. Ann. Conf., Vol. 10 (1978),
     p. 8 (in Japanese).

A COMPUTING SYSTEM ORGANIZATION FOR IMAGE DATA RETRIEVAL

T. Ichikawa and H. Aiso*

R&D Laboratories, Kokusai Denshin Denwa Co., Ltd.

*Dept. Elec. Engrg., Keio University, Keio, JAPAN

1.   INTRODUCTION

As observed in remotely sensed or medical image data processing, data bases for particular types of images are more and
more regionally integrated. For the incorporation of these data
bases in a distributed environment, the retrieval of this data to
make images available for local processing is important. What is
needed is an advanced facility for selecting related image data
through an estimation of the relevancy of their essential features
at a physical or conceptual level. This problem involves large-
scale semantic data base handling and the facility for extracting
image features for the embodiment of the data bases.

This chapter describes a computing system organization for
image data retrieval where a multi-microprocessor associative
relevancy estimator (ARES) performs associative searches of the
semantic data bases as originally described by Ichikawa et al.
(1978). The ARES is coupled with an augmented content-addressable
memory for extracting local features by pyramidal image data structures (ELPIS). The word "semantic" implies features which are
commonly or conceptually accepted. Each data item in a semantic
data base is coded categorically in such a manner that the relatedness of the symbolic partial representations for the categorical
blocks is measured in terms of the distances between the code
vectors assigned to them. Hence the blurred representation of each
data item is obtained systematically by the hierarchical application of error corrections to its original representation. Then, by
simply applying content addressing after blurring, a high-speed
search of related data items is accomplished based on their deep-

structured relatedness. This is the association principle behind
the ARES.

The ELPIS which stores physical features of images using a
pyramidal data structure makes rapid access and selective use of
particular areas feasible with local control of detailed fine structure. Content-addressing for local feature extraction is discussed
extensively in this chapter together with the method of communicating with the ARES on the organization and development of semantic
data bases.

2. SYSTEM ORGANIZATION

Figure 1 illustrates the total system organization for image
data retrieval. The ARES multi-microprocessor which performs high-
speed searchs of the semantic is based on associating stored information through relevancy estimates. The following is a brief explanation of the association principle as well as a discussion of the
feature extraction methods employed by the ELPIS.

2.1 The ARES for Associative Search of Semantic Data Bases

A semantic data base is defined as a collection of data items
which are symbolic representations of image features. The word
"semantic" implies features which are commonly or conceptually
accepted. Each data item D_0 is coded categorically as shown by
equation (1) in such a manner that the relatedness of the symbolic
representations for the categorical blocks d_{si} is measured in terms
of the distance of the code vectors assigned to them.

$$D_0 = (d_{s1}, d_{s2}, \ldots, d_{si}, \ldots, d_{sp}), \ 1 \leq i \leq p. \tag{1}$$

As categories we adopt the classification measures which are
consciously or unconsciously used for discriminating objects through
observed features. Therefore, the lengths of the categorical blocks
d_{si} are not necessarily the same depending upon the numbers of
classes which belong to the categories. Now divide D_0 into several
blocks of equal length n_1 as shown by equation (2), and apply an
error correction scheme to the code of length n_1 with t_1-error correction capability for the blocks d_{bj}.

$$D_0 = (d_{b1}, d_{b2}, \ldots, d_{bj}, \ldots, d_{bq}), \ 1 \leq j \leq q. \tag{2}$$

The data representation thus derived from D_0 is denoted by equation (3).

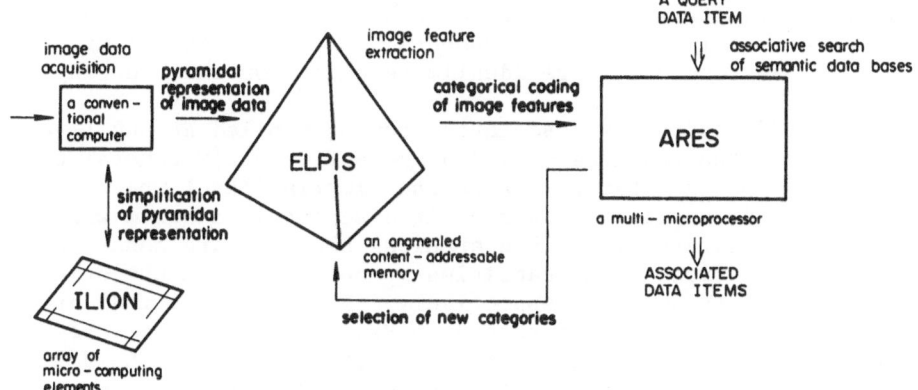

Fig. 1. Organization of the image data retrieval system.

$$e_1(D_0) = (e(d_{b1}), e(d_{b2}), \ldots, e(d_{bj}), \ldots, e(d_{bq})),$$

$$1 \leq j \leq q. \tag{3}$$

The error corrected code vectors $e(d_{bj})$ of length n_1 represents blurred features of the partial representations d_{bj} of the original data item. The block of length n_1 for error correction is then called a blurring block. The value of n_1 is selected to cover several categorical blocks so as to blur the influence of artificial categorization of the semantic data items. Thus $e_1(D_0)$ is comprised of a global structure of data items which are inherently related to each other within the distance t_1. Data items related to a query item on their global structure are specified simply by applying content addressing to the blurred data representations without executing complex relatedness calculations.

The application of content addressing after blurring is referred to as "association." Content addressing is accomplished as follows: Let $e_1(Q_0)$ denote the blurred representation of a query item Q_0. Then D_0's, whose $e_1(D_0)$'s coincide with $e_1(D_0)$ by an amount equal to or greater than θ $(0 \leq \theta \leq q)$ blurring blocks, are selected.

Now suppose we want to have a more general structure of data items. We apply the code e_2 of length n_2 with t_2-error correction capability to $D_1 = e(D_0)$ by dividing it into blurring blocks of length n_2. Thus we can continue the blurring as indicated by equation (4) until we reach the blurredness required for the association, where the relation $t_s \geq 2t_{s-1}+1$ should necessarily be satisfied when $n_s \geq n_{s-1}$.

$$D_0 \to D_1 = e(D_0) \to D_2 = e_2(D_1) \to \cdots \to D_s = e_s(D_{s-1}) \to \cdots \tag{4}$$

Figure 2 illustrates the hierarchical blurring of a categorically coded original data item by successive applications of error corrections. Perfect codes are ideally selected for the blurring.

Content addressing for association can be applied at any level of blurring when the features provided are sufficiently general to meet the requirements. Let s* denote the blurring level where content-addressing is applied. Figure 3 is a schematic diagram explaining how content addressing and hierarchical blurring are connected to perform association. The quantities B_0 and B_s denote the original and blurred data spaces, respectively. The hit marks indicate the D_{s*}'s which are content addressed in B_{s*}. When we have a restriction on the number of associated data items, the threshold θ for content addressing is heuristically selected so that hit marks do not exceed a certain limit δ.

The ARES has also a facility for data modification. Suppose $Q_0(\epsilon B_0)$ is given as a query item. Assume that some D_0's in B_0 are specified by Q_0 through the content addressing at the blurring level s*. These D_0's and related D_s's $(0 < s \leq s*)$ are replaced by Q_0 and its blurred expressions, respectively. Otherwise, Q_s $(0 \leq s \leq s*)$ is registered, unmodified, in B_s $(0 \leq s \leq s*)$. The quantity Q_0 might also be given as a weighted average of two data items previously stored in B_0.

To summarize the control of the ARES, the parameters which affect association performance are the length n_s and the error correction capability t_s of the code applied for blurring at the level $s(s=1,2,...,s*)$. The final level of blurring s* for content addressing, the threshold θ on the number of coinciding blocks at content addressing, and the limit δ on the number of associated data items.

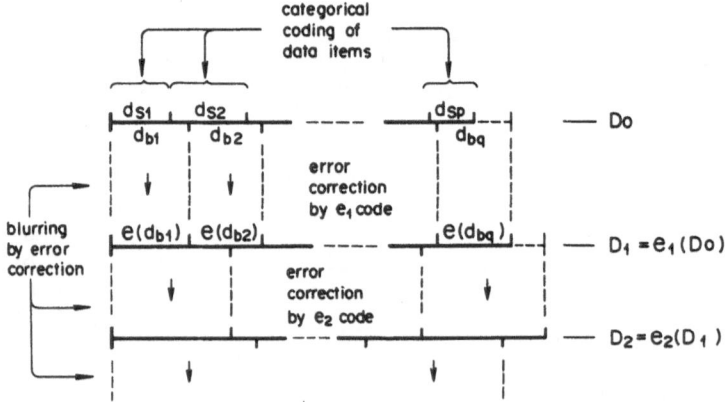

Fig. 2. Hierarchical blurring of a categorically coded data item.

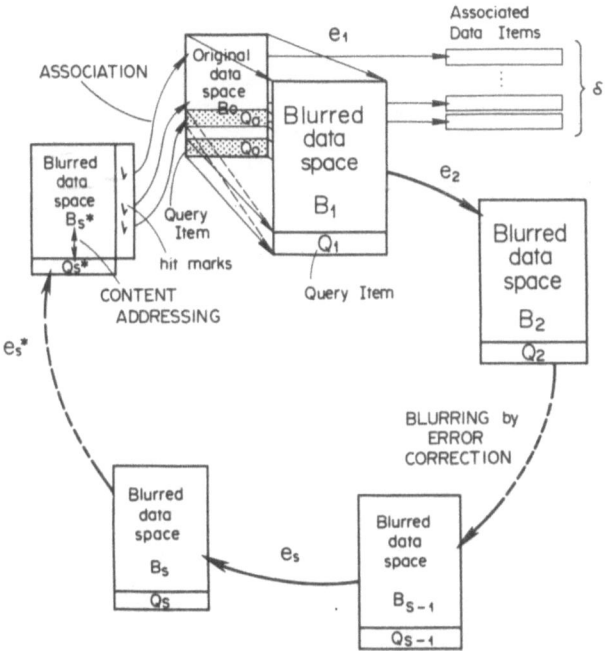

Fig. 3.   Association through blurring and content addressing.

A multi-microprocessor architecture for the ARES is also described in the chapter by Aiso and Sakamura.

2.2   The ELPIS for Image Feature Extraction

In semantic data base, each data item represents the essential
features of an image which are commonly or conceptually accepted.
As pointed out by Akin and Reddy (1977), these are, for example,
the shapes and textures which are observed with an arbitrarily selected resolution. For ease in extracting these features, we assume
a pyramidal data structure for representing images in a digitized
square of n^L x m^L pixels according to the methods of Hanson and
Riseman (1975). Information contained in an original image can be
used to form a partial image where pixels have binary values only;
nonzero pixels are called "informative." A hierarchical search of
these informative areas is accomplished by repetitive n x m regular
decompositions of the image space using a window of n x m elements
as shown in Figure 4. Each informative pixel is then specified by
a vector V in equation (5) below which is similar to the subpicture
notation of Klinger (1976) in that each element of the vector provides steering information down a path in the regular decomposition
tree.

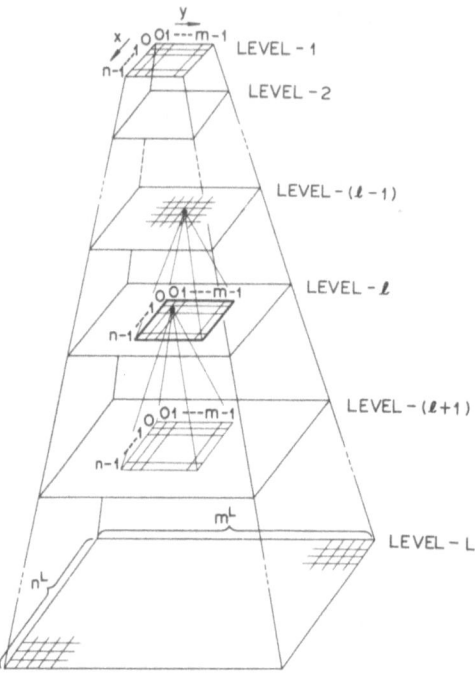

Fig. 4. A hierarchical structure of image coding (see Hanson and Riseman, 1975).

$$V = (v(1), v(2), \ldots, v(\ell), \ldots, v(L))$$

$$V(\ell) = (x_\ell, y_\ell), \; 1 \leq \ell \leq L \tag{5}$$

$$0 \leq x_\ell \leq n-1, \; 0 \leq y_\ell \leq m-1.$$

    Figure 5 illustrates how the specification of informative pixels works according to the pyramid structure of an L-step hierarchical decomposition of an image space for the case of $n=m=2$, $L=3$. The pixel k indicated by a bold-faced square in Figure 5(a) is observed in the upper right corner of the window space as shown in Figure 5(b) when the window is applied at the first level of pyramidal coding. Hence we have (01) as the coordinate $v(1)$ of equation (5) for representing the pixel. As is obvious from Figure 5(c) and 5(d), subsequent applications of the window give us the second and the third coordinates of equation (5); $v(2) = (00)$, $v(3) = (10)$. Thus we have $V = (01,00,10)$.

    A vector is called "primitive" when it corresponds to a single informative pixel as defined in equation (5), and can be converted from one-dimensionally scanned external image data in real time in the process of image data acquisition by providing 2L counters which

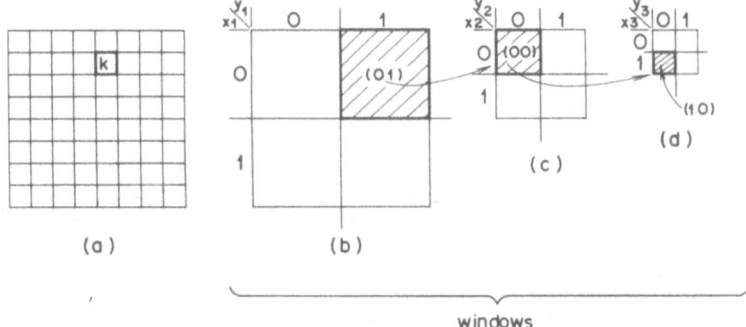

Fig. 5. Pyramidal representation of a pixel.

operate in parallel. The advantage of this vector representation of
informative pixels is that it facilitates global and local feature
extraction with arbitrarily selected resolution when a content-
addressable memory is provided for storing it.

However, the bottleneck problem here lies in the complexity of
the vector representation and wasteful use of memory space. One of
us has solved this problem as briefly described below and in Ishi-
kawa (1978). A novel scheme has been devised to reduce the number
of data words to be stored in a memory without distorting the abovementioned feature extraction facility inherently possessed by the
pyramidal representations. The simplification is essentially based
on the reduction of the number of informative elements observed in
a window space of n x m elements. Window patterns are simplified,
such that the original patterns are reproducible, by providing an
extended space of (n+1) x (m+1) elements. In the extended space,
each row or column takes a complementary appearance of informative
elements when a special mark is placed at or taken away from the
additional square of corresponding row or column as illustrated in
Figure 6.

This window pattern simplification scheme is called "cross-
shooting" of flip-flop targets and is applicable to the window space
at any level of the pyramid throughout the simplification process.
Thus the 52 nonzero pixels whose primitive vectors represent the
sample image pattern of Figure 7, for example, are reduced to $V_1'=$
$(22,22,22)$, $V_2'=(00,01,22)$, $V_3'=(01,00,22)$, $V_4'=(10,11,22)$, $V_5'=$
$(01,00,10)$, and $V_6'=(11,10,11)$ by using a two-by-two three-step decomposition of the image space, where the value 2 implies 0 and 1.
Details are provided in Ichikawa (1978).

Cross-shooting can be accomplished by locally communicating
with a parallel spatial array of micro-computing elements while con-
verting the one-dimensionally scanned external image data into the
pyramidal representation.

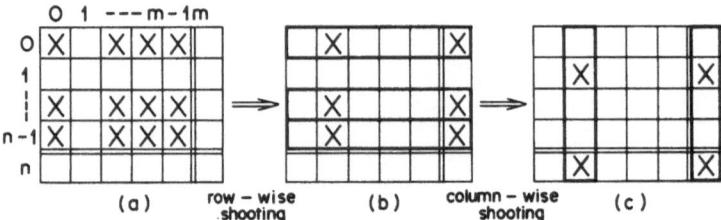

Fig. 6.   Cross-shooting of targets.

Fig. 7. A sample image pattern
(from Tanimoto, Proc. 4th Conf.
Pattern Recog. Image Proc., 1977).

3.   IMAGE FEATURE EXTRACTION FACILITY

The ELPIS, which works with the simplified pyramidal representa-
tation of images, provides a facility for selecting particular areas
of primary interest and detecting the spatial distribution of particular information for the purpose of analyzing global and local
features. Feature extraction is accomplished by content addressing.

3.1 Feature Extraction by Content Addressing

In the ELPIS each vector of the simplified pyramidal representation of an image appears in one word as a stored item within the
data structure specified in Figure 8. Images are not necessarily
binary. The rightmost block marked by "inf" in the figure indicates
the particular information contained in a pixel or a cluster of
pixels which is specified by $v(1)$, $v(2)$, ..., $v(L)$ in the word.

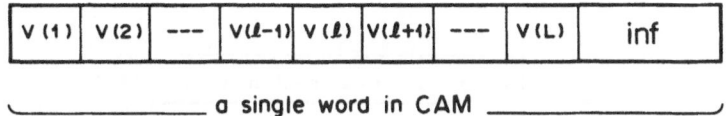

Fig. 8. Information assignment to a word in the Content Addressable Memory (CAM).

        The following modes are provided for content addressing.

Mode 1: Vector representation of pixels with the particular infor-
         mation specified by "inf."

Mode 2: Vector representation of informative pixels contained in
         a particular subarea.

Mode 3: Vector representation necessary for describing an image
         with the fineness specified at the ℓ-th level ($1 \leq \ell \leq L$)
         of the pyramid.

        Mode 1 is accomplished by simply masking the first to the L-th
block associated with the coordinate expressions of the pixels.
Mode 2 is accomplished by applying the vectors which specify the
subarea to the ELPIS as query items, where the function of checking
blockwise coincidence should be augmented since the vectors for the
simplified representation of the images are no longer primitive;
the values n and m, respectively, are permitted for x_ℓ and y_ℓ in
equation (5). Mode 3 is discussed in Section 3.2.

        Augmentation is generally described as follows: Let α and β
denote the values of a single digit of the query and of the stored
items, respectively. Then the value of the corresponding digit of
an associated item is specified by Figure 9, where the entry "–"
denotes the case of no associated items.

        The following example, although trivial, explains how coinci-
dence checking should be functionally augmented for the extraction
of image features. Take the binary pattern given in Figure 7 using
Mode 2 with ℓ=L. Suppose we want to get the informative pixels in
the subarea indicated by a bold-faced square in the Figure 7. As
query items we provide the vectors Q_1=(01,00,12) and Q_2=(01,10,02).
Content addressing is then carried out as illustrated in Figure 10,

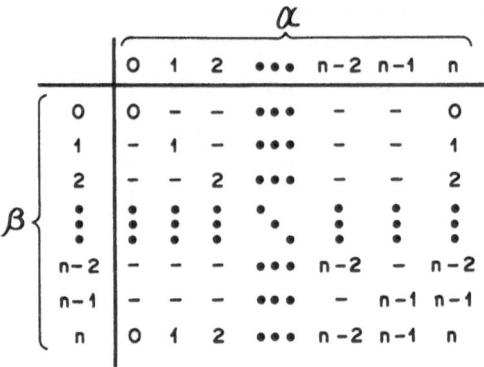

Fig. 9.   Augmentation diagram.

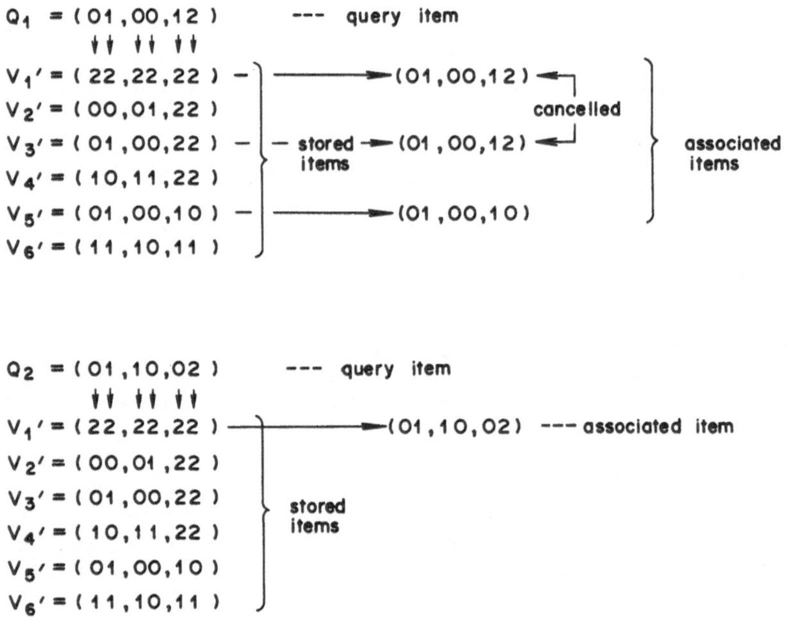

$Q_1 = (01,00,12)$ --- query item

$V_1' = (22,22,22)$ —⌉   →(01,00,12)←⌐
$V_2' = (00,01,22)$ | cancelled
$V_3' = (01,00,22)$ — |— stored →(01,00,12)←⌐   } associated items
$V_4' = (10,11,22)$ | items
$V_5' = (01,00,10)$ —| →(01,00,10)
$V_6' = (11,10,11)$ —⌋

$Q_2 = (01,10,02)$ --- query item

$V_1' = (22,22,22)$ —⌉   →(01,10,02) --- associated item
$V_2' = (00,01,22)$ |
$V_3' = (01,00,22)$ | stored
$V_4' = (10,11,22)$ | items
$V_5' = (01,00,10)$ |
$V_6' = (11,10,11)$ —⌋

Fig. 10.   Augmented content addressing.

where the vector (01,00,12) which appears as two associated items cancels out. Thus we have vectors (01,00,10) and (01,10,02) indicating the informative pixels in the specified subarea.

3.2   Control of Detailed Fineness

The following is a procedure for content addressing using Mode 3 with an arbitrary ℓ such that $1 \leq \ell < L$. Let V_ℓ ($1 \leq \ell \leq L$) denote a set of vectors describing an image with the fineness specified by ℓ. Assuming a set V_L of vectors which contain particular information already detected and mask "inf" throughout the processing hereafter.

(i) Mask the first to the ℓ-th blocks and apply $V_0 = ((n,m), (n,m), \ldots, (n,m))$ as a query item. Let a vector in V_L thus detected be denoted as V^ℓ_{pi} ($i=1,2,\ldots$). Find all V^ℓ_{pi}'s and let $V^\ell_{pi} \epsilon V_\ell$.

(ii) Apply V_0 as a query item with all blocks unmasked and find V_0 among V_L. Let $V_0 = V^\ell_0$ when V_0 is detected and next do the processing described in (iii). Otherwise do the processing given in (iv).

(iii) Mask the (ℓ+1)-th to the L-th blocks and apply each vector in $V_L - V_\ell$ as a query item. Delete V^ℓ_{pi}'s from V_ℓ when we have associated items whose first to ℓ-th blocks coincide with those of $V_L - V_\ell$. Thus we have $V_\ell = \{V^\ell_0, V^\ell_{pi}\}$.

(iv) For each vector in $V_L - V_\ell$, replace the (ℓ+1)-th to the L-th blocks by (n,m)'s and denote it by V^ℓ_{qj} ($j=1,2,\ldots$). Then we have $V_\ell = \{V^\ell_{pi}, V^\ell_{qj}\}$.

The V_ℓ thus obtained gives the image S_ℓ with the fineness specified by ℓ. Let the spaces S^ℓ_0, S^ℓ_{pi}, and S^ℓ_{qj} correspond to V^ℓ_0, V^ℓ_{pi}, and V^ℓ_{qj}, respectively. S_ℓ is then obtained by using equation (6) or equation (7), where $\oplus$ denotes an exclusive-OR of the specified spaces and + denotes the ordinary logical OR

$$S_\ell = S^\ell_0 \oplus \sum_i S^\ell_{pi}, \text{ when } V_0 \text{ exists.} \qquad (6)$$

$$S_\ell = \sum_i S^\ell_{pi} + \sum_j S^\ell_{qj}, \text{ when } V_0 \text{ does not exist.} \qquad (7)$$

Figure 11 is an example which demonstrates the case where process (iii) is adopted. Suppose we want to find S_2 for the sample image of Figure 7 where $V_L = \{V_1', V_2', \ldots, V_6'\}$, (L=3). We get

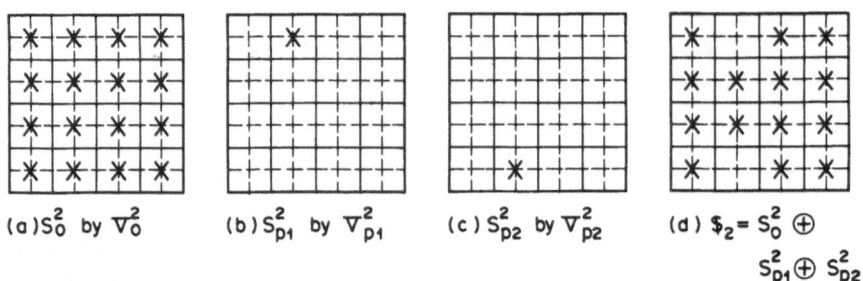

$(a) S_0^2$ by ∇_0^2 $(b) S_{p1}^2$ by ∇_{p1}^2 $(c) S_{p2}^2$ by ∇_{p2}^2 $(d) S_2 = S_0^2 \oplus$
$$S_{p1}^2 \oplus S_{p2}^2$$

Fig. 11. Restoration of an image with the fineness specified by $\ell(1 \leq \ell \leq L)$.

$V_{p1}^2 = (10,11,22)$, $V_{p2}^2 = (00,01,22)$, $V_{p3}^2 = (01,00,22)$ as V_ℓ, $(\ell = 2)$ by using (i), and $V_0^2 = (22,22,22)$ by using (ii). Then, by using the process (iii), V_{p3}^2 is deleted from V_ℓ. Figure 11(d) shows S_2.

So far we have explained the fundamental modes of content addressing. In addition, the blocks "inf" in the vectors detected by Mode 2 will provide information which characterizes the corresponding subareas. Selection of informative subareas which contain particular information is accomplished by successive applications of Mode 1 and Mode 2. Local control of detailed fineness is accomplished by the application of Mode 3 to particular subareas already specified by Mode 2.

This demonstrates the validity of the ELPIS for the organization of an image data retrieval system, clearly showing the flexibility of feature extraction and characterization of images. Figure 12 is a simulated result which demonstrates the local control of detailed fineness. The original image shown in Figure 12(a) consists of 256x256 pixels requiring a 4-step decomposition with a window of 4x4 elements. Figure 12(b) is that observed at the third level of the pyramid. In Figure 12(c), the fourth-level decomposition is partially adopted in the image of the third level.

4. CONCLUDING REMARKS

In the above we have investigated the fundamental functions required for implementing an advanced image data retrieval system. We now discuss briefly the communications between the ELPIS and the ARES for the organization and development of semantic data bases.

In a semantic data base, categories are selected so as to make the global and local description of features feasible at an arbitrarily selected resolution. For practical use, however, the

 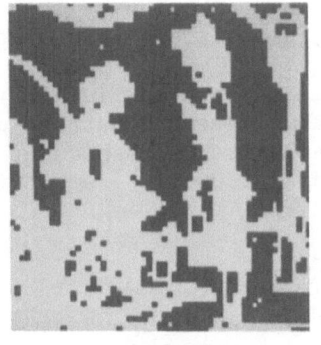 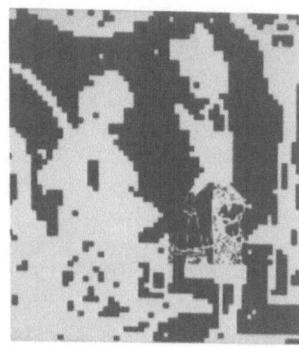

(a) Original Image   (b) Rough Image   (c) Partially Fine Image

Fig. 12. Simulated results of local feature extraction.

ability to adapt to the growth of semantic data bases should be provided. We sometimes need to add new categories to maintain reasonable association performance. New categories are selected on the
basis of a relevancy estimation of the associated data items.
Stored data items have to be partially updated in connection with
the addition of these new categories. We can expect the best performance when the updating is done for all data items already registered in the data base whenever new categories are adopted. This
may cause a serious degradation of system efficiency especially as
regards on-line use. Hence we limit the execution of updating to
the associated data items which are not completely coded. The
problem then becomes how to find the best compromise between association performance and the cost of updating. Optimization is
carried out by controlling the quantity δ which defines the limit
of the number of associated data items obtainable at every step of
the association.

The validity of the categorical coding of features for attain-
ing improved association performance and developing semantic data
bases has been demonstrated through the simulation of the ARES
applied to semantic word retrieval using the model of word meaning
proposed by Haralick and Ripken (1975). (See Ichikawa and Konishi,
1978.)

A system which incorporates multi-level knowledge sources
normally requires sophisticated communications because of the
necessity of communicating with knowledge sources at every stage
of feature extraction and analysis by going up and down a hierarchical level of knowledge as pointed out by Williams and Lawrence
(1977). However, as observed in Figure 1, the communication link-
age in the proposed system will be simplified to some extent. This

is due to the categorical organization of the data base which works
on a variety of features; global/local and fine/course, in parallel.

In conclusion, the image data retrieval system proposed in this
chapter provides a real-time advanced pattern processing capability
supported by a fully parallelized mechanism of extracting features
and associating related data items in a semantic data base.

5.   ACKNOWLEDGEMENTS

The authors gratefully acknowledge Professor Ramamoorthy of
the University of California at Berkeley and Professor Fu of Purdue
University for their encouragement and valuable advice. Professor
Klinger of the University of California at Los Angeles, Professor
Tanimoto of the University of Washington, Professor Riseman of the
University of Massachusetts, Professor Hanson of Hampshire College,
and Professor Kitahashi of Toyohashi University of Technology stimu-
lated the study on the pyramid a great deal through many discussions.
Thanks are also due to them.

6.   REFERENCES

Akin, O., and Reddy, R., "Knowledge Acquisition for Image Under-
     standing Research," Comput. Graph. Image Proc. 6:307-334
     (1977).

Hanson, A. R., and Riseman, E. M., "The Design of a Semantically
     Directed Vision Processor (Revised and Updated)," COINS Tech.
     Rpt. 75C-1, University of Massachusetts, Amherst, February
     1975.

Haralick, R. M., and Ripken, K., "An Associative-Categorical Model
     of Word Meaning," Artificial Intell. 6:75-99 (1975).

Ichikawa, T., "A Pyramidal Representation of Images," Proc. 4th
     Internat'l Joint Conf. Pattern Recog. (Kyoto, November 1978).

Ichikawa, T., and Konishi, Y., "A Behavioral Observation of ARES
     Through Semantic Word Retrieval," Tech. Rpt. AL78-50, Inst.
     Electronics Comm. Engrs. Japan, October 1978 (in Japanese).

Ichikawa, T., Sakamura, K., and Aiso, H., "A Multi-Microprocessor
     ARES with Associative Processing Capability on Semantic Data
     Bases," AFIPS Conf. Proc. 47:1033-1039 (1978).

Klinger, A., and Dyer, C., "Experiments on Picture Representation
     Using Regular Decomposition," Comput. Graph. Image Proc. $\underline{5}$(1):
     68-105 (1976).

Williams, T. D., and Lawrance, J. D., "Model-Building in the VISIONS
     High Level System," COINS Tech. Rpt. 77-1, University of Massa-
     chusetts, Amherst, January 1977.

A COMPOUND COMPUTER SYSTEM FOR IMAGE DATA PROCESSING

J. Iisaka, S. Ito, T. Fujisaki and Y. Takao

IBM Japan Ltd.

Tokyo Scientific Center, Tokyo, JAPAN

1.   INTRODUCTION

Practical image data processing (IDP) is comprised of a sequence of functions, i.e., image data acquisition, correction, editing, storing, analysis and feature extraction, display and output/dissemination. The major application areas are environment control, medicine, document processing, etc. Some users handle multiple frames for reference or comparison as well as a single frame at a time. A single CPU cannot afford to deal with various kinds of image data at various processing phases in an interactive multiuser environment.

In order to achieve the objective of better cost/performance, a Compound Computer System (CCS) is proposed to distribute specialized processing capabilities. The CCS is a computer complex composed of a general purpose computer system as a host, a slave computer as peripheral I/O controller and a special purpose parallel processor having a multi-microprocessor architecture. The host system performs such functions as image data base management, which is the unified handling of image and coded data, interactive operation by a query language through graphic and color displays, slave control and multi-microprocessor management in a virtual machine (VM) environment. The slave computer controls various special I/O devices such as a scanner, dot matrix printer, photoprinter, etc. The multi-microprocessor performs arithmetic operations on image data in parallel; typical IDP involves handling a large amount of data and performing repetitive operations on picture elements.

257

2.   SHARED INTELLIGENT I/O CONTROLLER (ETOILE)

In spite of recent progress in software and hardware technologies, computer applications which handle non-digital information sources such as maps, blueprints, voices, pictures, etc. are
difficult to implement on large-scale computers because few currently available operating systems can support inputs and outputs
from these types of information sources. One of the conventional
solutions to this problem is to incorporate a mini-computer in the
system configuration so as to control the peripherals which are
required for the handling of such information sources. However,
the involvement of such mini-computers usually makes the implementation costly and complex. This is because not only are the codes
for applications very complex but also the codes for communication
control between host and satellite computers. Another problem with
this solution is its high cost.

For these reasons, a general system, called ETOILE, was developed as a complement to the currently available operating system. ETOILE creates an operating system which is open-ended as
regards the architecture of the connectable I/O peripherals. Any
I/O peripherals, including laboratory instruments, can be incorporated in the configuration and, moreover, it is possible to ex-
pand the configuration by adding other peripherals in the future.
Another aspect of ETOILE is that it incorporates a programmable
I/O controller as a part of its nucleus so that interfaces to each
I/O peripheral can be logically defined in the most suitable way.
This logical interface concept makes it possible to interface
applications peripherals independently. Multiple unrelated users
on large computers can access special I/O peripherals easily and
concurrently. Thus each application will be provided with its own
high-speed, sophisticated data acquisition and data representation
capability.

Figure 1 schematically shows the architecture of ETOILE system.
The paragraphs below introduce the major components of the system.

2.1 Programmable I/O Controller

An intermediate computer equipped with I/O and program interrupt features is incorporated as a part of the ETOILE supervisor.
It enables a control program in the I/O controller to support multiple peripherals in a multi-programming environment.

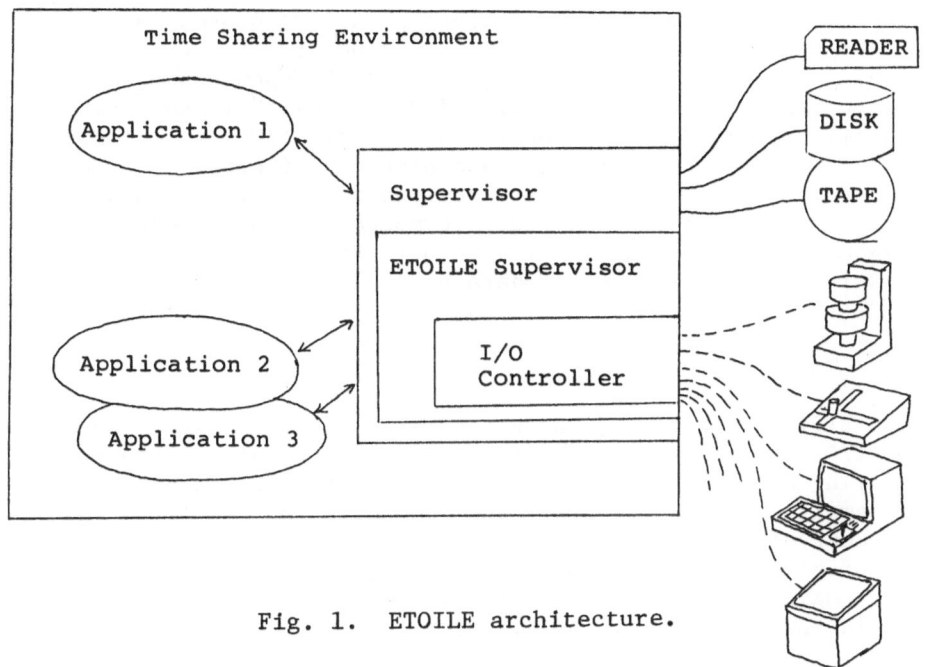

Fig. 1. ETOILE architecture.

2.2 Host-Controller Communication Facility

A facility which provides signal transfer and high-speed data transmission between the host computer and the I/O controller is used for host-controller communication.

2.3 I/O Peripherals

I/O peripherals are connected to the input and output registers in the I/O controller. In order to control these peripherals, the I/O controller provides I/O instructions to read and write the contents of the input/output registers. A change in the contents of any input register will cause an I/O interrupt to the I/O controller.

2.4 User-Friendly Interface (UFI)

In order to relax the complexity of the communication between users and the host supervisor, a user-friendly interface, referred to as the UFI, is incorporated through which the communications with the host supervisor can be easily performed. This interface can accept two types of ETOILE command requests from users:

(1) conversational command with prompting and (2) subroutine calls
in a high-level programming language environment. Functions supported by this interface are "link," "detach," "open subsystem,"
"close subsystem," "start peripheral I/O operation" (read, write
and control types), "halt I/O operation," "wait I/O completion,"
"query buffer parameters" (number and size of buffers), and "set
buffer parameters." For the start-peripheral I/O operation commands, four variations are provided (combinations of synchronous
and asynchronous transmission from/to buffers in main storage and
transmission from/to the user's data set).

2.5 The ETOILE Supervisor

The ETOILE supervisor, which resides both on the host computer
and the I/O controller, has the following management functions:
(1) handling and queuing of commands from users, (2) data transmission among the host computer, the communication facility, and
the I/O controller, (3) task management in the I/O controller,
(4) resource management of the I/O controller (resources include
CPU time, main storage, interruption level, I/O peripherals, I/O
buffers in the controller, etc.), (5) abnormal condition handling.

2.6 Subsystems

The ETOILE subsystems are user-written codes which map real
I/O operations from each of the I/O requests issued by the applications software. These subsystems are normally kept in the
library of the host computer. Whenever an open-subsystem request
from an application is issued, the relevant subsystem will be
fetched from the library and will be sent into the main storage of
the I/O controller. After this point, the subsystem code will
give the actual protocol for each of the I/O requests from that
application. The role of the mapping from requests to real I/O
operations is to act as an intermediary between a logical interface and a real I/O peripheral, i.e., to control and to perform
real I/O operations with the peripheral pursuant to the specifications given for that interface. This intermediate service of the
subsystem enables applications to be designed as peripheral-
independent. It also enables application software writers to
create code for performing peripheral I/O operations without full
knowledge of the peripheral.

3. IMAGE DATA BASE

One of the key aspects of image handling technology is to
store and retrieve both coded data and image data in a systematic

way. To facilitate this an interactive database system has been implemented as a research prototype. The basic idea of this system is that "image" is regarded as one of the data types. Each image is treated as a data element just like a number or a character string within the framework of a relational database. This approach enables the usage of a simple but powerful relational data language for storage and retrieval of both coded data and image data in a unified way.

3.1 System Overview

As shown in Figure 2, the prototype system now under development consists of four subsystems: (1) the interaction subsystem, (2) the edit/process subsystem, (3) the workspace subsystem and (4) the database subsystem.

The *interaction subsystem* controls man-machine interaction via a workstation having character and image display. The interface provided here is command-oriented. This subsystem analyzes the command entered through the keyboard and then dispatches the appropriate other subsystems. The screen management of both character and image display is done by this subsystem. This facilitates the scrolling of each screen.

The *edit/process subsystem* consists of an image editor and a user function manager. The image editor supports basic image editing functions such as extraction, overlay, magnification, etc. which are common to all image-oriented applications. To meet the diverse requirements of image processing, the user function manager is provided with which users can add their own commands and processing functions.

The *workspace subsystem* is a collection of named temporary image data kept only during a processing session. Image data generated by editing/processing or scanning are first stored in the workspace and then, with related coded information, stored in the database and retrieved from the database into the workspace as required. This approach simplifies the user commands because database commands and edit/process commands can be completely separated. It also cuts the database access overhead if the same image data are accessed many times during a session.

The *database subsystem* controls database accesses and provides an easy but flexible interface to access both coded data and image data in a unified way. From the user's viewpoint, a database is a collection of tables (a relational database) and each image is treated as a data element. Image data are further categorized into two sub-types: binary image and halftone image. Binary image data

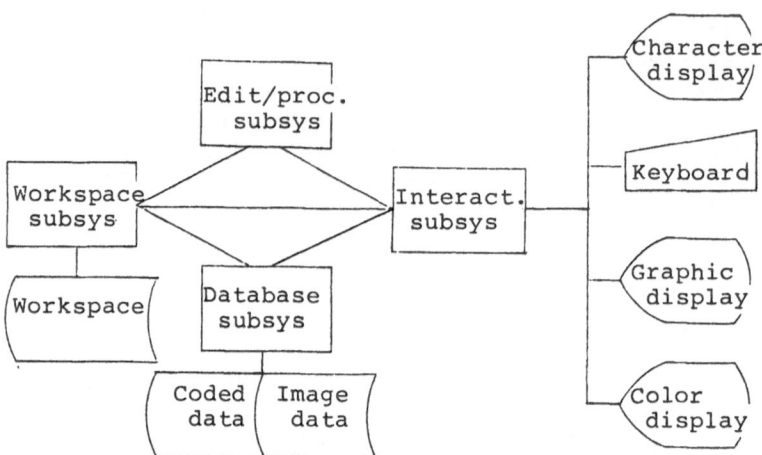

Fig. 2.   Functional system structure.

are compressed with 1-dimensional run-length coding using 4, 8, 12, and 16 bit code words. This fairly simple algorithm is selected as a tradeoff between the compression time and the processing time. Halftone image data are currently restricted to only 4 bits/pel (16 gray levels) and stored in the database without any compression.

Although color images are not supported as a special data-type, we can handle them by interpreting the pel intensities of halftone images as color codes (in this case we can get only 16 colors) or combining three halftone images as the red, green and blue components. Within the system, coded data and image data are managed separately because of their different characteristics. Each image is given a unique internal identifier and the association between coded data and image data is established by storing these image internal identifiers in a coded database. As a coded data manager, SYSTEM-R (an experimental relational data base management system developed at IBM San Jose Research Laboratory) is used. Hence SEQUEL (Structured English Query Language) is used as a data manipulation and definition language with slight modifications for image data handling.

4.   SPECIAL-PURPOSE PROCESSOR

When using a conventional general-purpose computer (GPC) for image data processing (IDP), a heavy CPU load is always one of the most critical problems. For this reason, better algorithms for IDP and hardware technology to implement them should be developed in order to make IDP practical in an interactive multi-user environment. From the parallel processing viewpoint, IDP is characterized

by handling a large amount of data and performing repetitive operations. Considering other IDP aspects such as image data base management, interactive operations using display devices, etc., in a
multi-user environment, IDP needs tremendous computing power. Thus,
IDP applications need a more powerful computer system than a conventional one to attain the objective of better cost/performance.

4.1 Structure of Special-Purpose Processor

A Special-Purpose Processor (SPP) should have a multi-microprocessor structure for processing image data. The multi-microproc-
cessor consists of a control processor (CP), arithmetic processors
(AP's), and a common memory (CM). The GPC has such functions as
loading a microprogram in the SPP, initiating SPP tasks, image data
transmittion, gathering calculated results, and supporting microprogram development. On the other hand, the SPP has such functions
as loading arithmetic programs into the AP's, arithmetic processing
of image data by the SP's and dynamic control of each arithmetic
operation by the CP.

The CP can dynamically arrange the microprogram modules of each
AP, at an appropriate time, from the host computer's disk. In addition, the CP can supply partitioned data to each AP.

Figure 3 shows a block diagram of an experimental special purpose processor architecture. The interconnection between the AO's
is through the common memory under the control of the CP's commands
according to each AP's interrupts. Processed data are transmitted
to the host computer through the I/O buffer directly. Microprogram
modules are stored in the host computer's disk storage.

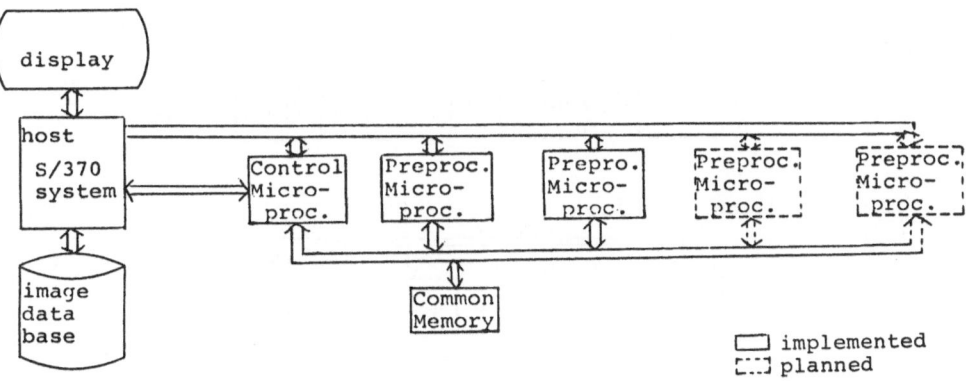

Fig. 3. Experimental special purpose processor.

4.2 Parallel Processing Schemes

Parallel processing schemes in the multi-microprocessor architecture are proposed for the purpose of increasing the processing
power of the special purpose processor. Two types of parallel processing schemes (Figure 4) at the task level may be realized in the
multi-microprocessor architecture.

First, parallel processing may be performed on the continuous
data stream in a pipeline mode. In this mode, after the CP acquires
data, a block of data is transmitted to the common memory by the CP.
The first AP processes the data and its output data is written to
another area of the common memory to continue with the next AP's
processing of the output data of first AP. At this time the first
AP is processing the next block of data.

Second, parallel processing is performed on each segment of
data by the same/different algorithm in a simultaneous mode. In
this mode the image data are segmented into several blocks by the
host computer and transmitted to the I/O buffer of each AO. The
data are processed by each AP simultaneously. In parallel processing it is important to balance the timing between loading of a program module and transmission of a data block. Both the data stream
and the program stream should be strategically planned according to
the processing sequence.

A user can select appropriate concurrent operation modes (pipeline, parallel, or mixed mode) at a task level in accordance with
the requirements of the application. Therefore, any application
whose algorithm and data have a well-defined structure so as to be
easily decomposed in a task-level parallel processing organization,

1. Pipeline Mode

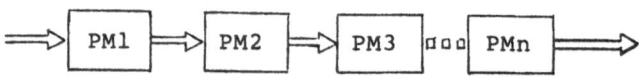

2. Simultaneous Mode

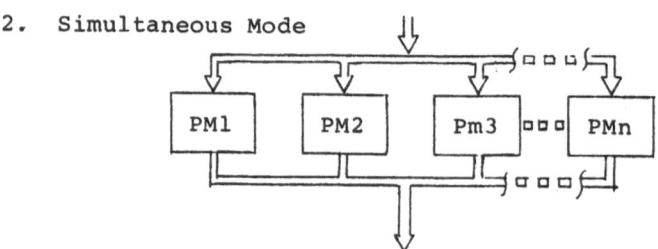

Fig. 4. Parallel processing modes.

will be able to get good cost/performance in the CCS environment. System throughput is one of the major objectives of the CCS and is achieved with a high degree of parallelism between the SPP and the host system, and between data transmission, the CP's and the AP's.

5.   CONCLUSION

The whole compound computer system described in this chapter operates under IBM VM/CMS. The user oriented image processing commands are defined and executed as CMS commands. The system described here is presently used by many users. The users interact with the system through comprehensive commands utilizing various types of input-output according to their wishes. The user evaluation is very good especially as regards easy access, data transferability, and easy upgrading.

6.   ACKNOWLEDGEMENTS

The authors wish to express thanks to Mr. M. Ohkohchi, Mr. M. Morohashi, and Mr. H. Mashita who did the design and implementation of the ETOILE system.

INTERACTIVE TECHNIQUES FOR PRODUCING AND ENCODING COLOR GRAPHICS

T. Kamae, T. Hoshino, M. Okada and M. Nagura

Visual Communication Development Division

Yokosuka Electrical Communication Laboratory, NTT, JAPAN

1. INTRODUCTION

The Video Response System (VRS) with which the Nippon Tele-graph and Telephone Public Corporation (NTT) is now experimenting is a public visual information service using home TV sets as described by Nakajima (1977). The VRS stores visual information in a large capacity video store. The stored video information is converted into a TV signal by using a video codec and is tempo-rarily stored in the video frame memory (Figure 1). The TV signal is read out of the video frame memory at a rate of 30 frames/sec and is transmitted to a subscriber's video adapter through a video transmission line. Among the most important factors in the VRS is the video store and the visual information stored therein.

Color graphics, consisting of simple lines and uniform colors covering closed areas surrounded by the lines, are frequently employed to represent various information in the form of visual images. Production of these color graphics usually needs skilled labor and is, therefore, costly. The first part of this chapter deals with computerized techniques for producing color graphics.

Short access time to the stored information is one of the most important points in a centralized system such as VRS. Another im-portant factor is re-writability since it is necessary to update stored information. Digital mass storage meets this requirement. The encoding of color graphics which is necessary for digital storage of visual images is discussed in the second part of the chapter.

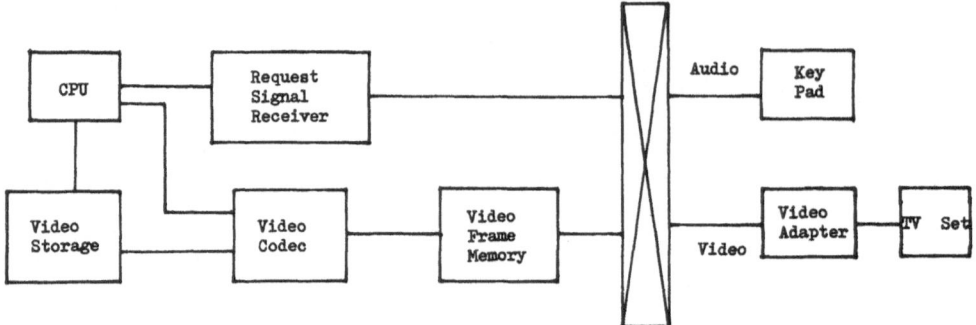

Fig. 1. Block diagram of the video response system.

2.   INTERACTIVE PRODUCTION OF COLOR GRAPHICS

This section discusses computer techniques for producing color
graphics.

2.1   Input of Line Drawings

There are many ways to input line drawings. Although a data
tablet can be employed as a direct input tool, a facsimile transmitter is a more practical tool. Line drawings on paper are
scanned by a facsimile transmitter and digitized as binary data.
The binary data are encoded line-by-line in the form of run-length
codes. These run-length coded data are decoded into a TV signal at
a rate of 30 times/sec and displayed on a CRT.

2.2   Trimming Lines

Line patterns taken from a facsimile transmitter may not be
uniform and may contain notches. A line thinning algorithm is useful for trimming these patterns. This also may be done by man-
machine interactive trimming. As shown in Figure 2, picture elements which should be changed from black to white or vice versa are
pointed to by using addresses in an enlarged portion of the original
line drawings.

2.3   Detection of Uncolored Closed Areas

Line drawings represented as a series of run-length codes are
processed to detect closed areas. Figure 3 shows a portion of a
run-length coded picture using the method of Hoshino et al. (1978).
We first scan the run-length codes and find an uncolored code, C_{11}
for example. We next scan codes on the second line and find codes

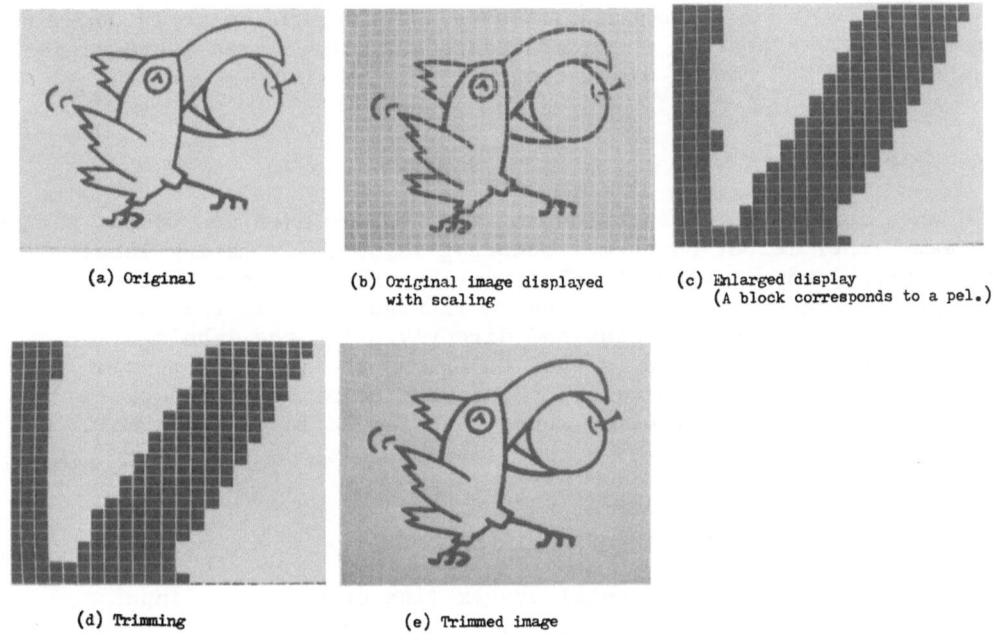

(a) Original

(b) Original image displayed with scaling

(c) Enlarged display (A block corresponds to a pel.)

(d) Trimming

(e) Trimmed image

Fig. 2. Man-machine interactive trimming.

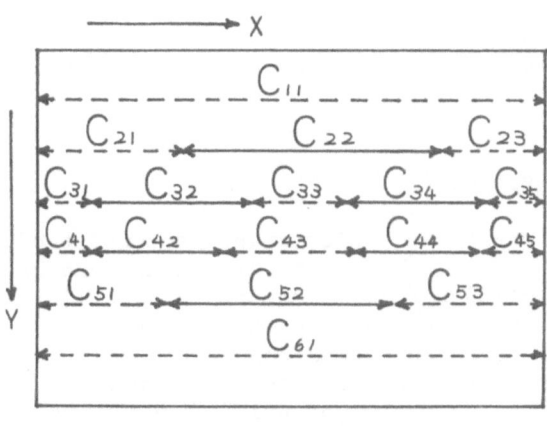

C_{mn}: nth RL code on the mth line

Fig. 3. Portion of run-length-coded picture.

C_{21} and C_{23} which have some X addresses and the same color in common with C_{11}. Repeating these procedures, a closed area containing
C_{11}, C_{21}, and C_{23} can be disclosed.

2.4 Coloring

When run-length codes belonging to the same area are found, the
the same label is put on them. Coloring is actually making labels
which correspond to colors. Color run-length codes used in the
present experimental system are shown in Figure 4. In this system
64 kinds of colors can be assigned directly using the labels R, G
and B. The extension code is used to extend the lengths of color
runs without changing the color. Hence the coloring procedure is
to replace white runs by color runs. An example is given in Color
Plate 16*.

2.5 Outline of the Experimental System

In the present experimental system line drawings are input
from a flying spot scanner and coded in the run-length form shown
in Figure 4. Lines can be trimmed as shown in Figure 2. Then,
detecting a closed area, the system modifies it by coloring it with
a particular color. When an operator inputs a color number from a
typewriter, then the area is colored. When the color is satisfactory, the operator inputs "Y" to accept the color. The system
then detects the second closed area and the same procedure is repeated.

3.   ENCODING OF COLOR GRAPHICS

In this section we discuss procedures for encoding color
graphics for digital storage.

3.1 Color Run-Length Coding

As stated in the above, color graphics consist of lines and
regions of uniform color. This kind of picture can be effectively
encoded in the run-length form as demonstrated by Ishii et al.
(1977). Figure 4 shows the run-length codes employed in the
system. The color code represents continuation of the same color
for up to (2^8-1) picture elements (pels). Since the extension
code can represent a longer run of pels of the same color, any run
whose length is smaller than $(2^{15}-1)$ can be represented by the
three byte code. The monochrome code is used to represent a line,
which usually separates different color areas.

*The color plates will be found following page 46.

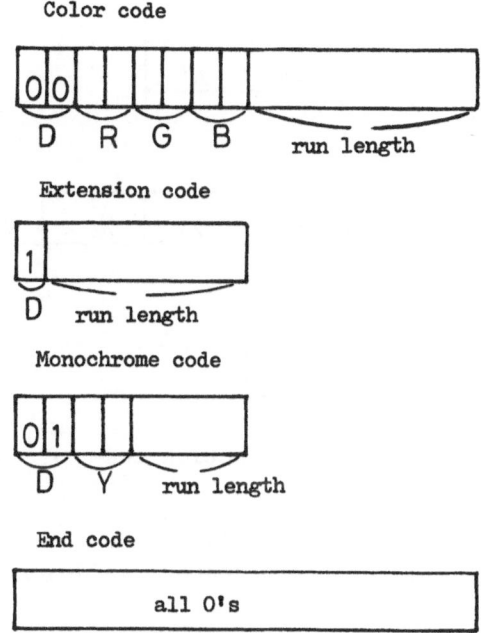

Fig. 4. Color run-length code
(D = distinction code).

Using this color run-length coding scheme, color graphics corresponding to a standard TV frame can be represented by an amount
of data equal to only 20K to 25K bytes. (Note that DPCM coding
needs 100K to 150K bytes to encode a TV frame.)

3.2 Boundary Coding

Since lines usually have nearly uniform width in color graphics,
they can be handled as line drawings as in the computer graphics
field. Lines are approximated by straight line and circular arc segments. Circular arcs are found to be very effective in approximating
curvilinear parts of line drawings smoothly. Thus line drawings can
be represented as a series of end points plus a distinction code
which distinguishes a line segment and a circular arc segment.

To encode colors, one point contained in each closed area is
selected. A color code is attached to the address of this point.
Thus color graphics are represented by line segments, arc segments,
and/or colors attached to each point address, as shown in Figure 5.

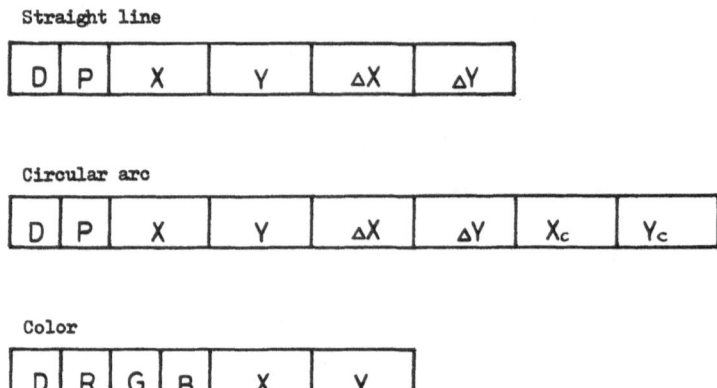

Fig. 5. Boundary code (D = distinction code; P = parameter).

To approximate line drawings by line and arc segments, thinning and fitting of lines are necessary as described below.

3.2.1 *Thinning of Lines*

Line drawings are stored in a TV frame memory. An unscanned boundary is detected by consecutively scanning picture elements in the raster scanning manner. The thinning procedures proceed as follows:

(1)   All boundaries are found and labeled by repeating the raster scan and the boundary scan alternately. When raster scanning detects the 0 to 1 transition belonging to a boundary a_i, boundary scanning starts from the transition point as shown in Figure 6. In boundary scanning, deletable pels on the border a_i are labeled i+2 and undeletable pels are labeled 2. Here deletable pels are 8-deletable pels as defined by Rosenfeld (1970). Undeletable pels are actually a part of the skeleton. In this way all boundary and outmost pels on the boundaries are found.

(2)   Boundary scanning goes deeper into interior points. Here pels whose label is i+2 are regarded as "0" pels. In the boundary scanning process, the eight pels X_0 through X_7 are considered in the neighborhood of the present pel X as shown in Figure 7. When X_0 is 0 and X_2 nonzero, X proceeds to X_2.

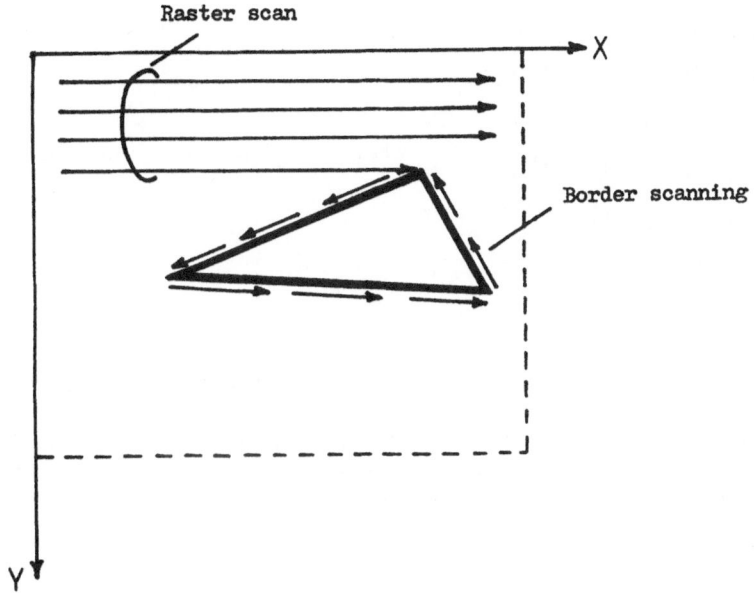

Fig. 6. Boundary following method.

X_3	X_2	X_1
X_4	X	X_0
X_5	X_6	X_7

Fig. 7. Picture element (X) and
neighborhood $(X_1, \ldots, X_8)$.

When both X_0 and X_2 are zero, and X_1 nonzero, X proceeds to X_1. If X_0, X_1, and X_2 are all zero, the
scanning direction turns to the left and the matrix
shown in Figure 7 is reorganized. If X_0 is nonzero,
the scanning direction turns to the right. An illustration is provided in Figure 8.

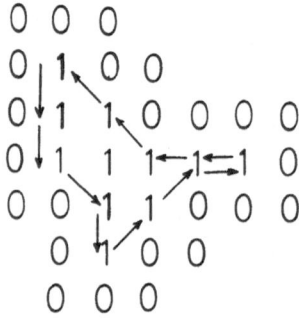

Fig. 8.   Example of scanning method.

By repeating this scanning procedure along the boundary the skeleton can be obtained. Figure 9 shows an example of an original figure and its skeleton. According to this algorithm, most picture elements are scanned only once and hence the time required for scanning can be reduced substantially.

3.2.2 *Approximation of Lines*

In the next stage, the skeleton of the line drawing is approximated by straight line and circular arc segments. For approximation, it is necessary

(1)   to extract bends or nodes, which are end-points of line segments,

(2)   to separate linear components and curvilinear components, and

(3)   to approximate curvilinear components by a series of circular arcs.

Following the method of Rosenfeld and Johnson (1973), the bend index $t_{i,d}$ at pel D_i is defined to be $\cos \theta_{i,d} + 1$, where $\theta_{i,d}$ is the angle from vector $\overline{D_i\,D_{i-d}}$ to $\overline{D_i\,D_{i+d}}$ in Figure 10. The quantities D_{i-d} and D_{i+d} are the dth pels from pel D_i. The bend index has the following properties:

(1)   When line D is straight, the bend index is nearly zero.

(2)   When line D is circular, the bend index is almost constant for fixed d.

Fig. 9.   Original line drawing (left) and its skeleton (right).

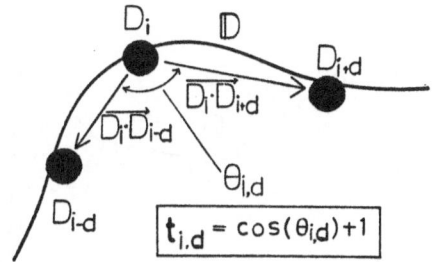

Fig. 10.   Graph defining the bend index.

    To distinguish curvilinear portions from straight portions,
several values of d are employed. Let $d_1 < d_2 < \ldots < d_n$. Bend
indices t_{i,d_1} are calculated for every pel D_i in D. Let D_1 be a
set of pels whose bend index t_{i,d_1} is smaller than a threshold T_1.
Then bend indices t_{i,d_2} are calculated for every D_i in $D - D_1$.
Similar procedures are repeated. Thus a portion of D which has
the larger bend index is selected at an earlier stage. There are
various ways to determine thresholds T_k and the values of d_1
through d_n.

Line D consists finally of curvilinear portions D^C and straight portions D^L as shown in Figure 11. Straight portions with a small number of pels are integrated into neighboring curvilinear portions. Next the algorithm examines D^C with a small number of pels. A curvilinear portion with fewer pels than a threshold is recognized as a node, as shown in Figure 11.

In the final stage curvilinear portions are approximated by circular arcs. The bi-arc approximation method used is based on the work of Kosugi and Teranishi (1977). This method can be applied as shown in Figure 12 where the line drawing of an elephant is approximated by using 51 straight lines and 81 circular arcs.

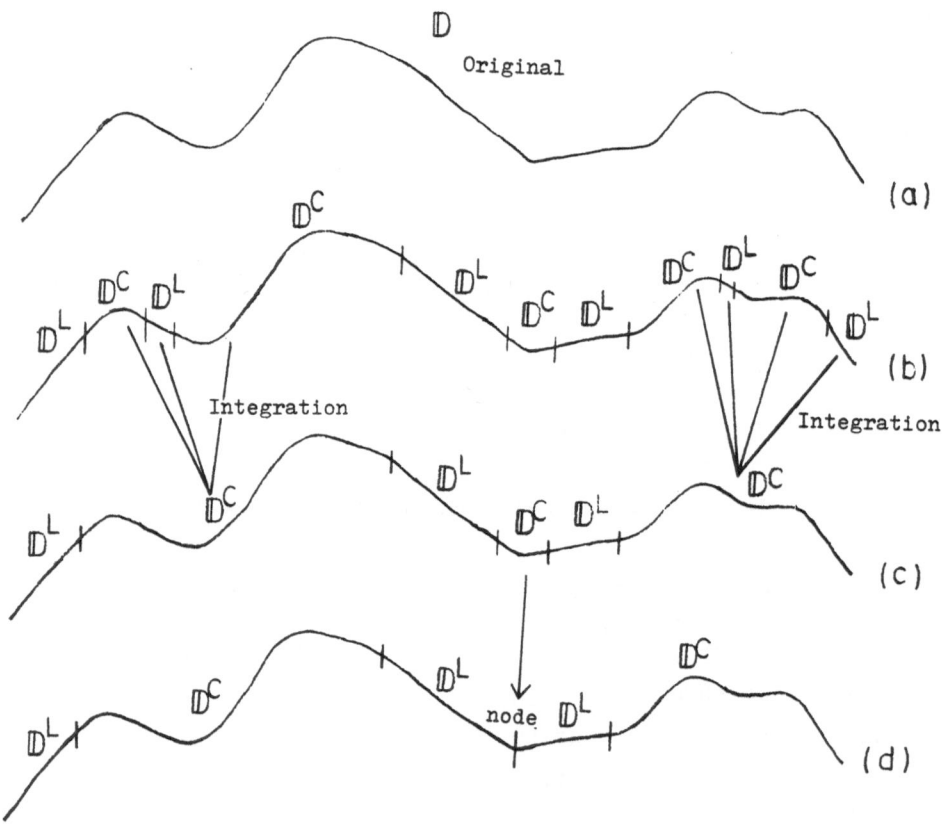

Fig. 11. Integration of segments of a line and detection of a node.

Fig. 12. Example of encoding method (left to right): original
skeleton; approximated skeleton; superposition of original and
approximated skeleton; end points of all circular arcs and straight
lines found by encoding algorithm.

3.2.3 *Merits of Boundary Coding*

Boundary coded color graphics are represented only by the
addresses of pels. Hence color graphics encoded in this manner
can be enlarged or reduced easily at any time. This is one of
the strong points of this coding scheme and may lead to the development of a pictorial data base using this technique.

4.   CONCLUDING REMARKS

Interactive techniques for producing and encoding color graphics have been outlined in this chapter. In the production process
difficulty still exists in line trimming or thinning. In the
coding process further work is needed on this problem.

5.   REFERENCES

Hoshino, T., Kawakubo, S., and Yonezawa, S., "On a Simple Coloring
    Equipment," Proc. Image Engrg. Mtg. Inst. Electronics Comm.
    Engrs. Japan (February 1978) (in Japanese).

Ishii, A., Kishino, F., and Kawakubo, S., "Run-length Coding Tech-
    nique for Color Video Graphics Filing System," Proc. 1977
    Picture Coding Symposium (Tokyo, August 1977).

Kosugi, M. and Teranishi, T., "Construction of a Curve Segment with
    Two Circular Arcs," Trans. Inst. Electronics Comm. Engrs.
    Japan 60-D(11):944-951 (1977) (in Japanese).

Nakajima, H., "An Experimental Interactive Video-Audio Information
    System," Proc. Comm. Systems Mtg. Inst. Electronics Comm.
    Engrs. Japan (July 1977) (in Japanese).

Rosenfeld, A., "Connectivity in Digital Pictures," J. Assoc. Com-
    put. Mach. 17(1):146-160 (1970).

Rosenfeld, A. and Johnston, E., "Angle Detection on Digital Curves,"
    IEEE Trans. Comput. C-22(5):875-878 (1973).

IMAGE PROCESSING UNIT HARDWARE IMPLEMENTATION

M. Kidode, H. Asada, H. Shinoda and S. Watanabe

Toshiba Research and Development Center

Kawasaki, JAPAN

1. INTRODUCTION

Conventional digital computers are inefficient for even simple local image processing operations in terms of cost-effectiveness. Image processing by serial computers requires a large amount of computing time. Programs and data are stored in the same memory and all operations are serially executed in the so-called "von Neuman" architecture, even though image processing could usually be performed mostly in parallel. Two-dimensional image data require a large amount of memory storage and usually exceed the main memory capacity. This results in much overhead time transferring image data between the main memory and secondary storage, e.g., magnetic disk. Therefore, we have elected to implement cost-effective image processing by developing special hardware optimized to overcome the above problems.

Several image processing machines have been proposed and some of them have already been realized as described by Fu (1978). It is becoming possible to implement such special hardware units at a modest cost with modern digital technologies in memory chips, arithmetic circuits and advanced digital control circuits. The problem now to be solved is to determine what image processor architecture is superior in terms of cost-effectiveness, or cost-performance ratio. Based on trade-off studies in speed, economy, and generality, a microprogrammable image processing unit (IPU) has been developed by us to improve the functional capabilities of another machine (a local parallel pattern processor-PPP) described by Mori et al. (1978).

In Section 2 the major concepts for the IPU design are discussed. In Section 3 the hardware implementation details are described. Software implementation techniques are listed in Section
4, with several applications.

The IPU is presently playing an important role as a high speed
image processor in an interactive image processing system as described by Asada et al. (1978). Basic image processing functions,
such as the two-dimensional FFT, vector operations, spatial filtering, affine transformations, data conversion, logical filtering, etc.
have been implemented using relatively simple hardware.

2.   DESIGN CONCEPTS

The IPU has employed characteristic concepts described below
in its design so as to improve image processing cost-effectiveness.
These ideas contribute to rapid execution time and to reduced hardware cost and complexity.

2.1   Image Data Hierarchies

The IPU is directly connected to an image memory unit in an
interactive image processing system as shown in Figure 1. This
configuration minimizes the "idle" data transfers among hierarchical levels of the memory. The time for these data transfers, which
are not necessary for the essential calculations in image processing,
generally consume a large fraction of the total throughput time in
conventional digital computers.

The image memory consists of four 512x512 pixel blocks, each
having 8-bit gray levels and four 512x512 planes for binary graphics.
Advanced memory fabrication technology makes it possible to build
such a memory storage unit large enough to contain standard images.

2.2   Important Image Processing Functions

There are several types of frequently or commonly encountered
image processing functions, listed by Hunt (1976), most of which
take a large amount of execution time. The most important of these
functions are discussed as follows.

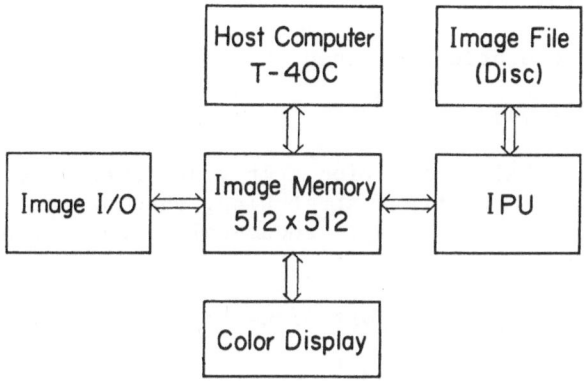

Fig. 1. Interactive image processing system.

2.2.1 *Sum-of-Products--Two Dimensional Convolution*

This operation is expressed by the equation

$$G(x,y) = \sum_i \sum_j W_{ij} F_{ij}(x,y) \tag{1}$$

where $F_{ij}(x,y)$ are neighboring pixels around the point (x,y) in the
input image and W_{ij} values are weighting coefficients. Depending
upon the values of the coefficients this computation is frequently
used in image enhancement, averaging, differentiation, correlation,
matching, etc.

2.2.2 *Fast Transforms*

The general transform considered here is given by

$$(u,v) = \sum_x \sum_y F(x,y) W^{ux+vy} \tag{2}$$

where W is the transform kernel. The most commonly used fast transform is the Fourier transform (W = exp(-jz)). Fast Fourier Transform is frequently used in restoration, enhancement, reconstruction,
coding, compression, etc.

2.2.3 *Point Mapping--Data Conversion*

This type of operation is given by

$$G(x,y) = \phi\{F(x,y)\} \tag{3}$$

where ϕ is a single-valued function. It represents a common technique which is used in contrast stretching, shading-correction, gray-scale mapping, thresholding, pseudo-color generation, etc.

2.2.4 *Linear Coordinate Transformation--Affine Transformation*

This transform is expressed by the following two equations

$$\begin{cases} G(x,y) = AF(x,y) \\[2mm] A: \begin{pmatrix} X \\ Y \end{pmatrix} = \begin{pmatrix} a & b \\ c & d \end{pmatrix} \cdot \begin{pmatrix} x \\ y \end{pmatrix} + \begin{pmatrix} e \\ f \end{pmatrix} \end{cases} \tag{4}$$

where A is an affine transformation specified by a linear transformation matrix (a,...,d) and shift values (e and f). This operation is used in magnification, shrinking, rotation, shifting, etc.

2.2.5 *Vector Operations*

Some vector operations are frequently used for color or multi-spectral image data where each pixel value is considered as a vector with three or more components. For example, the maximum likelihood pixel classification algorithm can be realized by such a vector operation.

2.2.6 *Summary*

In addition to the above operations logical filtering and region labeling for binary pictures, histogram computation, and pixelwise arithmetic/logical operations are frequently used. These basic image processing functions have all been implemented by specially designed circuits whose common implementation techniques are embedded in local operations combined with the raster scan regime.

2.3 Local Parallel Operations

There are two types of hardware implementation to achieve parallelism in image processing: locally parallel and fully parallel. The most intuitive idea is a parallel array of identical processing elements corresponding to each pixel which works simultaneously on all of the image data. Image processors of this fully parallel type, however, face several problems in hardware realization and processing capability. These problems involve the very large number of circuit cells required as well as interconnection wiring and data communication among neighbors.

These considerations have led to an IPU design which carries out image processings on a locally parallel basis. Specially designed circuits accomplish the local parallel operations (parallel data access and computation in the local window) which sequentially scan the entire image in the raster mode. The combination of locally parallel operations and sequential scanning fits the serial data transfer characteristic of an image memory and also reduces hardware cost and complexity.

2.4 Flexible Control

Even with the above basic important functions realized in special purpose modules, it is still necessary to perform more global and complex control functions. For this purpose the IPU has a dynamically rewritable microprogram control architecture capable of combining basic functions or using a microprocessor which has been included in the system.

2.5 Word Length

The two-dimensional FFT and vector operations must be performed precisely in the sense of the number of bits per word, while other basic functions are computed with 8-bit integers. Using a small number of bits and fixed point data has always restricted arithmetic processings (due to overflow and underflow) as well as the available dynamic range. From a trade-off study considering hardware complexity and computational precision, we chose a 24-bit word length using floating data (8-bit exponent and 16-bit mantissa) to be fed through the corresponding circuits in the IPU. According to these design considerations, the IPU has been developed to perform major image processing functions approximately 100 times faster than by conventional digital computers.

3. HARDWARE IMPLEMENTATION

An overall block diagram of the IPU is depicted by Figure 2. The host computer is the TOSBAC-40C minicomputer with 64 KB memory. The host computer controls the operational modes of the IPU and the image memory. Each image processing program assigns the image

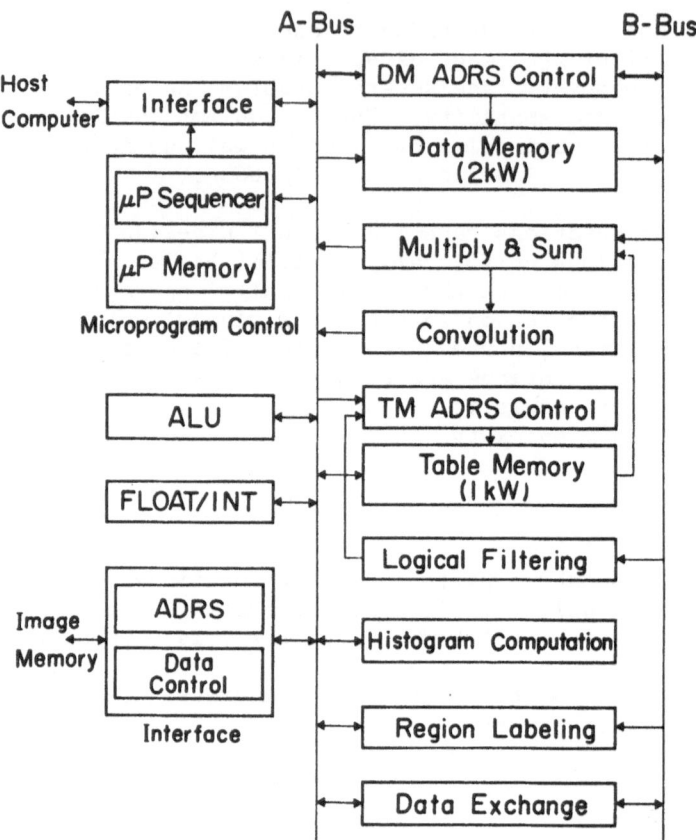

Fig. 2. Block diagram of image processing unit.

memory to input and output, sets parameters in the IPU, specifies the function using the execution command, and then senses the end of execution from the IPU. Another important role of the host computer is to transfer microprogram data to the IPU. Details on these software implementations are presented in Section 4. The interface from the IPU to the host computer controls the transfer of microprograms, commands, command parameters, and generates an interrupt signal at the end of each operation.

The image memory is directly connected to the high speed data bus, which has a transfer rate of 1M words (36 bits per word) per second. The interface to the image memory has an address calculator and data controller. The address calculator independently controls the input and output addresses, while the data controller

transfers image data simultaneously to or from any block or plane
of the image memory.

There are specially designed modules for executing all basic
functions: (1) two dimensional convolution, (2) logical filtering,
(3) region labeling, (4) histogram computation, multiplication and
addition with 24-bit floating point data, etc. The data memory and
table memory are provided for using a scratch-pad memory, table
look-up memory, and so forth. Two sets of 48-bit buses transmit
the internal data between functional modules.

The IPU has a microprogram control architecture, where all the
functions are instructed by the microprogram controller (μC). The
μC consists of the program sequencer and the microprogram memory
and has a 24-bit arithmetic/logic unit (ALU) and associated data
conversion circuit (FLOAT/INT).

3.1 Microprogram Control and ALU

The μC plays an important role in the IPU as a central brain.
The μC functional roles are communication to the host computer, interface control to the image memory, internal data bus control,
control of functional modules and pixelwise ALU operations. Programs for executing all basic functions are stored in the read-only
microprogram memory, while any user-defined programs can be transferred from the host computer to the random access memory and then
executed. The word length of a microprogram instruction is 48 bits
with a cycle time of 300 ns.

3.2 Address Control

The address calculator is capable of computing next addresses
automatically and checking the window for the image data to be processed using local operations. The addressing capability within
the local window is independent of the address calculator. The
input and output addresses each consist of 18-bit words (two 9-bit
addresses for the x and y coordinates). Each 9-bit address is independently calculated, incorporating a high speed ALU and counter
for input and output addresses, respectively. Initial address data
are set by the host computer at the beginning of an image processing operation.

The affine transformation is performed by the address calcu-
lators. Equation (4) in Section 2 can be rewritten as

$$\begin{cases} x = pX + qY + r \\[2em] y = sX + tY + u \end{cases} \tag{5}$$

where X and Y are the input addresses and x and y output addresses.
As the output address values of X and Y incremented in the raster
mode by the output address counters, the input address values are
calculated using equation (5) in the high-speed ALU. In digital
image processing, image data are expressed in the form of spatially
sampled quantities so that (x,y) and (X,Y) are integers. Since the
computed values (x,y) are not always integers, the input address
calculator should have a sufficient number of bits so as to perform
the accurate affine transformation and to interpolate the gray level
at (X, Y) from several surrounding image data points which have
integer values near the point (x, y).

In practice, all the values in the high-speed ALU are expressed
using 20 bits with a 9-bit fraction, so that an image of 512x512
pixels can be rotated with a precision of one part in 512. Three
interpolation algorithms can be employed by the IPU: (1) the nearest neighbor method, (2) the bi-linear interpolation method using
4 neighbors, and (3) the cubic spline function method with 16 neighboring pixels. In addition to these features the address controller
has additional registers to calculate the addresses of neighboring
pixels at high speed.

3.3 Data Memory and Table Memory

The IPU data memory and table memory are commonly used to implement several functions. Their data accesses are performed via
two sets of internal buses on either 8-bit, 24-bit, or 48-bit data.
The data memory consists of four sets of 512-word blocks (24 bits
per word). It serves as the line buffer memory for feeding image
data to local operation function modules for executing spatial filtering, logical filtering, and region labeling. It is also used
for storing complex data and vector data in FFT and vector operations, respectively. The table memory is comprised of four sets of
256-word blocks. It provides an important function in storing the
weighting coefficients (spatial filtering), the output code table
(logical filtering), the look-up table (data conversion), the kernel
sine and cosine data (FFT), and some parameter data (vector operations). It is also employed for computing the gray-level frequency
distribution (histogram computation) and for memorizing the con-
nected-region labels (region labeling).

Execution of equation (3) for point mapping is also accomplished
with the table memory. Here the conversion function ϕ is stored in

tabular form for table look-up. Since the gray levels of a given
input image are expressed by 8 bits, the output value corresponding
to any input gray level can be stored in 256 words in the table
memory. Histogram computation is also performed with the aid of
the table memory and histogram counter. The contents of the table
memory are incremented so as to count the frequency of points with
the same gray level in the input image.

3.4 Spatial Filtering

Rapid computation of the sum-of-products operation expressed
by equation (1) is realized in hardware by a sophisticated implementation method. In order to reduce the number of arithmetic elements required (multipliers and adders) a time-shared pipeline control with a partial sum-of-products circuit was successfully adopted
from the work of Mori et al. (1978). The refined configuration reduces the number of multipliers and adders from m^2 to m. This also
results in an execution time improvement in comparison with the
fully-parallel configuration.

In practice, the weighting matrix size was determined to be
12x12 as a result of a trade-off study between speed and economy
and the common utilization of arithmetic elements when performing
the FFT function. The data memory buffers twelve lines of image
data, while the weighting matrix is preset in the table memory.
Partial sums are accumulated in the spatial filtering module to
complete the total sum-of-products operation and then normalize the
results to 8-bit output data.

3.5 Fast Transform with Multiplier and Adder

The FFT computations are performed by iterative butterfly
block operations. Figure 3 shows the butterfly diagram for a complex signal with 8 samples. The Fourier transform of x(n) can be
expressed as

$$X(k) = \sum_{n=0}^{N-1} x(n)W_N^{kn} \quad (0 \le k \le N-1) \tag{6}$$

where N = 8 and the kernel $W_N = \exp(-j2\pi/N)$. The fast Fourier
transform (FFT) algorithm processes a signal of N samples through
$\log_2 N$ stages, in which N/2 sets of the butterfly operations are
computed. (See Gold and Rader, 1969.) The butterfly operation,
including multiplications and additions of complex signal data in

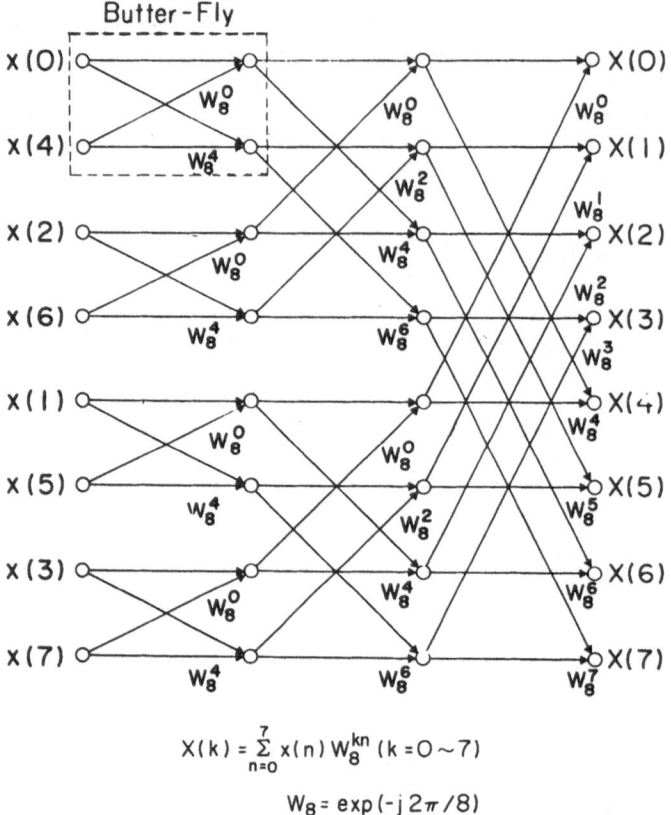

$$X(k) = \sum_{n=0}^{7} x(n) W_8^{kn} \ (k = 0 \sim 7)$$

$$W_8 = \exp(-j 2\pi/8)$$

Fig. 3. Flow chart of discrete FFT algorithm.

the form of the butterfly, can be realized with 4 multipliers and 6 adders, as depicted in Figure 4.

A butterfly computation block set was implemented with 4 multipliers and 4 adders in cooperation with the pipeline timing control. The input samples are stored in the data memory, while the kernel sine and cosine data are provided in the table memory from the host computer. All the intermediate values in each computing stage are successively stored into two data memory sets.

Two-dimensional FFT computations can be achieved by recursively performing one-dimensional computations line-by-line and column-by-column. The 8-bit image data are converted into 24-bit floating real data (imaginary part equal to zero) iteratively to

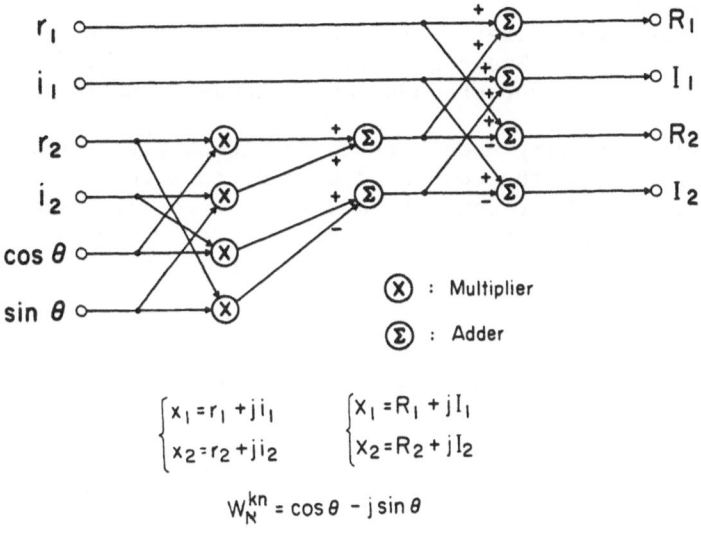

$$\begin{cases} x_1 = r_1 + j i_1 \\ x_2 = r_2 + j i_2 \end{cases} \qquad \begin{cases} x_1 = R_1 + j I_1 \\ x_2 = R_2 + j I_2 \end{cases}$$

$$W_N^{kn} = \cos \theta - j \sin \theta$$

Fig. 4. Flow chart of butterfly computation.

perform the FFT line-by-line and then are stored in the image memory. After the linewise FFT is performed, the columnwise FFT is accomplished and, finally, the transformed complex data is generated in the image memory. Power spectrum calculations and spatial frequency filtering can be performed on this complex data. Note that the multipliers and adders used for the butterfly calculator are also used in performing spatial convolution and vector operations.

3.6 Logical Filtering

Logical filtering works on binary image data for the purpose of thinning, boundary detection, noise cleaning, feature extraction, etc. The logical filtering module examines every 3x3 local window in an input binary image and converts the 9-bit binary data to the corresponding output code by consulting the look-up table stored in the table memory. The 9-bit binary data formed by a 3x3 window represents a unique binary address in the table memory. Any type of 8-bit output code is available by sending the desired table from the host computer via the interface.

3.7 Region Labeling

Region labeling is a local operation which labels the connected components of a binary image. The region labeling module constructs a transformed image where the "1" points have labeled numbers. Two points have the same number if they belong to the same connected region of "1" points. Two connectivity definitions are possible in the square array: 4-connectedness and 8-connectedness.

The region labeling module tests the connectivity with a 3x2 local mask and labels the connected components with the aid of the table memory, new number counter, and associated circuits. After obtaining the label number in the table memory, a microprogram edits the label numbers and performs point mapping to produce the final labeled image.

4. SOFTWARE IMPLEMENTATIONS AND APPLICATIONS

The IPU has been specifically designed to perform basic image processing functions at high speed. It is necessary to operate the IPU without complex programming procedures. The ability to program data transfer and function execution easily in the host computer allows efficient and flexible processing. All IPU operations are initiated by the host computer and then carried out independently, leaving the host computer free to carry out other tasks in the interim. Upon completion, the IPU generates an end-of-operation interrupt to the computer.

4.1 Functional Commands for the IPU

Table 1 shows the command repertoire for directing the IPU which resides in the host computer. These commands are grouped in three categories: initialization, data transfer, and function-execution commands.

An INT command initiates the internal status of the IPU: status registers, flip-flops, and so on.

An LMP command is used for transmitting user's microprograms into the random-access program memory.

A TDM data transfer command establishes data communication between the IPU and the host computer: data conversion table, output table code, weight matrix data, sine and cosine table, histogram data, etc.

A TIM command is used for data transfer between the host computer and image memory.

Table 1 - IPU Command Repertoire

Class	Command Code	Function
Initial	INT	Initialization
	LMP	Load Microprogram
Data Transfer	TDM	Transfer Data from/to Data Memory/Table Memory
	TIM	Transfer Data from/to Image Memory
Image Processing	MOV	Image Data Transfer
	FLT	Spatial Filtering
	AFN	Affine Transformation
	HST	Histogram Computation
	DCB	Data Conversion
	LFL	Logical Filtering
	LAB	Region Labeling
	FFT	Fourier Transform
	MLH	Maximum Likelihood Classification
	EXB	Execute Microprogram

Function execution commands initiate image processing functions. All functions, except the histogram computation HST, require paired-operand image data for input and output, address parameters for scanning and boundary checking, a set of parameters for mode selection, etc. When an operation has been completed, the status information is returned to the host computer via the interface.

A typical command, except for EXM, has the following program sequence in the host computer.

Step 1*   Transfer a set of parameters to the IPU using the TDM command (FLT, DCV, LFL, FFT, MLH).

Step 2   Issue a function execution command.
          Assign image memory to input and output.
          Set address parameters.
          Set parameters or select mode.
          Start the function.
          Sense the status information.

Step 3*   Receive data from the IPU using the TDM command.
          (HST)

(* Steps to be accomplished if necessary.)

4.2 Microprograms in the IPU

As described in the previous section, the IPU operates under
the microprogram control. The microprogram controller plays an important role in data communication, module control, and ALU operations. The word length of a microprogram instruction is 48 bits
with a cycle time of 300 ns. The microprogram instructions are designed to accomplish the above functional roles in one cycle as
simultaneously as possible, i.e., as much in parallel as possible.
The instruction word specifies the ALU operation, data transfers
from source to destination via the bus, control pulse generation,
status selection, sequence address, and so forth.

The ALU operates on up to 24-bit data with several arithmetic
and logic functions. Data transfers are provided with two sets of
48-bit data buses. Each 48-bit data bus carries data from one
module to another with source and destination specified by the
microprogram. The control pulse generator synchronizes the operation of the IPU under the microprogram control. The microprogram
sequence is controlled by the status selected and the next address
according to the program branch, jump-to-subroutine, and return-
from-subroutine conditions.

4.3 Applications

The IPU is an efficient tool for research experiments in image
processing, especially when used in an interactive image processing
system. It should be emphasized that the IPU has processing capabilities for binary and gray-scale images as well as for all types
of two-dimensional signal data. Table 2 shows a list for applicable
image processing operations using IPU's basic functions.

The gradient vector calculation is one of the applications
which exemplifies the use of an assortment of basic functions. The
gradient operation computes the magnitude and direction of the gradient vectors. If the input image is denoted by $F(x,y)$, then the
magnitude $G(x,y)$ and direction code $D(x,y)$ of the gradient vector
at point (x,y) are obtained from the expressions

Table 2 — List of Image Processing Applications

Basic Function	Applications
Two-Dimensional Convolution	Smoothing, Noise Averaging, Enhancement, Differentiation, Restoration, •••
Affine Coordinate Transform	Enlargement, Shrinking, Rotation, Shift, Geometric Correction, Mosaicing, •••
Logical Filter	Thinning, Noise Clearing, Gap Filling, Feature Detection, •••
Region Labeling	Region Separation, Region Coloring, Area Counting, •••
Data Conversion	γ-correction, log/exp, sin/cos, $x^2/\sqrt{x}$ Thresholding, Histogram, Equalization, •••
Histogram Generation	Gray Level Frequency Counting, Area Counting, •••
Pixel Operation	Pixelwise arithmetic and logical operations, Ratioing, Shading Correction, •••
Fourier Transform	Spatial Frequency Filtering, Image Enhancement, Restoration, Power Spectrum, Coding, Compression, •••
Vector Operation	Maximum Likelihood Classification, color Information (Hue, Saturation, Intensity), •••

$$\begin{cases} G(x,y) = \sqrt{(\partial F(x,y)/\partial x)^2 + (\partial F(x,y)/\partial y)^2} \\[2mm] D(x,y) = \Theta \left\{ \tan^{-1}\left[\frac{\partial F(x,y)}{\partial y} \Big/ \frac{\partial F(x,y)}{\partial x}\right] \right\} \end{cases} \qquad (7)$$

where $\partial F(x,y)/\partial x$ and $\partial F(x,y)/\partial y$ are output values of the sum-of-products with x and y derivative weights, respectively. The symbol

Θ indicates a multilevel thresholding function which, in this case, is used to encode an angle θ into one of 16 direction codes.

Let the input image be stored in an image memory by M1. It is assumed that the gradient operation program is able to make use of three other image memories M2, M3 and M4. The main procedures to calculate G(x,y) and D(x,y) using a 4-bit magnitude value and a 4-bit direction code packed into 8-bit data are as follows.

Main Step 1   Execute FLT with x-derivative weights and
              store the result in M2.

Main Step 2   Similarly, execute y-derivative FLT and
              store the result in M3.

Main Step 3   Execute EXM for the square, sum, and square
              root pointwise with the magnitude stored in
              M4.

Main Step 4   Execute a division using EXM and store the
              result in M1. (The original image is
              destroyed here.)

Main Step 5   Execute DCV with Θ conversion table and
              store the direction code in M2.

Main Step 6   Execute pointwise operation EXM to pack
              4-bit magnitude (M4) and 4-bit direction
              code (M2). The final output is stored in
              M3.

Other extended programming examples can be found in the case of spatial frequency data processing. After the Fourier transform is computed for a given image, the low-pass, high-pass, and band-pass filters can be performed on the transformed data. These spatial frequency filtering operations can be implemented in the microprogram by using the multipliers, adders, and the ALU. Then, the inverse FFT function can be used to compute the averaged image, edge-element images, and enhanced images, respectively. A power spectrum display of the result can also be obtained by a very simple pixelwise EXM function command used to calculate the square root of the real and imaginary data and, if desired, using logarithmic mapping.

The IPU shows an ability to perform parallel pipelined classi-fication, compute color information in terms of intensity, hue, and saturation, etc. The concepts and algorithms of image processing programs for the parallel machine may differ from those of conven-tional programs implemented in serial computers. It is still an unsolved problem now to develop analytically new image processing solutions suitable to all high-speed image processing problems.

5.   CONCLUSION

A microprogrammable image processing unit (IPU) has been proposed and realized using simple hardware. The IPU has improved the
cost-effectiveness of high-speed digital image processings. This
system can perform those basic image processing functions which are
frequently used, time-consuming operations. Its speed is two orders
of magnitude faster than the speed of conventional digital computers.
The IPU has the capability of being programmed to conduct local
parallel operations, coordinate transformations, fast image transforms such as the FFT, vector operations, and pixelwise ALU operations. More complex functions can be microprogrammed in combination with these basic functions.

The IPU is an efficient tool for exploring novel applications
requiring quick analysis in the field of image understanding research for such as the analysis of remotely sensed data, medical
diagnosis, and industrial automation. A further expansion of the
IPU, which is currently being developed, is to design a more advanced image memory unit in terms of capacity and intelligent function as described by Akers et al. (1977). In addition, the development of application programs in the host computer, as well as micro-
programs in the IPU, is a critical task for enhancing the image processing capability and power of the IPU system.

6.   ACKNOWLEDGEMENTS

The research and development activities reported in this chapter are being performed under contract with the Ministry of International Trade and Industry as part of the Pattern Information Processing System (PIPS) project. The authors wish to express their
appreciation to Dr. H. Nishino and his research staff at the Electro-
Technical Laboratory. They also wish to extend their appreciation
to Drs. K. Kakizaki, H. Gench, and K. Mori who have encouraged them
and contributed many significant concepts to improving on the system
design.

7.   REFERENCES

Akers, A. E., Persoon, E., and Fu, K. S., "A Virtual Memory Com-
    puter for Image Processing," Proc. IEEE Comp. Software Appl.
    Conf. (1977).

Asada, H., Shinoda, H., Kidode, M., Yoneyama, T., Watanabe, S., and
    Mori, K., "Interactive Image Processing System with High Per-
    formance Special Processors," Proc. 4th Internat'l Joint Conf.
    Pattern Recog., Kyoto (1978).

Fu, K. S., "Special Computer Architectures for Pattern Recognition
    and Image Processing--An Overview," Proc. Amer. Fed. Info.
    Proc. Soc. 47:1003-1013 (1978).

Gold, B., and Rader, C. M., Digital Processing of Signals, New York,
    McGraw-Hill (1969).

Hunt, B. R., "Computers and Images," Proc. Soc. Photo. Ind. Engrs.
    74:3-9 (1976).

Mori, K., Kidode, M., Shinoda, H., and Asada, H., "Design of Local
    Parallel Pattern Processor for Image Processing," Proc. Amer.
    Fed. Info. Proc. Soc. 47:1025-1031 (1978).

MAN-MACHINE INTERACTIVE PROCESSING FOR EXTRACTING METEOROLOGICAL

INFORMATION FROM GMS IMAGES

N. Kodaira, K. Kato and T. Hamada

Meteorological Satellite Center

235, Nakakiyoto 3-chome, Kiyose-shi, Tokyo 180-04, JAPAN

1. INTRODUCTION

The GMS (Geostationary Meteorological Satellite) views the earth's disk via the VISSR (Visible and Infrared Spin Scan Radiometer). GMS is positioned at 140°E above the equator at an altitude of about 36000km. The VISSR provides concurrent observations in the infrared (IR) spectrum (10.5-12.5μm) and in the visible (VIS) spectrum (0.5-0.75μm). These observations are transmitted to the ground at periodic intervals; usually every three hours. About 25 minutes are required for the VISSR to produce the digital image of the full earth's disk.

The computer facility located at MSC (Meteorological Satellite Center) provides a large scale computer for image data processing. The configuration of the computer system consists of four FACOM 230-75 computers. The computer complex is mainly divided into two systems: (1) the "online system" and (2) the "batch system." The batch system has the responsibility for extracting meteorological information such as cloud drift wind vectors (CWV) and cloud-top heights (CTH). An overview on the utilization of the GMS satellite is provided by Murayama et al. (1978).

We have developed man-machine interactive processing methods for extracting CWV's and CTH's because there are some difficulties in selecting clouds and in assigning emissivities to target clouds. A skilled analyst inspects synoptic cloud patterns, selects target clouds, and assigns emissivities for CWV and CTH extraction. The Cloud Wind Estimation System (CWES) and Cloud-Top Height Estimation System (CTHES) are our approach to introducing human expertise into

an automatic procedure. The system for wind derivation and the
limitation of the method now used are described by Hamada and
Watanabe (1978A, 1978B). Detailed procedures for the CTHES system
and the CWES system are described in Sections 2 and 3 of this
chapter, respectively.

2. GMS CLOUD TOP HEIGHT ESTIMATION SYSTEM

The MMIPS (Man Machine Interactive Processing System) at NESS
(National Environmental Satellite Service) is described by Bristor
et al. (1975). Also the McIDAS (Man-Computer Interactive Data
System) at the University of Wisconsin is discussed in Suomi (1975).
These systems have as their purpose cloud top height estimation.

The total radiation sensed by the GMS represents the sum of
target cloud radiance and that from the underlying sea surface or
low level clouds. The Cloud Top Height Estimation System (CTHES)
was developed at MSC (Meteorological Satellite Center) to derive
cloud top heights (CTH) by means of man-machine interactive processing. Derived CTH's are the most basic information for interpreting cloud features such as the extension of cloud tops, their
flatness, and their convective activities. This information is
typically depicted on a "Neph-Analysis Chart" which is one of the
routine products at MSC.

2.1 Basic Equation

The blackbody temperature of a cloud equals the actual cloud
top temperature if the cloud is dense enough to shield the infrared sensor from radiance from below the cloud. In general, actual
clouds are not dense enough optically, so the radiance sensed by
the satellite represents the sum of both the cloud radiance and
that from the underlying surface. The relationship between the
cloud top temperature and the measured blackbody temperature can
be expressed by the following equation considering the atmospheric
attenuation correction (AAC).

$$N(Tbb) = eN(Tc) + (1-e)N(Ts-t+dt) \qquad (1)$$

where $N(Tbb)$ is the radiance from the cloud observed by satellite
and $N(Tc)$ is the radiance from the cloud.

The quantity $N(Ts-t+dt)$ is the radiance from the underlying
surface (sea or low-level cloud). Tbb is the representative blackbody temperature of the cloud observed by satellite; Tc, the blackbody temperature of the cloud; Ts, the blackbody temperature of
underlying surface. The quantity t is the total atmospheric

attenuation correction value (from the surface to the top of the atmosphere); dt, the atmospheric attenuation correction value (from the cloud top to the top of the atmosphere); e, the cloud emissivity estimated on the basis of empirical rules. The quantities related to atmospheric attenuation corrections can be estimated given the vertical profile of the atmosphere and the optical path length. The quantities N(Tbb) and N(Ts) are measurable but cloud emissivity is not measurable objectively. These three basic parameters are needed in order to solve the radiation equation. (Figure 1 illustrates equation (1) schematically.)

2.2 Basic Parameters

We next discuss methods for determining Tbb, e, and Ts.

2.2.1 *Representative Blackbody Temperature* (Tbb)

To estimate Tbb, a histogram analysis is performed within a target cloud area. This area contains typically 17 lines by 45 overlapped pixels along a scan line. This area is about 85Km on a side at the sub-satellite point (SSP). The histogram based on the samples of infrared (IR) image data is usually multimodal. The cold mode represents the temperature of a cloud-covered area and the warm mode yields the temperature of a cloud-free area. These two modes can be distinguished by comparison with a threshold value which is empirically defined as Ts-5 Kelvin. Each sample warmer than the threshold value is discarded. The retained samples are distributed in a histogram which contains a single mode associated with the cloud top temperature.

The following three methods are available for determining Tbb from this histogram: (1) the mode value is regarded as Tbb in the "Synoptic Scale Cloud Top Height Estimation System" described in Section 2.4.1; (2) the minimum value is defined as Tbb in the cloud drift wind height assignment and "Meso-Scale Cloud Top Height Estimation System" as described in Section 2.4.1; (3) the mean value is assigned to Tbb which is an optional method.

2.2.2 *Cloud Emissivity* (e)

The emissivity is primarily a function of the opacity of the cloud. Convective clouds tend to radiate almost as blackbodies (e=1) but other types of clouds display emissivities that range from less than 0.1 up to nearly unity. The emissivity is sensitive in the CTH estimation. To reduce uncertainty in emissivity estimates, the CTHES adopted an empirical relation (see Allen,

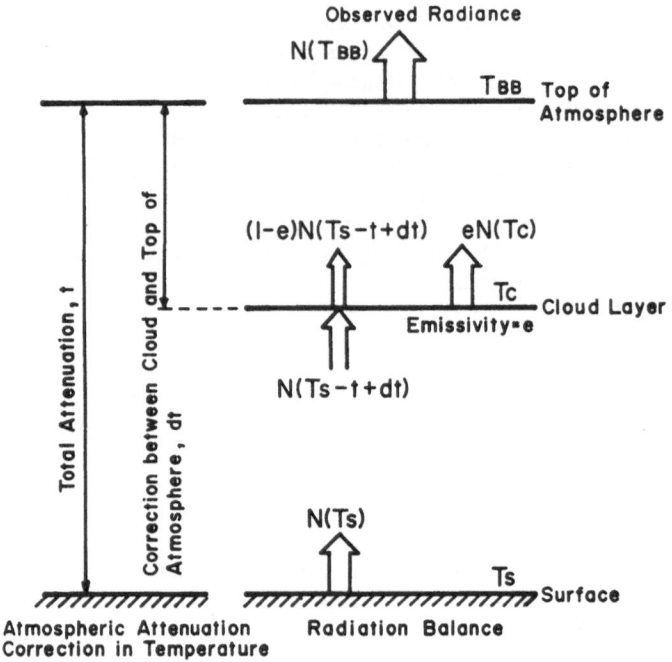

Fig. 1.   Schematic diagram of radiation balance.

1971) between emissivity and the cloud type and opacity (or cloud thickness). This is summarized in Table 1. An analyst makes a decision only as to cloud types and opacities, then emissivity is inferred from this table.

When a thin cirrus cloud occurs in a target cloud area, a tropopause level is accepted as the CTH because observational facts indicate that this type of cloud has almost the same height as the tropopause level.

2.2.3 *Underlying Surface Temperature* (Ts)

The radiance from the underlying surface is not inferred from the satellite when a target cloud is covered by an emissive cloud. This causes an underestimation of Ts. Instead of the direct measurement of Ts that is accomplished by MMIPS operation (see Bristor, 1975), the CTHES uses a satellite derived sea surface temperature (SST). This data set is routinely produced every ten days (at grid point intervals of one degree in latitude and longitude). If the SST data are not available, a 1000 mb temperature is obtained from climatological data as described in Section 3.

Table 1 – Empirical Cloud-Type Emissivities
as a Function of Cloud Thickness

Cloud types and thickness		Emissivity	Legend
Cirrus	Thin	—(%)	
	Medium	—	
	Dense	60	
Altostratus	Thin	—	
	Medium	70	
	Dense	90	
Altocumulus	Thin	50	
	Medium	80	
	Dense	90	
Cumulus	Thin	80	
	Medium	90	
	Dense	100	
Stratocumulus	Thin	30	
	Medium	70	
	Dense	90	
Stratus	Thin	60	
	Medium	80	
	Dense	100	
not specified		—	✳

Finally, the cloud temperature Tc can be derived from equation (1)
by the use of these three parameters: Tbb, Ts, and e.

2.3 Conversion to Cloud Top Height

The conversion of cloud top temperature (CTT) into cloud top
height (CTH) is performed based on vertical temperature profile
(VTP) data. The data set is reconstructed from the climatological
data (see NCAR, 1971, and U.S. Navy, 1970). This is provided in
the form of monthly mean grid point values at five degree intervals in latitude and longitude called "GMS Standard Atmosphere"
(GMSSA). As a daily operation, an actual VTP is simulated by the
GMSSA data referring to the results of air-mass analysis. Once
the location of a target cloud area is designated, the derived CTT
is referenced to the VTP data at the nearest grid point.

2.4 System Configuration and Function

The configuration of the CTHES soft-ware is summarized in
Table 2, and the functional block diagram is shown in Figure 2.

Table 2 – Job Group Organization of Software Programs Used in the
Cloud Top Height Estimation System (CTHES)

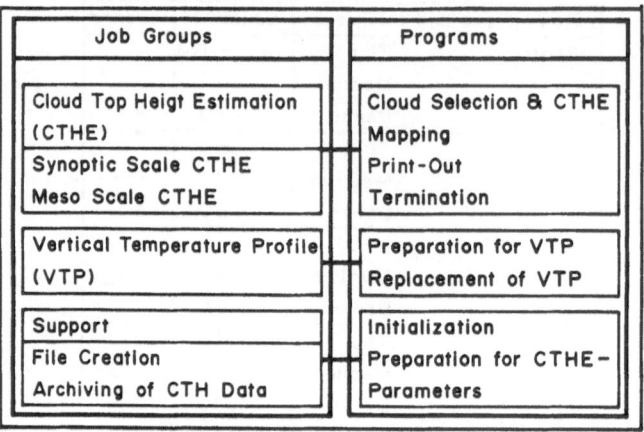

Job Groups	Programs
Cloud Top Heigt Estimation (CTHE) Synoptic Scale CTHE Meso Scale CTHE	Cloud Selection & CTHE Mapping Print-Out Termination
Vertical Temperature Profile (VTP)	Preparation for VTP Replacement of VTP
Support File Creation Archiving of CTH Data	Initialization Preparation for CTHE- Parameters

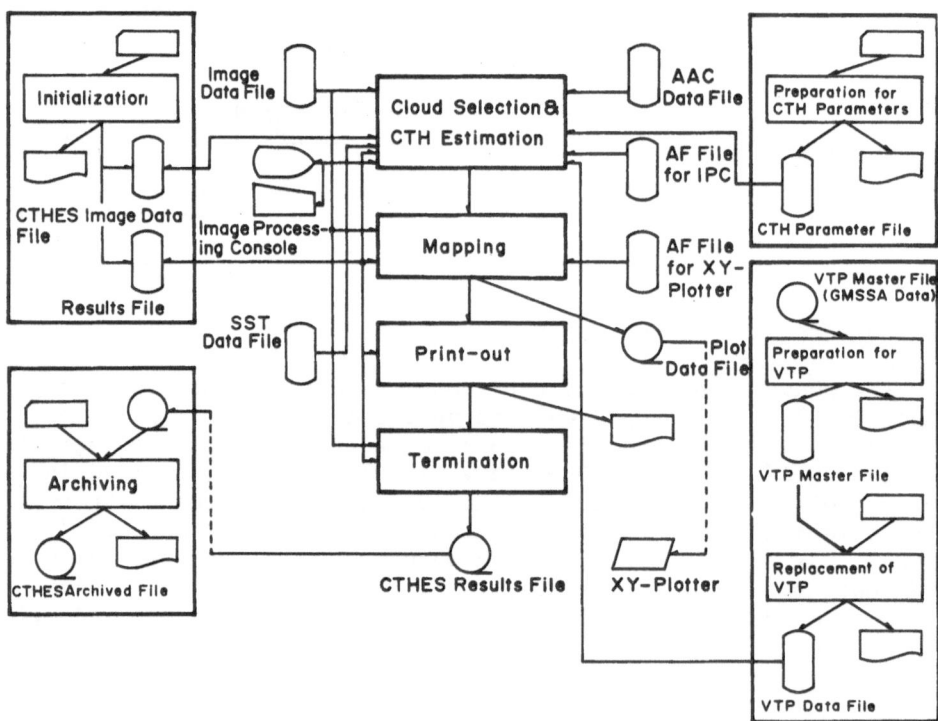

Fig. 2.   System configuration and functional block diagram of the
CTHES.

2.4.1 *Cloud Top Height Estimation Job Groups*

This is the core of the model. The job groups are divided into two parts. One is the Large Scale CTHE which aims to derive the CTH's associated with a large-scale meteorological disturbance. The area of interest is mainly restricted to the northern hemisphere. A target cloud area size is chosen normally as 17 lines by 45 pixels and is used to find the mode value for Tbb.

The other part is the "Meso Scale CTHE." It is intended to provide an effective means for finding the CTH's corresponding to a meso-scale disturbance. An area of interest is Japan and its vicinity. The target cloud area size is normally selected as 9 lines by 23 pixels and the minimum value is used for Tbb.

These job groups consist of the following programs.

(1) Target Cloud Selection and Cloud Top Height Estimation

This program allows an analyst to estimate CTH on the basis of a man-machine interactive procedure. For visual interaction the "Image Processing Console" (IPC) is equipped for displaying various kinds of image data in black and white (B/W) or color. The following operations can be selected on the basis of the analyst's command.

(a) Visible (VIS) and IR image data within the area of interest are displayed immediately on the IPC with either the original or reduced spatial resolution.

(b) VIS and IR image data which are independently colored are superimposed and displayed. The cloud type and thickness are distinguishable by inspecting this display.

(c) Pseudo-color IR image data are displayed to examine the "thermal slice profile" around the target cloud area.

(d) Gray-scale or color-bar scale/VIS-count or IR-count conversion curves are displayed. This permits the analyst to enhance a cloud feature.

(e) Target cloud selection is accomplished by moving a target selector with the "Positioner." After centering the target cloud, the analyst inputs the location of the target cloud area and its size to the computer via the "Send Position Key."

(f) Using a keyboard and/or a function key, the analyst issues commands to the computer on the basis of derived relevant information such as cloud type and thickness while inspecting the displayed image area. The derived CTH is displayed immediately on the IPC. The same procedure can be repeated if the estimation is not acceptable.

(2) Mapping

The derived CTH's are mapped on the X-Y plotter through the use of this program. The symbolized cloud types (Table 1) and other information, such as CTT, Tbb, and the standard deviation estimated from the target cloud area, are also added to the map. A sample of this is shown in Figure 3. The map is delivered to the neph-analysis section. In the case of the Meso-Scale CTHE, this program is skipped.

(3) Print-Out

All kinds of information needed for further investigations of the results are output to the line-printer. The frequency distribution derived from target cloud areas is available. When the Meso-Scale CTHE is performed, a temperature array display related to target cloud areas is available.

(4) Termination

Once the target cloud selection and cloud top height estimation procedure is completed, derived CTH's data are stored on magnetic tape. Disk files used for the CTHES are then deleted.

2.4.2 *Vertical Temperature Profile Job Group*

This job group provides the best VTP data on the disk file named "VTP Data File." The temperature/height conversion and atmospheric attenuation correction procedures access this file. The job group consists of the following programs.

(1) Preparation for VTP

The GMSSA data are stored on a magnetic tape. A disk file is generated from this magnetic tape through the program. This file, named "VTP Master File," is maintained as a permanent file.

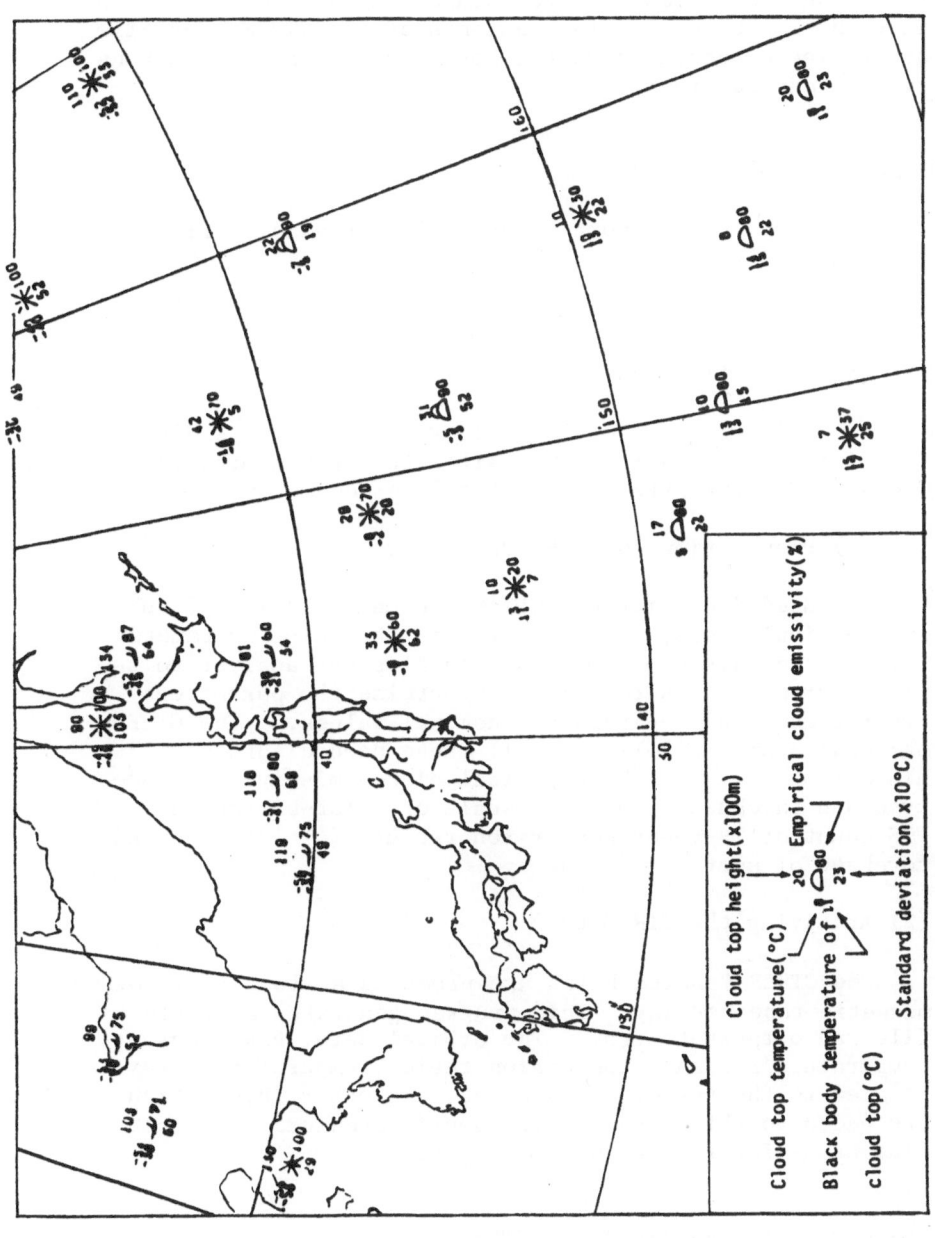

Fig. 3. A sample of cloud top height data display.

(2) Replacement of VTP

Once the air-mass analysis is completed, the results
are punched on cards. In this case, the name of the month
is used as a keyword. The contents of the VTP Master File
are replaced according to the keywords. This program runs
twice a day routinely.

2.4.3 *Support Job Group*

This provides the support programs for creating a file and
archiving the CTH's data.

(1) File Creation

(a) Initialization

This program sets up the system status. A disk
file initialization is done. After this procedure, the
CTHES image data file and results file become accessible.

(b) Preparation for CTHE Parameters

Needed parameters, which are mainly selected on
an empirical basis, are prestored in the "CTHES Parameter
File" using this program. This enables the analyst to replace parameters as required. Otherwise the parameters
are retained and regarded as "nominal values." The CTHE
parameters are as follows: (1) those needed for the
histogram analysis; (2) empirical cloud-emissivity/cloud-
type and thickness; (3) gray-scale or color-bar-scale/
VIS-count or IR-count conversion tables; (4) location and
coverage of sector image data.

(2) Archiving the CTH Data

The CTHES Results Files are produced at four reels of
magnetic tape per day. This program generates a single
file and outputs to tape. The orbital data, blackbody
temperature/radiance conversion table, temperature arrays
related to the target cloud areas, and other information
are added to this tape. This tape is archived on a rout-
tine basis for future investigation.

3.   GMS CLOUD WIND ESTIMATION SYSTEM

The technique for deriving cloud drift winds using an animated
film-loop was developed by Prof. Fujita of the University of Chicago

soon after the launch of the first Spin Scan Cloud Camera installed
on the Application Technology Satellite (ATS 1) late in 1966. Leese
et al. (1971) applied a cross-correlation technique to the derivation of cloud displacements for the first time. This made it possible to extract a large number of cloud drift winds routinely.

Now the European Space Agency (ESA), the National Environmental
Satellite Service (NESS) in the United States and the Meteorological
Satellite Center (MSC) of Japan produce cloud drift winds, reformat
them into a WMO code, and transmit them to world-wide users twice a
day. These data are expected to be valuable especially over data-
sparse areas, oceans, deserts and mountain areas, especially for
the input of numerical prediction processing.

The Japanese system, the GMS Cloud Wind Estimation System
(CWES), provides man-machine interactive processing and automatic
processing. A skilled analyst selects suitable target clouds interactively, the computer automatically tracks them to get the cloud
displacements, and the analyst performs quality control on the results interactively. Finally, reliable wind vectors are transmitted
to world-wide users by the Global Telecommunication System (GTS).

The processing systems operated by ESA, NESS, and MSC are to be
described by WMO in a Technical Note. Detailed processing operations at NESS are described by Bristor (1975). Similar details of
the CWES operations are described in this chapter.

3.1 GMS Cloud Wind Estimation System

The GMS Cloud Wind Estimation System (CWES) has three procedures, MM-1, MM-2, and Film-Loop (FL). The former two are man-
machine interactive procedures and the latter is film-loop procedure
as shown in Table 3. All vectors derived from these procedures are
quality-controlled and transmitted to world-wide users by the Global
Telecommunication System (GTS). The general flow of these procedures
is shown in Figure 4.

3.2 Registration of Images

Registration of the GMS VISSR images is performed using the coordinate transformation process from attitude and orbital predicted
data of GMS satellite. The orbit of the satellite is predicted
daily using Trilateration Range and Range Rate (TRRR) data which are
measured four times a day using three ranging stations at Hatoyama
and Ishigaki-jima, Japan, and Orroral Valley, Australia. The nominal error of the predicted satellite position in a day is about
100m. The error is not so large as to cause significant image misregistration.

Table 3 – The Procedures of Wind Derivation in
GMS Cloud Wind Estimation System

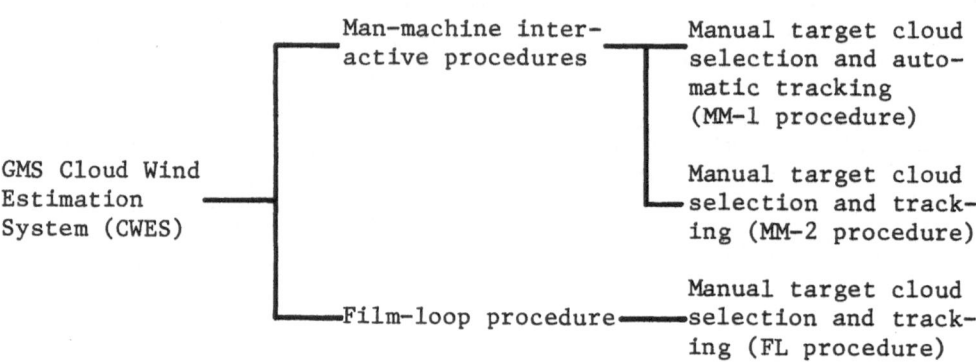

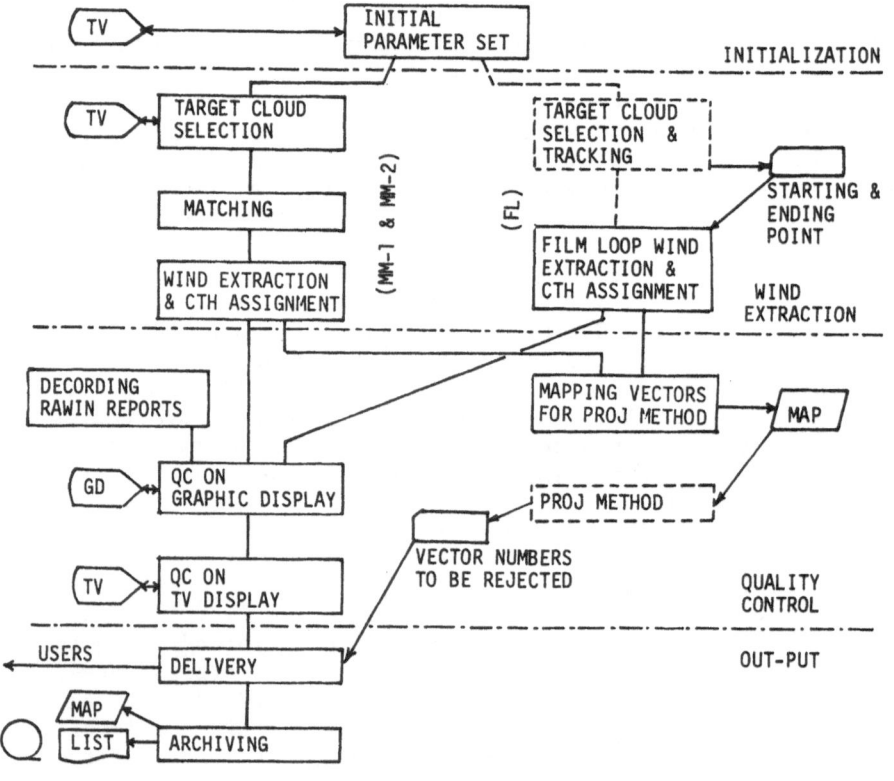

Fig. 4.   General flow of the CWES system.

The attitude of the satellite is also predicted daily on the basis of the results of man-machine interactive landmark matching procedures using several VISSR visible images. The nominal error of the predicted satellite attitude is less than 140μrad of the spin-axis direction. This causes misregistration of four visible pixels (picture elements) and 2.8m/s of wind error at the sub-satellite point (SSP). However, the actual error in the cloud drift winds calculated comes mainly from the relative misregistration of two images. Our investigations show that the misregistration between images used for deriving winds using CWES is less than one visible pixel (see Hamada and Watanabe, 1978B). Consequently the registration accuracy of VISSR images is accurate enough to derive cloud drift winds when using CWES.

3.3 Wind Derivation

This section outlines several methods employed in the cloud drift wind calculation procedure.

3.3.1 *Man-machine Interactive Procedures for Target Cloud Selection*

In this section the man-machine interactive procedures MM-1 and MM-2 are described. The Image Processing Console (IPC) used for man-machine interactive processing is equipped with a TV-screen, a cursor dial for positioning purposes, an alpha-numeric keyboard for commanding the computer, and function keys. Usually three TV-screens, two for black-and-white and one for color, are used for display. Images at the original spatial resolution are displayed on two TV-screens, black-and-white and color. The visible and/or infrared images covering the same area of approximately 600 x 600 km are displayed using different enhancement tables.

The black-and-white screen is used to display the black-and-white visible or infrared image to be used for target cloud selection and tracking. The color screen is used for three purposes: (1) pseudo-color visible or infrared image displayed by manual selection by the analyst; (2) both visible and infrared images simultaneously displayed in different colors so that the analyst can inspect the characteristics of the clouds; (3) time sequential display of two or three images in different colors in order that the analyst can inspect the development and displacement of clouds. Another black-and-white TV-screen is used for displaying the full disk image or a sampled image covering one-sixteenth of the full disk. This makes it possible for the analyst to inspect the synoptic pattern of the clouds to get information for cloud target selection.

(1) MM-1 procedure

Using the color or black-and-white screen display of
the image at its original spatial resolution, the analyst
moves the cursor to select target clouds for tracking.
In routine operations, sequential images are usually displayed on the color screen so that the analyst may classify usable "passive tracers" as shown by Hubert and Whitney (1971). After selecting target clouds, the analyst
assigns an emissivity value to the cloud so as to estimate the cloud top height from the infrared data. The
operation described above is repeated about 200 times.
Information on the location of selected points and
assigned emissivities is accumulated in the disk for
subsequent processing.

(2) MM-2 procedure

On the color screen where time-sequential images are
displayed in different colors, the analyst selects both
the starting point and the ending point of a target cloud
and assigns an emissivity value in the same manner as with
the MM-1 procedure. Information on the location of selected points and assigned emissivities is accumulated in
the disk for subsequent processing. This procedure is
followed for research purposes and is not used in routine
operations.

3.3.2 *Automatic Tracking*

Target clouds are tracked in the MM-2 man-machine interactive
procedure. The processing described in this section applies only to
MM-1 target clouds. The cloud pattern of each target cloud selected
in the man-machine interactive procedure is located in the next
image taken with a 30-minute interval using cross-correlation technique. Digital image data formated as 32 pixels by 32 lines, 16 by
16, or 8 by 8 and centered at a selected point in the initial image
is put into a computer as a template. In the next picture, taken
30 minutes later (or before), image data formated 64 by 64 is put
into the computer as the search area data. The correlation of
brightness between template and search area is calculated for different lag values and a cross-correlation coefficient matrix is obtained as the result. The correlation matrix is given by

$$C(p,q) = \frac{\sum_{i=1}^{N} \sum_{j=1}^{N} [T(i,j) - \overline{T}][S(i+p,j+q) - \overline{S(p,q)}]}{\sqrt{\sum_{i=1}^{N} \sum_{j=1}^{N} [T(i,j) - \overline{T}]^2} \sqrt{\sum_{i=1}^{N} \sum_{j=1}^{N} [S(i+p,j+q) - \overline{S(p,q)}]^2}}$$

where

$T(i,j)$ = Brightness level of template data

i,j = 1,2,3, . . ., N

$S(i+p,j+q)$ = Brightness level of search area data

i,j = 1,2,3, . . ., N

$$p,q = -\frac{N}{2}\;,\; -\frac{N}{2}+1\;,\; \ldots\;,\; \frac{N}{2}-1\;,\; \frac{N}{2}$$

(p,q) is lag-position on matching surface

$$\overline{T} = \frac{1}{N^2} \sum_{i=1}^{N} \sum_{j=1}^{N} T(i,j)$$

$$\overline{S(p,q)} = \frac{1}{N^2} \sum_{i=1}^{N} \sum_{j=1}^{N} S(i+p,j+q)$$

Figure 5 provides a schematic of the relation between the template and the search area used in calculating the cross-correlation matrix. Leese et al. (1971) as well as Smith and Phillips (1972) were the first to apply cross-correllation techniques to the derivation of cloud displacements. The cross-correlation coefficient matrix is depicted in three dimensions as shown in Figure 6. It is called a "matching surface" and is used for quality control as described in Section 3.4.

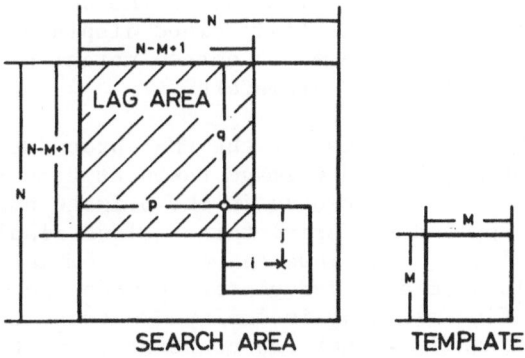

Fig. 5. Schematic showing the search area and the template area used for calculating the cross-correlation matrix.

Fig. 6. Matching surface displayed on the screen of a graphic
display.

A double-matching method (Figure 7) has been adopted for tracking target clouds when using the MM-1 procedure. First, using reduced resolution, each selected target cloud is tracked and a coarse
displacement vector is derived. This is called "coarse matching."
Next, using the image at the original spatial resolution, a correction vector is derived using processing similar to coarse matching.
This is called "fine matching." The sum of the coarse vector and
the correction vector yields the final cloud displacement. The reduced sampling interval to be used in coarse matching is given in
the process of setting initial parameters.

We usually use three images for the MM-1 procedure (Figure 8).
These images are the A-, B- and C-image taken at time intervals of
30 minutes. We select target clouds in the B-image and the double-
matching method is applied to derive target cloud displacement between the B- and C-image. The inverse vector from the B-image to
the C-image is defined as a coarse vector between the B-image and
the A-image. Only fine matching is employed between the B-image and
the A-image. Consequently we get two consecutive vectors, V_{AB} and
V_{BC}, corresponding to every selected target cloud in the B-image.

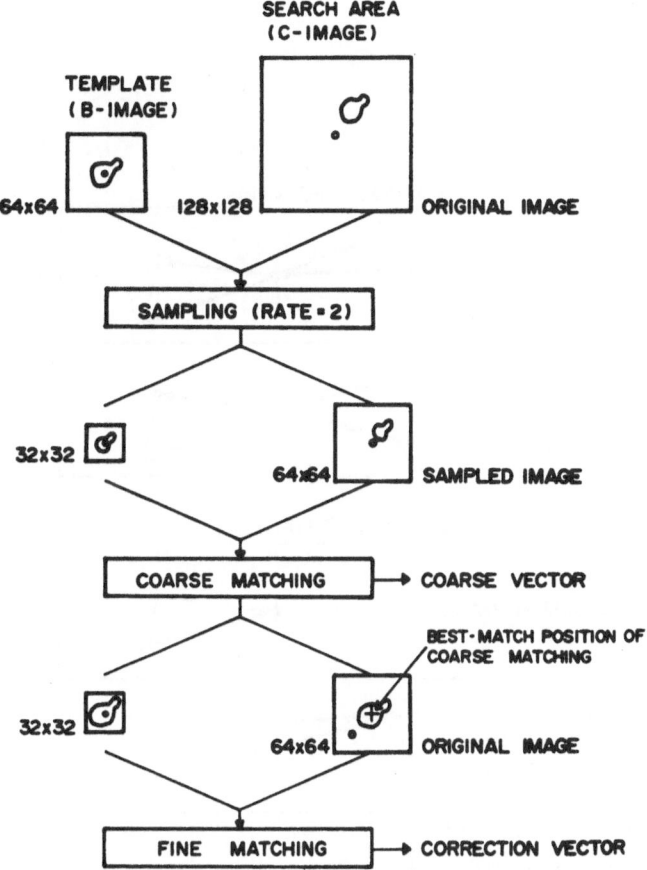

Fig. 7.   Schematic of double matching using the CWES.

The matching scheme mentioned above is used for routine operations. The two-image mode shown in Figure 8 is provided for the
case when one of these images is lacking.

The image data used for this matching process is composed of
picture elements taken at discrete positions which causes a truncation error in the calculation of cloud drift winds. In order to
eliminate the truncation error, interpolation of the position where
the maximum correlation value occurs is applied to the result of
fine matching. Interpolation substantially improves spatial resolution and is particularly effective when applied to the infrared
image because of the poor resolution in the infrared in comparison
with the visible. The procedure employed, when using CWES, to increase the resolution of the image match surface is to fit a

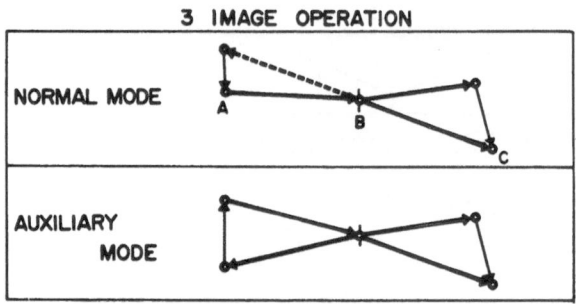

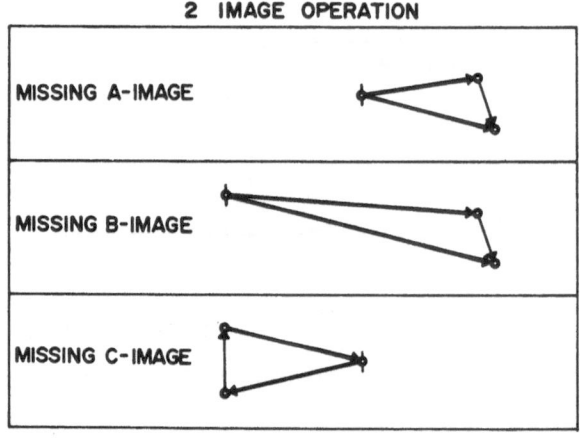

Fig. 8. Matching schemes for two or three images. Usually three
images (A, B and C) taken at 30-minute intervals are used for the
matching process. The mark "⧫" depicts the image used in target
cloud selection. Vectors depicted by double line are coarse vectors. Thin line vectors are correction vectors derived through fine
matching procedures. Heavy line vectors are the resultant vectors.
The broken line vector in the normal mode shown is an inverse vector
of the resultant vector between the B- and C-image and is replaced
with a coarse vector between the A- and B-image.

bi-directional quadratic in the neighborhood of the best match co-efficient. This method was developed at SSEC, University of Wisconsin, as described by Smith and Phillips (1973).

3.3.3 *Wind Derivation and Height Assignment*

The resultant vectors in the image coordinate system are transformed into wind vectors in the earth coordinate system on the basis of attitude and orbital predicted data. The vectors between the B- and the C-image are regarded as final winds. Vectors between the A- and the B-image are used for quality control.

Three cloud top heights (CTH's) are derived from three infrared images for each set of wind data. The CTH derived from the latest image is assigned to a current vector. A black body temperature, Tbb, is extracted from a small infrared image array centered at a selected point in the target cloud or at a matching point obtained in the matching procedure. Tbb is modified into a cloud top temperature, Tc, by considering atmospheric absorption. Using climatological data on vertical temperature profiles prepared on grid points (five degree interval in latitude and longitude), Tc is transformed into CTH both in pressure level and in geo-potential height.

3.3.4 *Wind Derivation Using Film-Loop*

A 35mm motion picture film, produced from four consecutive images taken at 30-minute intervals, is used for deriving winds. These pictures are called the Z-, A-, B- and C-image. The latter three images are the same as those used in the man-machine interactive procedures. The film-loop is back-projected on a digitizer board which provides a two-dimensional coordinate measuring system coupled with a card punch. The analyst selects and tracks benchmarks printed on each frame of the film-loop. The locations of the tracking points on the digitizer board are punched on data cards. The locations of tracking points for target clouds are also punched in the same way as above. Upon completion a control card and several identification cards (ID cards) are added to the card deck. The control card specifies the type of image used for target cloud tracking and the ID cards specify the mean value of the cloud top height assignment. The complete card deck contains the input data for the film-loop wind derivation and the cloud top height assignment program. In this program the digitizer coordinates are transformed into image coordinates referring to the nearest three benchmark locations. Final wind vectors are extracted in the same way as in the MM-1 procedure.

3.4   Quality Control in the CWES System

Quality control is divided into two stages as shown in Table 4.
The first stage is the automatic assessment of matching surfaces,
wind velocities, cloud top heights, and the missing line (line drop)
check. The second stage is man-machine interactive quality control
of the resultant winds, which is performed by using the graphic display, the TV display, and by means of film-loop projection.

Table 4 - Quality Control Steps Used in CWES

Automatic assessment

1) The features of matching surface

2) Picture-to-picture variation of cloud top heights

3) Wind acceleration

4) Checking on missing lines of the images used for
matching process

5) Checking on missing lines of IR images used for
cloud top height assignment

Manual quality control

- Using graphic display

1) Checking on horizontal consistency

2) Comparison with radiosonde winds

3) Checking on the features of matching surface

- Using TV display

1) Checking on reasonability of automatic tracking

- Using film-loop

1) Checking on identity of each result vector with
film-loop displacement (Projection Method)

3.4.1 *Automatic Assessment*

The resultant vectors are assessed automatically by checking
threshold values given by the initial parameter setting program.
The threshold values are pre-determined by another investigation.
Each resultant vector screened automatically on the basis of thresh-
old values and unreliable vectors is excluded in the final report.
Steps used in automatic assessment are as follows (see Table 4):

(1) The features of the matching surface

The parameterization of a matching surface is shown in Figure 9. (A sample of a matching surface is shown in Figure 6.) The threshold values are determined by another investigation and are improved empirically. The relationship between R (the difference between the dominant and second peak values) and D (the distance between the dominant and second peak positions) is assessed automatically. In the case that R is smaller than the threshold value and D is greater than the threshold value, the vector is sent to the manual check procedure as automatic assessment has failed. Relationships are shown in Table 5.

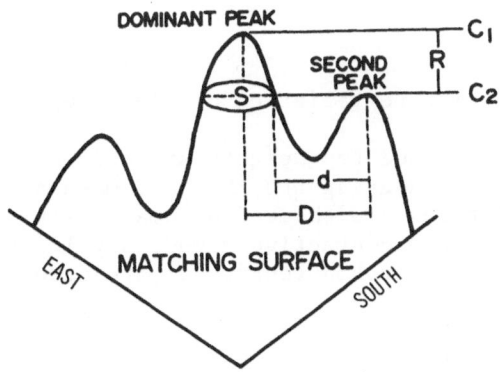

Fig. 9. Parameterization of a matching surface with multiple peaks. Parameters: C_1 and C_2 are peak values; S is the area dominated by the first peak with respect to the second peak; the distance d is used for searching second peak; the distance D is the distance between the dominant peak and the second peak.

Table 5. Automatic Assessment on the
Multiple Peaks of a Matching Surface

A view of matching surface with multiple peaks	Quality flag*		Result of assessment
	Distance (D)	Difference (R)	
	1	0	PASS
	0	0	PASS
	1	1	FAIL
	0	0	(Sent to manual judgment)**

* Quality flag: '1' or '0' is assigned to the flag in case the
parameter value is smaller or greater than threshold value
respectively.

**At the same time cross-correlation matrices are preserved in
DP for inspecting them in the process of interactive quality
check.

(2) Picture-to-picture variation of cloud top heights

Two or three cloud top heights are extracted from the
same target cloud in using the MM-1 and MM-2 procedures
and their variations are checked automatically. Unreliable
vectors derived from convective clouds developing or decay-
ing rapidly and mismatched vectors are rejected in this
check.

(3) Wind acceleration

The difference between two time-sequential vectors V_{AB}
and V_{BC} is considered as the acceleration of the wind.
This procedure is very effective in rejecting mismatched
vectors.

(4) Missing line check on the images used for matching process

When the template image data or search area image data used in the matching process include less than two/five missing lines in the infrared/visible image, the matching result may be slightly degraded. In case two/five lines or more are missing in the infrared/visible template or search area, the current vector is immediately rejected.

(5) Missing line check on infrared image used for CTH assignment

This is the missing line check on the infrared image which is used for cloud top height (CTH) extraction. In case there are more than two missing lines, cloud top height extraction is not made from the image.

The functions described as items (1) and (4) above are performed in the process of matching. Items (2), (3), and (5) are performed in the process of wind derivation and cloud top height assignment.

3.4.2 *Man-Machine Interactive Quality Control*

Manual quality control is performed as part of the interactive procedures.

(1) Using a graphic display

(a) Checking on horizontal consistency

Horizontal consistency of the resultant vectors is checked by depicting all resultant wind vectors on a graphic display. Wind vectors depicted on a graphic display are shown in Figure 10.

(b) Comparison with radiosonde winds

Resultant vectors are compared with radiosonde winds produced in real-time on a graphic display. Two kinds of comparisons are performed. One is a comparison of a resultant vector with a nearby radiosonde wind value gathered at the same altitude as the cloud top height assigned to the resultant wind. Another is a comparison of a resultant vector with the same radiosonde wind to find a level of best fit (LBF). Using these comparison procedures the analyst checks the vector in question.

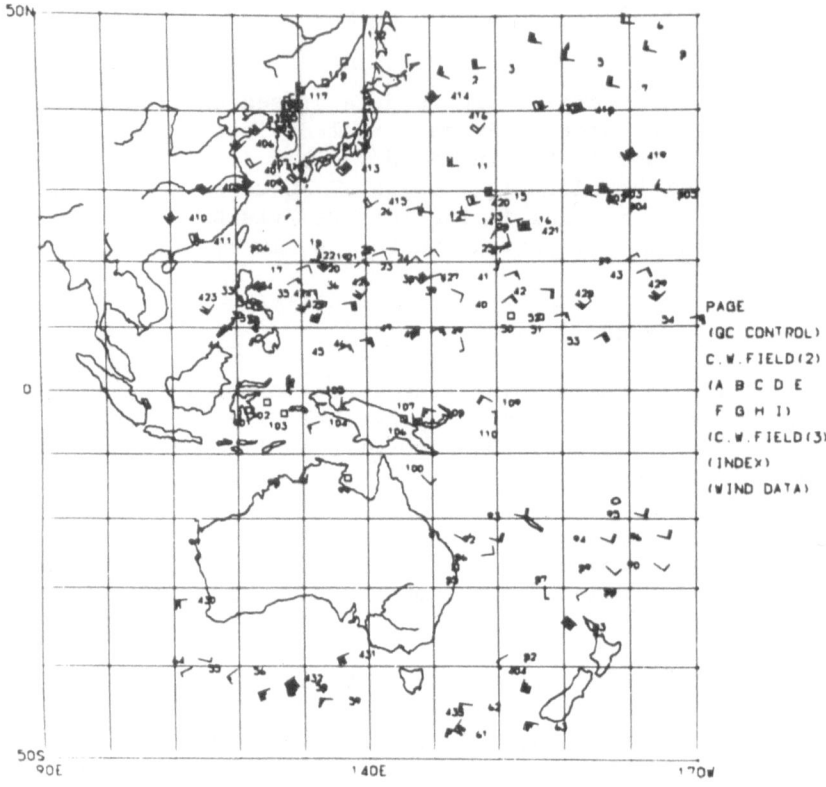

Fig. 10. Low cloud drift winds derived using the DWES at 00Z,
July 10, 1978, depicted on the screen of a graphic display for the
purpose of a horizontal consistency break.

(c) Checking on the features of a matching surface

A matching surface is checked on a graphic display by inspecting the three-dimensional features of the
surface. This function is provided in order to determine
the threshold values of the parameters characterizing a
matching surface. Usually this check is not used in routine operations.

(2) Using TV display

(a) Checking on the reasonableness of automatic tracking

Resultant vectors and three consecutive pictures
can be superimposed on the TV display. It is easy for the
analyst to check on reasonableness of automatic tracking.

(3) Using film-loop

    (a) Projection method

        The resultant vectors are plotted on a map of the
full disk projection as viewed by the GMS satellite.
The film-loop used for target cloud tracking is projected
on the map and the analyst checks on the identity of each
resultant vector along with the displacement of the cloud.
We call this procedure the Projection Method (PM).

    In these procedures, except for the matching surface check and
projection method, the analyst can reject unreliable vectors using
an alphanumeric keyboard at either the TV display or the graphic
display console.

3.5 Delivery and Archiving

    The product vectors considered reliable are reformatted by the
computer into WMO code for teletype transmission to world-wide users.
The transmission of these data is done within four hours from VISSR
observation.   These vectors are stored on magnetic tape to be archived, listed by a lineprinter, and plotted on the map using both
polar stereo and Mercator projection.   These wind data are included
in the "Monthly Report" issued by the Meteorological Satellite Center.   The resultant vectors derived in routine operations at 12Z,
Sept. 26, 1978, are shown in Figure 11.

4.   REFERENCES

Allen, J. R., "Measurements of Cloud Emissivity in the 8-13μ Wave-
    band," J. Appl. Meteor. 10:260-265 (1971).

Bristor, C. L. (ed.), "Central Processing and Analysis of Geosta-
    tionary Satellite Data," NOAA Technical Memorandum NESS 64
    (1975).

Hamada, T. and Watanabe, K., "Determination of Winds from Geosta-
    tionary Satellite Data - Present Techniques (Lecture 8A),"
    Paper presented at the WMO/UN Regional Training Seminar on the
    Interpretation, Analysis and Use of Meteorological Satellite
    Data, Tokyo (23 October - 2 November 1978A).

Hamada, T. and Watanabe, K., "Determination of Winds from Geosta-
    tionary Satellite Data - Limitation of Current Method (Lecture
    8B)," Paper presented at the WMO/UN Regional Training Seminar
    on the Interpretation, Analysis and Use of Meteorological Sat-
    ellite Data, Tokyo (23 October - 2 November 1978B).

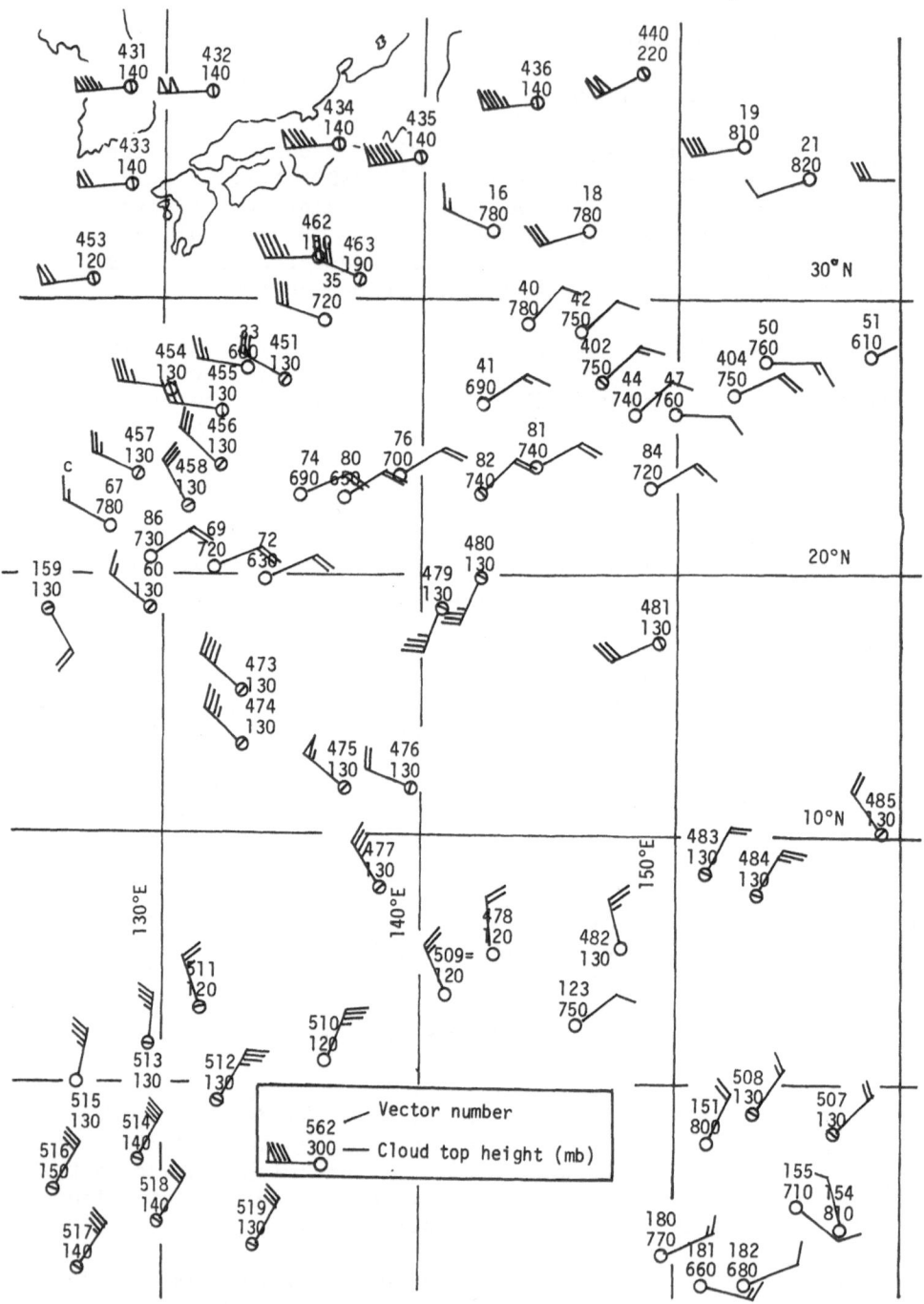

Fig. 11. Sample of wind data derived using the CWES at 12Z, September 26, 1978.

Hubert, L. F. and Whitney, L. F., "Wind Estimation from Geostationary Satellite Pictures," Mon. Weather Rev. 99:665-672 (1971).

Leese, J. A., Novak, C. S., and Clark, B. B., "An Automated Technique for Obtaining Cloud Motion from Geosynchronous Satellite Data Using Cross Correlation," J. Appl. Meteor. 10:118-132 (1971).

Murayama, N., Kato, K., and Kodaira, N., "An Overview on the Utilization of the Japanese Geostationary Meteorological Satellite," Paper presented at the Twelfth International Symposium on Remote Sensing of the Environment, Manila (April 20-26, 1978).

NCAR, "Climate of the Upper Air: Southern Hemisphere Vol. 1," National Center of Atmospheric Research, NCAR Tech. Notes STR-58 (1971).

Smith, E. A. and Phillips, D. R., "Automated Cloud Tracking Using Precisely Aligned Digital ATS Pictures," IEEE Trans. Comput. 21:715-729 (1972).

Smith, E. A. and Phillips, D. R., "McIDAS Cloud Tracking System," Internal SSEC Report, University of Wisconsin (1973).

Suomi, V. E., "Man-computer Interactive Data System (McIDAS)," SSEC, University of Wisconsin, Final Rpt. contract NAS 5-23296 (1975).

U. S. Navy, "Selected Level Heights, Temperature, and Dew Points for the Northern Hemisphere," NAVIR 50-1C-52 (1970).

AN ARRAY PROCESSOR FOR IMAGE PROCESSING

H. Matsushima, T. Uno and M. Ejiri

Central Research Laboratory, Hitachi Ltd.

Kokubunji, Tokyo 185, JAPAN

1. INTRODUCTION

Some early developments in pattern processing are reviewed in Unger (1958), McCormick (1963), and Murtha (1966). Some practical implementations are described by Golay (1969) and by Kruse (1976). For certain industrial and medical applications special processors for image analysis are in practical use such as in computed tomography (CT). Also, large scale pattern processors have become available in special fields.

However, image processors for multi-level image data are still under development and some problems remain. For example, the variety of objects in images is too diverse to be handled by most image processors. Therefore, it is difficult to settle on the target specifications for processor development. As a result, such processors tend to become too big and too expensive, representing a big investment risk. However, recent developments in commercially available IC's have had a profound effect on the size, speed, and cost of hardware. In addition, the need for general purpose image processors is getting stronger. The same situation exists both in 2-dimensional and 3-dimensional object recognition.

The IP (Image Processor) constructed in our laboratory is a processor developed against this background. Its biggest feature is its parallel processing ability which is due to an array structure of special processing elements. This chapter describes the design principles and hardware of the IP and gives actual examples of image processing.

2. DESIGN PRINCIPLES

    This section outlines the design philosophy which has been
taken in configuring the IP.

2.1 The Memory Unit

    The information included in the original image depends on the
two-dimensional arrangement of the pixel values. Therefore, two-
dimensional memory access is employed for data utilization. On
the other hand, information in the form of serial or sequential
data is different from image data. These differences require different memory organizations which makes it hard to implement such
a memory. Therefore, separation of the memory unit into two parts
was considered. One half is used for serial or sequential data and
the other half is used for two-dimensional data. These two parts
are closely connected by registers. By adoption of this method,
it becomes easy to access two-dimensional data which ordinary computers cannot do effectively.

2.2 The Processing Unit

    In image processing, the required precision is not as severe
as in performing scientific calculations. Another characteristic
of image processing is that the same operations are repeated many
times. Thus it is not efficient to increase processing speed by
using ordinary computer architecture. Therefore in the IP the execution part was separated into two parts: (1) parallel processing
and (2) serial or sequential processing. The former usually uses
a memory for two-dimensional data. The latter memory is designed
like that of a mini-computer and it mainly accesses data in the
serial or sequential data memory and controls the whole system.

2.3 The Array

    The major reason for using an array structure is to increase
parallel processing speed. In pre-processing or feature-extraction
the same operator is used for every pixel. For such position-
invariant processes an array structure is very useful. Many pixel
values are input to many processing elements at once. Then, all
processing elements execute the same operation simultaneously.
Ideally, it is possible to increase processing speed by an amount
equal to the number of processing elements.

2.4   The Processing Element

One operation which reduces the effectiveness of an array
structure is the occurrence of irregular data flow such as the
execution of a conditional jump. In such cases, all processing
elements cannot be controlled by one sequence. Thus, a special
processing element (PE) has been developed for image processing.
This PE can handle irregularities in the data flow.

2.5   Programming

As explained above, our IP permits parallel processing and
has plural processing elements which run simultaneously. It is
difficult to control such hardware by ordinary programs. Consequently programs which use mixed micro-instructions and macro-
instructions are permitted. Programmers trying new operations
experimentally can make direct use of micro-instructions and control the array. To facilitate this an assembler language with
mixed instructions is provided. These facilities permit easy use
by many researchers.

3.   SYSTEM CONSTRUCTION

This section describes the overall system and major subsystems.

3.1   System Outline

The overall IP configuration is shown in Figure 1. It consists of the following major subsystems:

(A)   A serial processing part

(1)   CM ; control memory

(2)   CPU ; central processing unit

(3)   SC ; sequence controller

(B)   A parallel processing part

(1)   MC ; micro instruction controller

(2)   PM ; picture memory

(3)   PEA ; processing element array

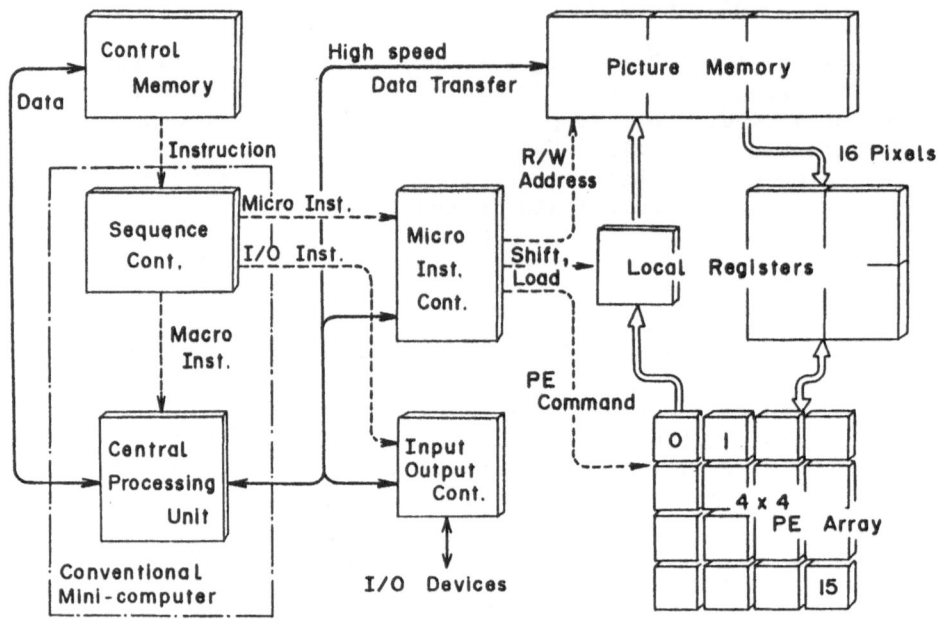

Fig. 1.   Image processor block diagram.

(4)   LR ; local register

(5)   IOC ; input output controller

Images are input by an industrial TV camera. They are converted to digital data by an AD converter in the I/O interface.
These data are transmitted serially to the picture memory plane
and stored. Each time they are processed they flow from the PM to
the PEA via the local registers and back again. When the process
is completed the processed image is transferred serially from the
PM to the output display.

When serial image processing is performed, data in the PM are
transferred to the CPU and the CM directly or through the LR and
are processed in the CPU.

3.2 Control Memory

The control memory (CM) is a random access memory which stores
programs and, sometimes, picture data. One word is 32 bits long
and the CM stores 16K words. Words having the full 32-bit length

are sometimes too long for image processing so that 16-bit half-
words are made accessible by use of special instructions.

3.3 Central Processing Unit

The CPU consists of two parts. The sequence controller (SC)
separates instructions into micro-instructions and macro-instructions. The micro-instructions are transferred to the MC and the
macro-instructions are processed in the CPU. In addition, the SC
generates timing pulses which are sent to all parts of the IP.

The CPU is an ordinary mini-computer so that the use of registers, the addressing methods, overall operations, etc. are not particularly unusual. However, in order to make use of micro-instruction subroutines, functions such as stacking of the subroutine jump
addresses or index modifications are used more frequently. Integer
arithmetic is utilized and all numerical data are considered as
integral numbers. Negative numbers are given by the 2's complement representation.

Between the CPU and the MC there is a 16-bit bus. Data transfer between the general registers of the CPU and the MC takes place
in 3.85 μs. Transfer times between the CPU and the other parts in
our IP are shown in Figure 2.

3.4 Micro Instruction Controller

The micro instruction controller (MC) controls the parallel
processing part of the IP, i.e., the PEA, LR, PM, and IOC. The
micro-instructions can be classified into two types. One type
consists of a PE command and other micro commands to the PM or the
LR. In this case, the MC transfers the PE command to the PEA immediately and controls the clock or the execution signal to the PEA.
The rest of this type of micro-instruction is interpreted in the
MC and used in order to control the PM and data shifts in the local
registers. The other type of micro-instruction is used to control
the I/O required for data tranfer. This micro-instruction is transferred to the IOC.

3.5 Picture Memory

The picture memory (PM) is a memory unit for images and consists of three 256x256 banks. These three banks may be considered
as storage for an input, an output, and an intermediate image.
However, bank usage is not fixed, so the programmer can designate

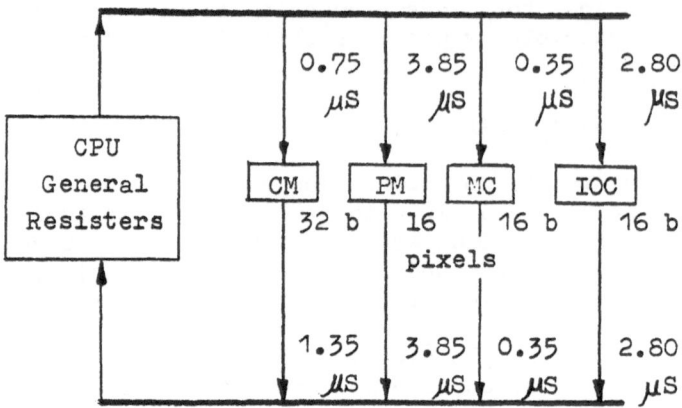

Fig. 2.   Data transfer rates.

the function of each bank freely. Each bank has two transfer modes
which are independent of those of the other banks. These modes are:

  (1) *Serial mode*: mode used for data transfer between
       the PM and general registers in the CPU or between
       the PM and the IOC. In this mode, 8 words (16 bits
       + 2 parity bits) are transferred successively.

  (2) *Parallel mode*: mode used for parallel processing
       where the unit of transfer is 16 bytes, i.e., 4x4
       pixels.

3.6   Processing Element Array

     The processing element array (PEA) is a parallel execution
unit consisting of 16 processing elements. A detailed description
of the PEA is given in Section 4.

3.7   Local Register

     The local register (LR) consists of a set of buffer registers
used in transferring data between the PM, the PEA, and the CPU
general registers. There are five registers, each register being
4x4. Four of the five registers are used to transfer data from
the PM to 16 processing elements. These registers act like one
8x8 register. The final register in the set is used for data
transfer from the PEA to the PM.

The LR can be used for shifting data upwards or to the left in parallel. The LR has a 16-byte bus connecting to the PM and the PEA. There also is a 1-byte bus connecting with registers in the MC or the CPU. A single processing element can have access to only one fixed point in a local register.

3.8 Input Output Controller

The input output controller (IOC) controls data transfer between the IP and input-output devices. There are two data transfer modes. One mode is programmed I/O; the other mode is direct memory access. The latter can fetch images at about 6 megabytes per second which is the speed necessary for real-time data acquisition from an ITV camera.

4. PARALLEL PROCESSING

This section provides further information on the processing element array and its use.

4.1 Processing Element

A special processor was developed as a basic element of the PEA. This processor is called the "PE" whose design principles are as follows:

(1) Execution of basic operations in the image processing field by a single PE command.

(2) To be able to absorb irregularity in program sequences caused by testing.

(3) To perform the functions of a stored-program computer.

(4) To be controlled easily from outside as a simple ALU unit.

The PE's block diagram is shown in Figure 3. The PE consists of four units: (1) control unit, (2) register unit, (3) execution unit and (4) scratch pad memory (SPM).

The control unit controls the PE by interpreting PE commands. The PE commands come from the MC or the SPM. The control unit furnishes local timing by synchronizing with the original clock given from the SC or the MC. Thus all PE's in the PEA can run simultaneously.

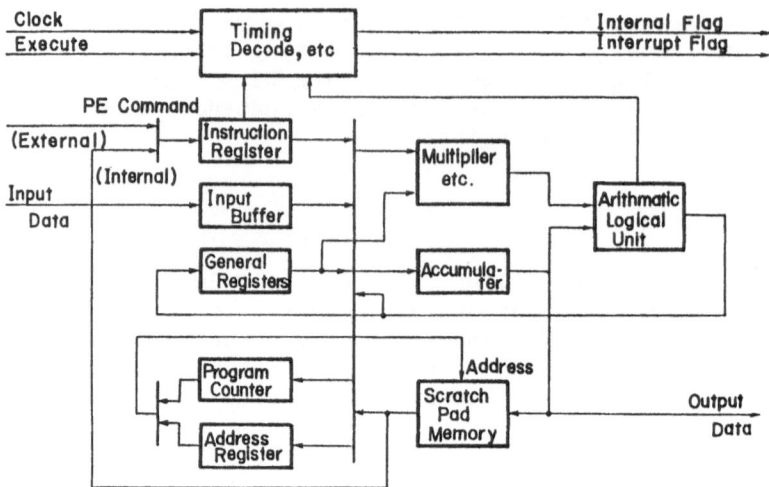

Fig. 3. Processing element block diagram.

A PE has four operating modes:

(1) *External mode*: operation by PE commands sent from
     the MC.

(2) *Internal mode*: operation by PE commands in the SPM.

(3) *Interrupt mode*: waiting for special PE commands
     from the MC. No-operation for ordinary PE commands.

(4) *Wait mode*: waiting for PE commands.

The relationships among these modes are shown in Figure 4. First,
the PE is cleared and stays in the wait mode. When a PE command
is transferred from the MC, the PE shifts to the external mode.
All PE's have their own Execute Signal inputs and can execute commands only if these are ON. These signals are controlled by the
MC. If they are OFF, the PE is not influenced by commands given
from outside.

When commands such as SET INTERNAL MODE and JUMP are input, a
PE shifts to the internal mode. Once in the mode and starting
with the next execution cycle, it reads out PE commands from its
SPM. Ordinary PE commands from the MC are ignored even if they
are sent in the Execute Signal ON-state. If a PE is set by a test
command (SET INTERRUPT), it shifts to the interrupt mode. Then,
the MC examines it and transmits a special PE command, such as
STATUS CONTROL and the PE shifts to the wait mode and can receive
commands.

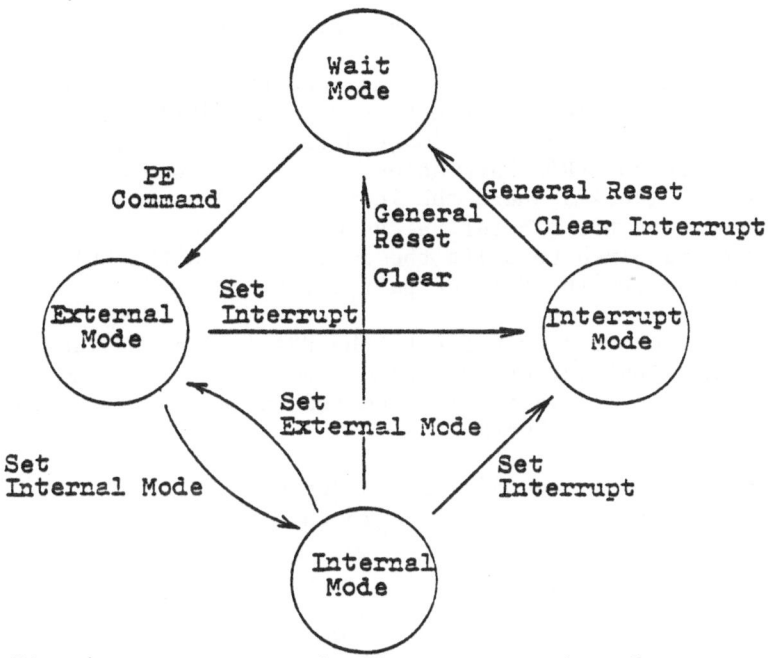

Fig. 4. Execution modes for a processing element.

ıne main registers in a register unit are general registers, an accumulator, an input buffer, a program counter, an instruction register, and the SPM address register. The program counter is used only in the internal mode.

The execution unit consists of a multiplier, a binary matching circuit, an ordinary ALU, etc. Operations in a PE require 12 bits. Image data are given from the LR and at 8 bits per pixel. The 4 bits in the upper third of these 12 bits are spread by the value of the most significant bit of the image data. Image data are input from the LR to the input buffer in a PE at the start of each PE execution cycle. Representative commands to the PE's are SUM OF PRODUCTS, SUM OF BINARY MATCHING, TABLE LOOKUP, etc. with execution times of 350ns.

The SPM stores PE programs, data, and some table information. It contains 256 16-bit words. Loading of a program to the SPM is done by two special PE commands: SET PROGRAM LOADING MODE and END OF PROGRAM LOADING. The commands to a PE which arrive between these two commands are loaded in the SPM.

When the PE is in the external mode, it is like an ALU and is directly controlled by the MC.. After shifting into the internal mode, it can run like a microcomputer. The internal mode and the interrupt mode are useful in adapting to data flow irregularities.

4.2   Data Supply to the Processing Element Array

Image data supply to the PEA is done through the local registers. Pixels readable from the PM are formated in a 4x4 square window as shown in Figure 5. Image data are not always immediately available since the LR's have access to windows whose upper-left corners are at (4n,4m) where the letters n and m stand for integer numbers from 0 to 63. Therefore, as many as two or four read operations may be required to shift the image data in the LR's and then input them into the 16 PE input buffers.

Broadcasting data transfer to the PEA is done by inserting data into the displacement portion of some of the PE commands. This insertion is automatically executed in the MC by setting one flag bit in the IP instruction.

Data transfer among PE's is done by using the LR. Data are output from one PE output, shifted in the LR, and input into another PE input buffer.

4.3   Processing Element Array Execution Mode

Several execution modes can be selected by means of PE and MC functions. These modes are:

(1)   Identical operations in all PE's. All Execute Signals are ON.

(2)   Identical operations in part of the PEA. In this case, PE's in the other part are in the wait mode. The Execute Signals of the designated PE's are ON; otherwise they are OFF.

(3)   Operation in only one PE as selected by a four-bit pointer. The Execute Signal is ON for this PE only.

(4)   Identical operations in PE's which satisfy some designated condition. The other PE's are in the interrupt mode. It is not always necessary for the CPU or the MC to know which PE's are running. All Execute Signals are ON.

(5)   Many operations. Some PE's are in the internal mode with each PE executing the program in its own SPM. The SPU knows when a program is completed by examining the PE's Internal Flag. In this case the CPU is free to perform other program routines.

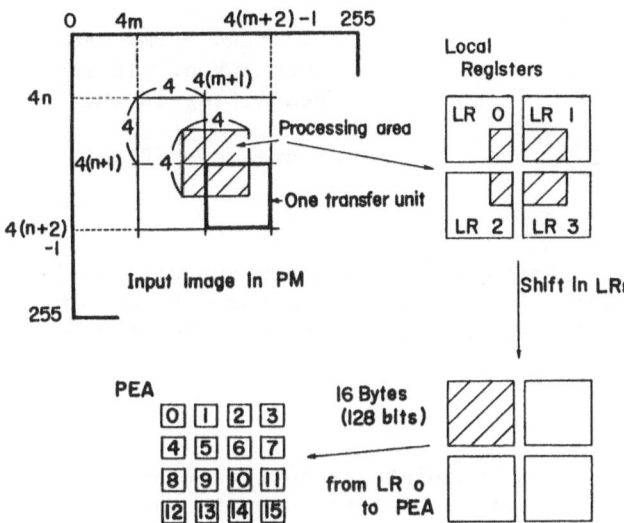

Fig. 5. Method of data transfer to the processing element array.

5.   EXAMPLES

This section gives examples of two representative image operations, i.e., thresholding and filtering.

5.1 Thresholding

Thresholding of images can be executed by using only the external mode with all PE's performing identical PE commands. Only one pixel is necessary per PE to obtain the result. Only 16 pixels are transferred from the PM to the PEA at step in the operation. The main part of this program is as follows:

```
L1   READ PM(0)/LOAD LRO           L1; label.  PM(0); bank 0.

     PE COMPARE I,/30 A            I; input.  A; All PE's run.

     PE LABEL 1,0,S   A            S; sign flag.  If S=ON(I>/30),
                                   one is output; otherwise zero.
```

WRITE PM(1) AND UPDATE
        /LOAD LR4   PEA → LR4 → PM. Address is
                                     incremented. If area is over,
                                     then A flag is ON.

TEST AND SKIP A   A; area over flag.

BRANCH L1

This routine is repeated 4,096 times for one 256x256 image and
takes about 11.5ms. By exchanging the above PE commands, image
inversion, table transformation, clamping, etc. may be performed.

5.2 Filtering

     Filtering by a 4x4 template is another example. Sixteen pixels
must be supplied to each PE per each step in the operation. Therefore, the PEA requires the 7x7 data array shown in Figure 6. At
first there is no image data in the LR's. For this case, PM access
must be repeated four times. However, for subsequent operations a
4x8 data window is available in the LR's. Therefore, only the remainder, i.e., 4x8 more data, is brought in from the PM. The main
part of the program is as follows:

     READ PM(0)/ LOAD LR0

     READ PM(0)/ LOAD LR2

L1   READ PM(0)/ LOAD LR1

     READ PM(0)/ LOAD LR3/ PE ACC CLEAR A

     SHIFT LR U/ PE SUM OF PRODUCT /** A   U; upwards. /**; template.

         ...   ...

     SHIFT LR U/ PE SUM OF PRODUCT /** A

     SHIFT LR L/ LOAD LR4

     WRITE PM(1) AND UPDATE

     TEST AND SKIP A

     BRANCH L1

The routine starting at L1 is repeated 4,096 times. It takes about
34.4ms. The processing time depends on the template width.

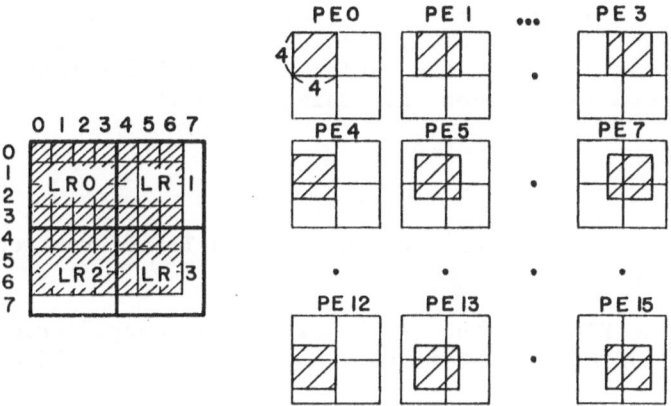

Fig. 6. Data flow in processing element array when executing 4x4 template filtering.

6. CONCLUSION

The IP has five design features: (1) memory unit separation into a PM for two-dimensional data and a CM for serial or sequential data, (2) processing unit separation into a parallel processing part and a serial or sequential one, (3) parallel processing using an array, (4) use of a special processing element (PE), (5) programming by use of micro-instructions.

First the PE was developed. It required some 200 TTL IC's. Next, the parallel processing array was implemented. Then the total hardware system and an assembly language were completed.

The usefulness of the IP in image processing has been established through many experiments. The PE and the PEA control method are especially useful for high-speed processing.

At present the IP uses a 4x4 array. This is adequate by today's hardware technology standards. However, in the near future, PE's may have one-chip LSI configurations. This will require development of still larger-scale processors.

7. ACKNOWLEDGEMENTS

The author wishes to thank Dr. Jun Kawasaki, Hirohide Endo, Yoshiji Fujimoto, and Haruo Yoda for constructive criticism regarding the machine design. Thanks are also due to Mitsuo Ohyama and Yoshihiko Kaido for development of the hardware.

This work was performed under contract to the Agency of Industrial Science and Technology, Ministry of International Trade and Industry, as a part of the National Research and Development Program "Pattern Information Processing System."

8. REFERENCES

Golay, M. J. E., "Hexagonal Parallel Pattern Transformations," IEEE Trans. Comput. C-18:733-740 (1969)

Kruse, B., "The PICAP Picture Processing Laboratory," Proc. 3rd Intern'l Joint Conf. Pattern Recog. (1976), pp. 875-881.

McCormick, B. H., "The Illinois Pattern Recognition Computer," IEEE Trans. Elect. Comput. EC-12:791-813 (1963).

Murtha, J. C., "Highly Parallel Information Processing Systems," Adv. Comput. 7:2-116 (1966).

Unger, S. H., "A Computer Oriented Towards Spatial Problems," Proc. IRE 46:1744-1750 (1958).

REAL-TIME SHADING CORRECTOR FOR A TELEVISION CAMERA USING A

MICROPROCESSOR

M. Onoe, M. Ishizuka and K. Tsuboi

Institute of Industrial Science, University of Tokyo

Roppongi, Minato-ku, Tokyo 106, JAPAN

1. INTRODUCTION

Digital image processing by computer has become practical in various fields. Video systems are particularly useful because of their high cost-effectiveness due to mass production. However, as they are basically analog devices, some attention is required when using them in digital systems. The television camera is comparitively compact and is a useful image input device for the practical image processing systems. However, there are problems with regard to data accuracy; as pointed out in Sawchuk (1977), for example, the shading phenomenon presents us with one of the most crucial problems.

An analog shading correction circuit may be added to the television camera system in many cases but this is not sufficient for digital processing. The upper waveform in Figure 1 shows the typical signal obtained from three horizontal scanning lines from a closed-circuit television camera whose input is uniform white image. The left image in Figure 2 shows the binary image obtained by simple thresholding of the image of some text produced by this television camera. This is one of the undesirable examples. Besides the television camera, the storage tube, which provides an economical frame memory, also requires the same kind of shading correction.

Although the shading problem has been conventionally solved by software processing in main computer, this takes so much time that it has been inadequate in practical systems. In order to cope with the shading problem, the authors have produced an image input system which has real-time shading correction function as initially described by Onoe et al. (1978A, 1978B). In this system a

339

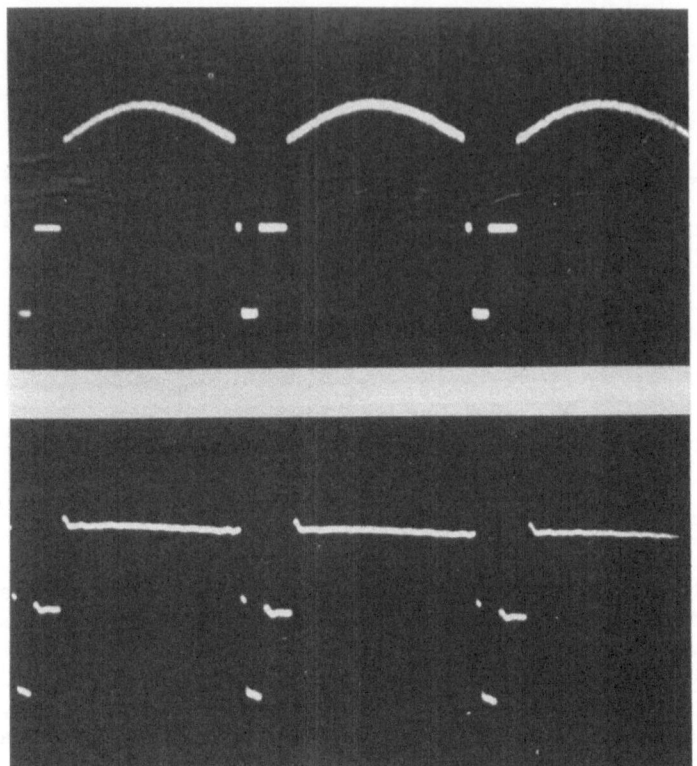

Fig. 1. Horizontal television scanning line waveforms: (upper)
with shading; (lower) after shading correction.

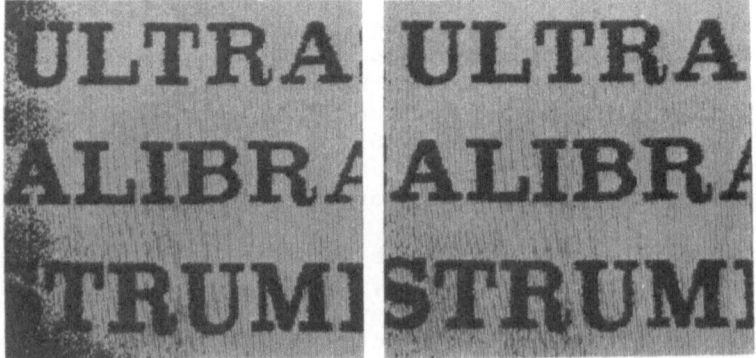

Fig. 2. Binary television image of text obtained by simple
thresholding: (left) with shading; (right) after shading correction.

microprocessor is employed to supervise the measurement of the
degree of shading and the production of a correction data table.
The shading correction is realized by using a wideband analog multiplier and it is effective for producing a corrected analog television signal as well as for digital input. The resultant signal
can be observed on the television monitor.

2.   HARDWARE CONFIGURATION

The overall hardware configuration of the equipment is shown
in Figure 3. Since the shading results mainly from image deterioration in the fringe part of the television picture, the system is
designed to compensate this deterioration by amplifying the signal
level. The microprocessor (Intel 8080A) supervises the measurement
mode, in which the degree of shading is measured, according to the
program stored in the ROM. It produces a correction data table in
a predetermined RAM area. To match the digitizer speed to the
operating speed of the computer, one sample is taken per scanning
line.

In the subsequent correction mode, the correction data table
stored in the RAM is read out repeatedly in DMA mode and converted
into an analog voltage. A wideband analog multiplier is employed
to amplify the television signal and the desired multiplication coefficient is applied to this voltage. The reason for employing an
analog multiplier is to permit shading correction for the analog
television signal itself for real-time display. While a digital
multiplier may be used for the digital input, it is not fast enough
to deal with real-time television signal.

2.1 Operational Details

Timing signals are generated from the vertical and horizontal
synchronization signals extracted from input television signal by
the synch separator. In addition, using a PLL circuit, a fast
clock is generated at a frequency 700 times that of the horizontal
synchronization signal (15.75 KHz). The period of this fast clock
corresponds to one pixel time of the television signal. For a
camera which does not generate complementary pulses in the vertical
synchronization period, such pulses are forced to be inserted for
the proper operation of the PLL.

The field F/F and the Y (vertical address) counter are driven
by the clock pulses generated from the vertical and horizontal
synchronization signals, respectively. The X (horizontal address)
counter is driven by the abovementioned fast clock. On the other
hand, the field index F/F and the Y and X index counters are set

by commands sent from the microprocessor or main computer via the
command decoder. The coincidence detector responds to the coincidence between the states of the abovementioned counters and flip-
flops to generate a pulse, which starts the operations of the S&H
circuit and the AD converter. The amplitude data at desired points
are obtained as explained above and the microprocessor or the main
computer can get them through their buses.

As the pixel time is about 90 ns ($63.5\mu s/700$), the S&H circuit
must be fast. Therefore, an S&H circuit with a settling time of
50 ns is employed. On the other hand, the AD converter does not
need to be so fast because of the abovementioned input scheme.
Accordingly, an AD converter whose conversion time is $2\mu s$ is used.
This AD converter employs 10-bit accuracy. Only the upper 8 bits
are entered into the microprocessor.

The block size in which the same multiplication coefficient is
applied is set to 16x16 pixels. Consequently, $(700/16)(525/16)$
$= (44)(33) = 1452$ (bytes) are required for the RAM to store the
correction data table. (Here the numbers 700 and 525 are the
number of pixels in one horizontal scanning line and the number of
scanning lines, respectively.) This equipment employs 2K-byte
RAM's. Although it is possible to set the block size to 8x8
pixels by increasing the capacity of the RAM, we have found experimentally that a block size of 16x16 pixels is sufficient.

2.2 Correction Mode

In the correction mode, the DMA controller sets the upper bits
of the X and Y counters on the address bus and sends a read strobe
pulse to the memory. After the access time of the memory, the data
which has appeared on the data bus is latched with a pulse sent
from the DMA controller. The DA converter converts this latched
data into an analog voltage which is supplied to one input of the
analog multiplier. A low-pass filter is inserted between the DA
converter and the multiplier to smooth the stepwise waveform. The
television signal is connected to another input of the multiplier,
where the pedestal level of the signal has been adjusted to 0 V
prior to the operation. The bandwidth of this analog multiplier
is 10 MHz to process the wideband television signal. The output
of the multiplier is applied to the S&H circuit and to the output
terminal via a buffer amplifier. The television monitor or digi-
tizers may be connected to this output terminal. Switches and
lamps (Figure 3) are provided to control and monitor the operation.

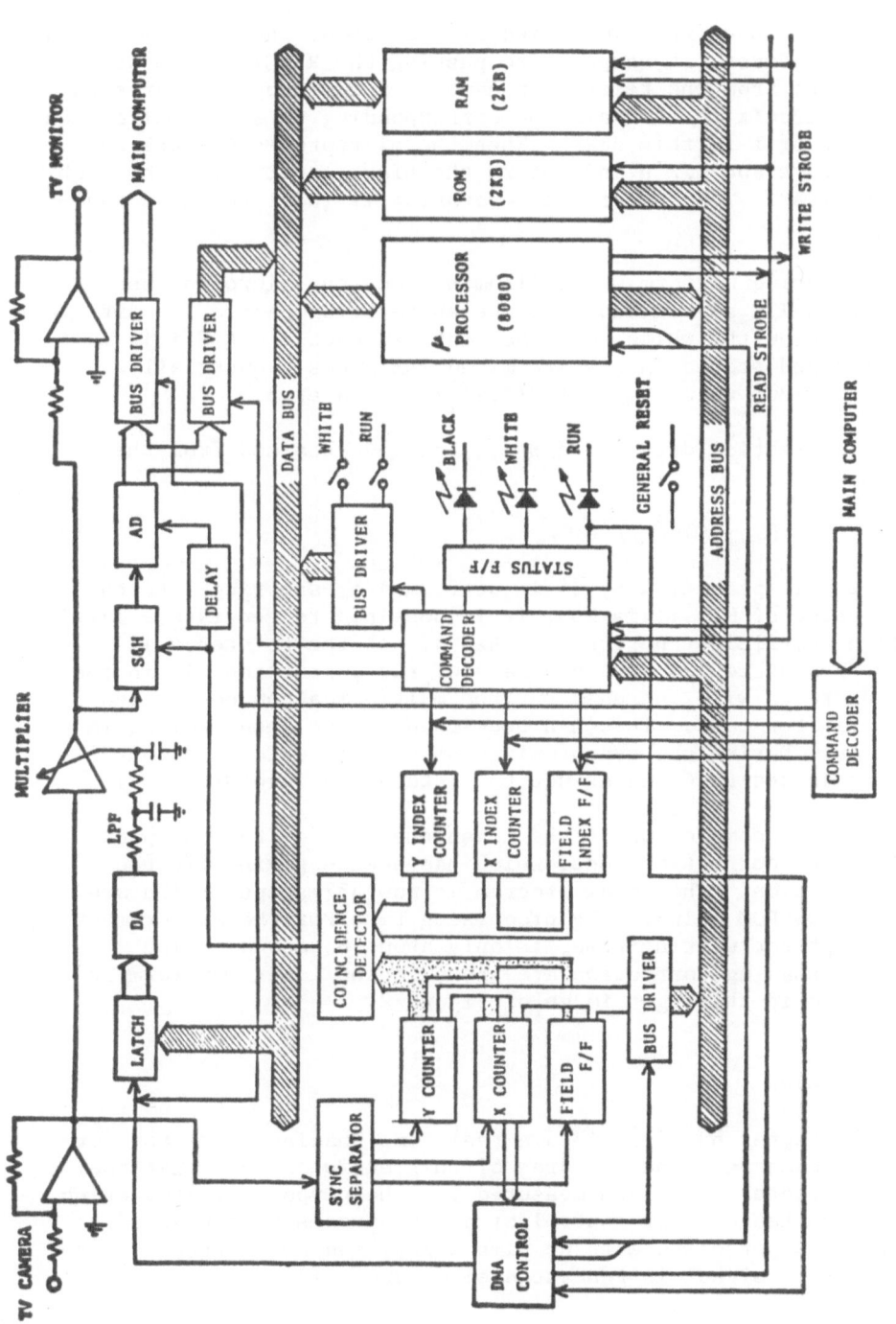

Fig. 3.   Hardware configuration.

3. OPERATION FLOW

    Figure 4 shows the simplified flow chart of the program stored
in the ROM. First, in response to pushing the RESET switch with
light blocked from the television camera, the microprocessor sets
the latch circuit to the value 64 corresponding to a multiplication
coefficient of 1 in this case. Then the microprocessor sets the
index counters and F/F's and enters the black level data from each
block into RAM. The end of this operation is indicated by a monitor lamp.

    Next, using uniform white illumination, the microprocessor
senses the WHITE switch and gets the white level data. The difference between the white and black data in each block $S(i,j)$ is
calculated and stored in predetermined RAM areas sequentially.
Then the maximum value S_{max} of $S(i,j)$ is calculated.

    The correction data table $M(i,j)$ can be obtained from the
equation

$$M(i,j) = (S_{max}/S(i,j))(64) \tag{1}$$

where the multiplication by 64 is performed by shifting. As the
maximum value of $M(i,j)$ is 255, it is possible to generate a multiplication coefficient no greater than 4. If the reference is
changed from 64 to 128, the accuracy of the correction within the
operating range will increase but the multiplication coefficient
will be limited to 2. This has been found to be insufficient in
some cases. During the synchronization signal period the correction data is set to 64 so as not to disturb the waveform.

    After the completion of the measurement mode, the equipment
goes into the correction mode, which has been explained in the
previous section. The total program is now 337 steps, which are
stored in two ROM chips. The processing time for the measurement
mode, except for waiting time, is only about 6 seconds. The
results of shading correction are shown in the lower waveform in
Figure 1 and in the right image of Figure 2.

4. CONCLUSION

    This chapter has described a real-time shading corrector for
television cameras. The features of this equipment are that the
shading characteristics are measured for the purpose of generating
a correction table under control of a microprocessor, the shading
correction is performed with no time delay, and the corrected analog television signal is generated simultaneously.

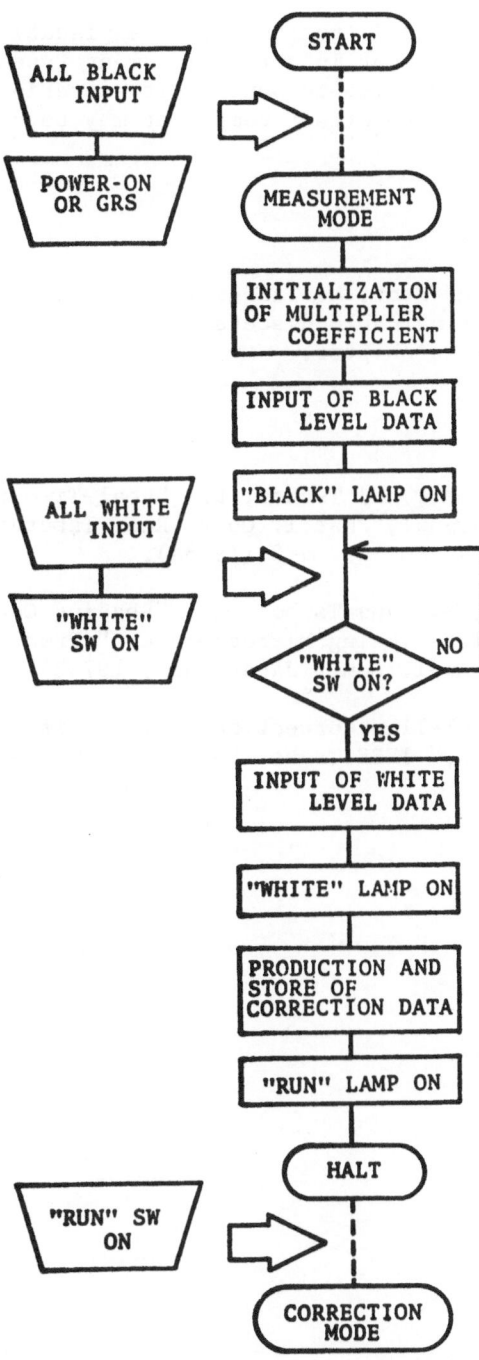

Fig. 4.   Operational flow for measurement mode.

As can be appreciated from the above explanation, this equipment is also efficient in correcting many types of illumination in addition to that due to sensitivity deterioration. For example, variation in the offset component may be corrected if necessary.

5.   ACKNOWLEDGEMENT

We wish to thank Prof. M. Takagi and the members of our laboratory for their valuable discussions and help in the experiments reported in this chapter.

6.   REFERENCES

Onoe, M., Ishizuka, M., and Tsuboi, K., "Real-Time Shading Corrector of TV Signal," Nat'l. Conv. Rec. Internat'l. Television Engrs. Japan, No. 9-2 (1978A).

Onoe, M., Ishizuka, M., and Tsuboi, K., "Shading Correction Equipment for TV Signal Using Microcomputer," Proc. Internat'l. Microcomputer Appl. Conf. Japan (Nov. 1978B).

Sawchuk, A. A., "Real-Time Correction of Intensity Nonlinearity in Imaging Systems," IEEE Trans. Comput. C-26:34-39 (1977).

PARALLEL ARCHITECTURES FOR IMAGE PROCESSING

S. R. Sternberg

Environmental Research Institute of Michigan

Ann Arbor, Michigan, USA

1. INTRODUCTION

Conventional computers do not readily lend themselves to picture processing. Digital image manipulation by conventional computer is accomplished only at a tremendous cost in time and conceptual distraction. Computer image processing is the activity of
modifying a picture such that retrieval of relevant pictorially encoded information becomes trivial. Algorithm development for image
processing is an alternating sequence of inspired creative visuali-
zations of desired processed results and the formal procedures implementing the desired process on a particular image processing
system. But our process of creative visualization is of pictures
as a whole. Implementation of the visualized image manipulation by
conventional computer requires fragmentation of the pictorial concept into information units matched to the word oriented capabilities of general purpose machines. Conventional computer image processing could be broadly categorized as manipulation of pixel
states rather than pictorial content.

The image processing approach presented in this chapter differs from conventional methods in that the basic manipulative in-
formational unit is pictorial. Image processing is treated as a
computation involving images as variables in algebraic expressions.
These expressions may combine several images through both logical
and geometric relationships. Finally, highly efficient parallel
computer architectures are proposed for implementing the computation automatically.

2.   THE BINARY IMAGE

The binary image is the fundamental unit of pictorial information. A binary image is a picture which is partitioned into regions of foreground and background. By convention, foreground regions are colored black, background regions are colored white. A binary image in Euclidean 2-space is an unbounded planar composition of interlocking black and white regions.

A binary image formulation is a rule for assigning the colors black and white to every point x,y in the plane. A binary image is normally described by a formulation which specifies the location of all of its black points. A binary image is fully specified by the set of its black points.

3.   BINARY IMAGE TRANSFORMS

A binary image transformation is a rule for mapping one binary image into another binary image. A binary image operation is a rule for mapping an ordered pair of binary images into a binary image. Binary image operations combine binary images.

Binary image transformations and operations are either logical or geometric. If the color of a point P in the resultant image can be formulated from the colors of point P in the transformed or combining images, then the transformation or operation is logical. Otherwise, it is geometric.

3.1   Geometric Transforms

Geometric operations on binary images add and subtract points vectorially. A point P in the plane is the sum or difference of the points (x_1, y_1) and (x_2, y_2) if $P = (x_1 + x_2, y_1 + y_2)$ or $P = (x_1 - x_2, y_1 - y_2)$ respectively.

Reflection is the geometric transformation which reverses the orientation of a binary image by assigning the color black to a point P in the reflected image, if and only if the point -P is black in the binary image undergoing reflection. Reflection of planar binary images rotates them 180 degrees about their origin or turns them upside down.

3.2   Logical Transforms

Complement is the logical transformation of a binary image which interchanges the roles of foreground and background by exchanging the colors black and white.

Union and intersection are the logical operations on binary images. A point is black in the union of a pair of binary images, if and only if it is black in either. A point is black in the intersection of two binary images, if and only if it is black in both.

3.3 Dilation and Erosion

Dilation and erosion are geometric operations on binary images. A point P is black in the dilation of an ordered pair of binary images, if and only if there exists a black point in the first and a black point in the second whose sum is P. A point P is black in the erosion of an ordered pair of binary images, if and only if for every black point in the second there exists a black point in the first, such that their difference is P.

Translating a first binary image to a point P may be accomplished by either dilating the first binary image with a second binary image whose foreground contains only a single black point P or eroding the first binary image with a second binary image whose only black point is at -P.

A first alternative definition for the geometric operations dilation and erosion are expressions involving translations of a first binary image by the black points of a second binary image. The dilation of two binary images is the union of translations of the first binary image by black points of the second binary image. Dilation is commutative, therefore, the dilation of a pair of binary images is also the union of translation of the second by the black points of the first. The erosion of two binary images is the intersection of translations of the first binary image by the black points of the reflected second binary image. Erosion is not commutative. (See Figures 1 and 2.)

$$A \oplus B = \bigcup_{b \in B} A \oplus b = \bigcup_{a \in A} a \oplus B \qquad (1)$$

$$A \ominus B = \bigcup_{b \in B} A \ominus b \qquad (2)$$

A second alternative definition of dilation and erosion expresses the resultant binary image in terms of those translations of a second binary image whose black regions either overlap or are completely contained in the black regions of a first binary image. A point P is black in the dilation of two binary images, if and only if there exists at least one black point in the translation

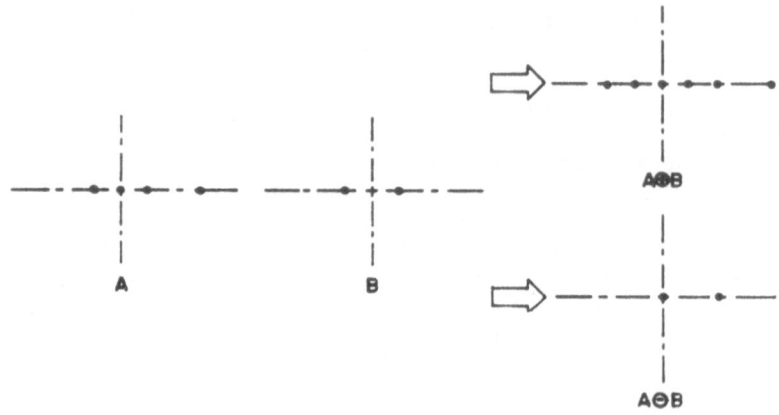

Fig. 1. Illustrating the definitions of dilation ⊕ and erosion ⊖.

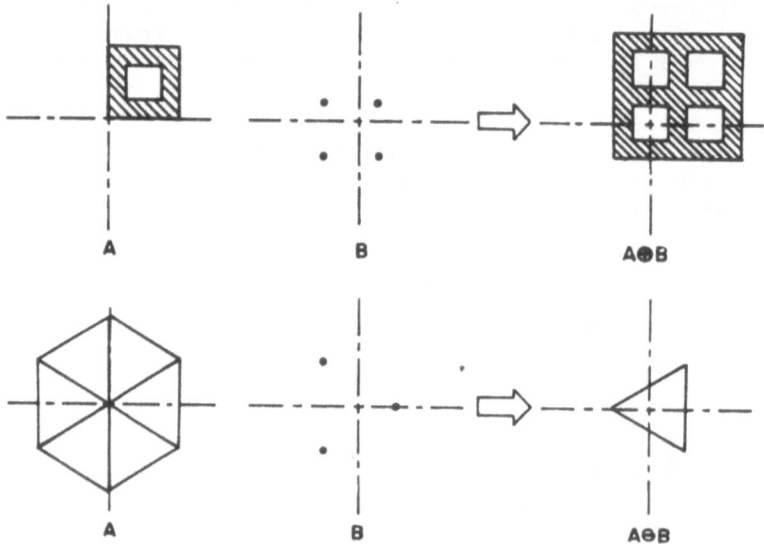

Fig. 2. Illustrating the first alternative definition of the geometric operations on binary images.

of the reflected second binary image to P which is also black in
the first binary image. A point P is black in the erosion of two
binary images, if and only if every black point in the translation
of the second binary image to P is also black in the first binary
image. (See Figure 3.)

$$A \oplus B = \left\{ p: \underset{b \in B}{\exists} \; p - b \in A \right\} \tag{3}$$

$$A \ominus B = \left\{ p: \underset{b \in B}{\forall} \; p + b \in A \right\} \tag{4}$$

It is the second alternative definition of the geometric operations dilation and erosion which are most suggestive of their utility. The first binary image of a paired relationship is considered
to be the image undergoing analysis, called the active image. The
second binary image of the paired relationship is the image to
which the active image will be compared, called the reference image.
The basis of comparison is overlapping or containment of the reference image foreground with respect to the active image foreground.
The reference image serves as a probe, labeling the resultant output image black at those points where, in the case of erosion, the
foreground of the probe is completely contained in the foreground
of the active image or, in the case of dilation, where the foreground of the reflected probe overlaps the foreground of the active
image.

3.4   General Comparison

Complement and reflection, union, intersection, dilation and
erosion are the primative binary image manipulations on which a
complete image processing system including a formal language and
a processor architecture can be based. The general utility of
these operations, particularly the geometric operations, dilation
and erosion, cannot be overstated. The obviousness of the utility
of dilation and erosion, as generalized image processing procedures, depends upon their interpretation in the somewhat altered
but equivalent definitions presented above.

Many frequently used image processing transformations may be
conveniently expressed in terms of sequences of logical and geometric operations combining the image undergoing analysis with
stored or specially constructed reference images. The medial axis
transformation, convex hull, size discrimination and template
matching are but a few of the commonly employed image analysis

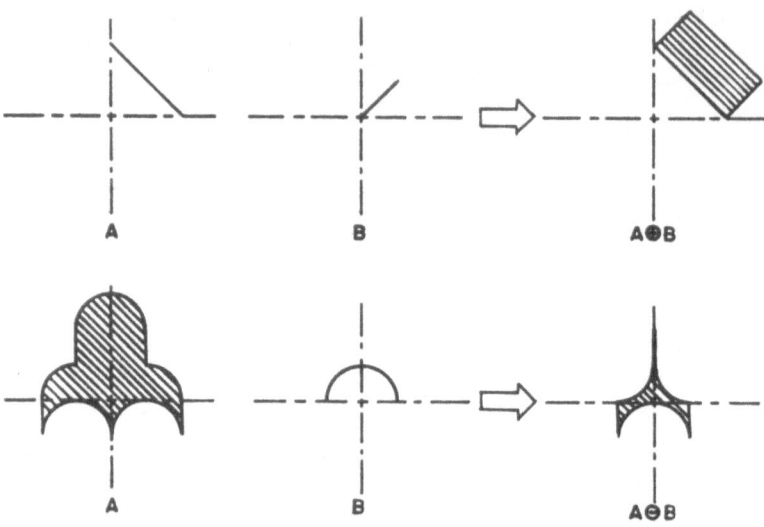

Fig. 3. Illustrating the second alternative definition of the geometric operations on binary images.

procedures which may be expressed as algebraic formulations involving logical and geometric binary image relationships. Nor are
the logical and geometric operations restricted in their utility to
planar binary images. The definitions quite naturally extend to
binary images in higher dimensional spaces and so to planar grey-
scale images which may be interpreted as binary images in three-
space. Indeed, a variety of edge detection and noise removal processes on planar grey-scale images may be formulated as sequences
of logical and geometric operations on three dimensional binary
images.

4.   HARDWARE IMPLEMENTATION

The logical and geometric operations on binary images are
easily comprehended as paired relationships between whole images
rather than manipulations of individual pixels. Successful employ-
ment of these operations in image processing systems would seem to
require that their implementation in hardware permits programming
by specifying the images and relationships involved without the
need to microprogram at a pixel level. Machine architectures specifically designed for computing the logical and geometric operations are highly parallel structures which efficiently minimize the
time and conceptual distraction implicit in modeling parallel processes on sequential machines.

Parallel implementation of the logical operations on pairs of binary images presents no conceptual difficulties. Computer architectures for parallel implementation of the geometric operations are suggested by the first alternative definition of dilation and erosion. The dilation of an active binary image by a reference binary image is computed by iteratively forming the union of the translations of the active image by the black points of the reference image. The eroded active image is computed by iteratively forming the intersection of its translations by the black points of the reflected reference image.

4.1 Optomechanical Model

An optomechanical model of a parallel mechanism for computing the geometric operations on pairs of binary images consists of a rigid transparent arm, one end of which scans the reference image while simultaneously scanning a shadow mask attached at its opposite end. A collimated light source is selectively switched on whenever the scanning arm encounters a black point in the reference image (Figure 4).

For implementing the dilation operations, the shadow mask is opaque in background regions of the active image. The active image shadow mask interrupts the light falling on a negative film whose exposure characteristics are determined such that the film remains white only at those points which remain entirely within shadow. For implementing the erosion operation, the procedure is modified by reflecting the reference image, constructing the shadow mask such that its opaque regions correspond to foreground region of the active image and employing a positive film which remains black at only those points which remain entirely within shadow.

4.2 Binary Cell Arrays

Digital electronic implementation of geometric and logical operation computations is strongly suggested by analogy with the optomechanical model but requires the introduction of the concept of a digital binary image. A digital binary image is the intersection of a binary image with a binary lattice. A binary lattice is a binary image which is black only at periodically spaced points in the plane. A binary lattice may be formulated as the dilation of two noncolinear binary ladders, a binary ladder being the binary image which is black only at periodically spaced points along a straight line, including the origin. Logical and geometric operations on pairs of digital binary images yield digital binary images.

A parallel electronic architecture for computing geometric operations on digital binary images consists of two planar arrays

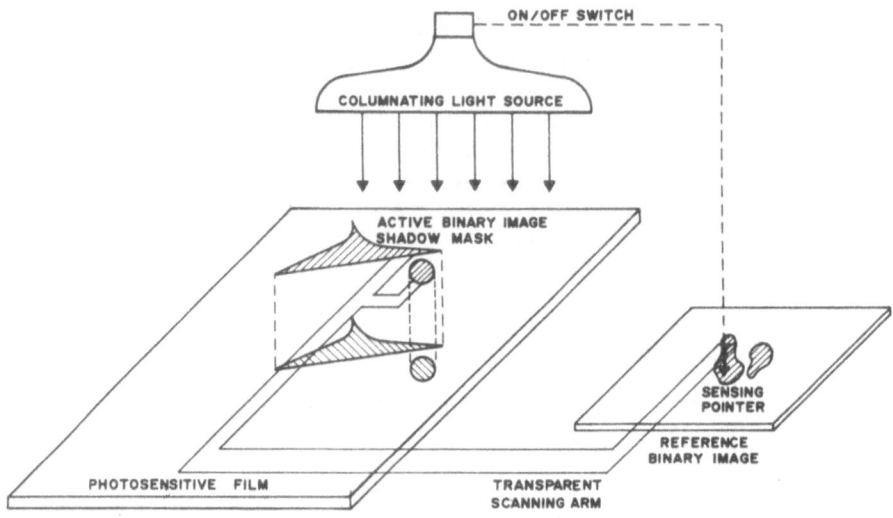

Fig. 4. Optomechanical model of a device for computing the geometric operations on binary images.

of identical binary cells arranged on binary lattice centers. The
cells of the first array store the active digital binary image.
The cells of the second array receives the iteratively computed
result. For dilation, the cells of the second array are initially
loaded with zeros representing white points; for erosion, the second
array is initialized with ones representing black points.

    The basic computational step consists of shifting in parallel
the contents of the first array by a displacement given as the
position of a pointer scanning an externally stored digital reference image and logically combining the shifted active image with
the contents of the stationary second array, the intermediate result of the basic computational step being stored in the second
array. (See Figure 5.) Dilation requires that the active image be
shifted to positions determined by the black points of the digital
reference image and the shifted active image logically OR'ed with
and replacing the contents of the stationary array. Erosion similarly requires the active image to be positioned in the shifting
array in accordance with the black points of the reflected digital
reference image with the AND function determining the logical relationship between the first and second arrays.

4.3 Computational Efficiency

    The total number of computational steps required to compute
the geometric operations on pairs of digital binary images in
shifting array architectures is equal to the number of black points

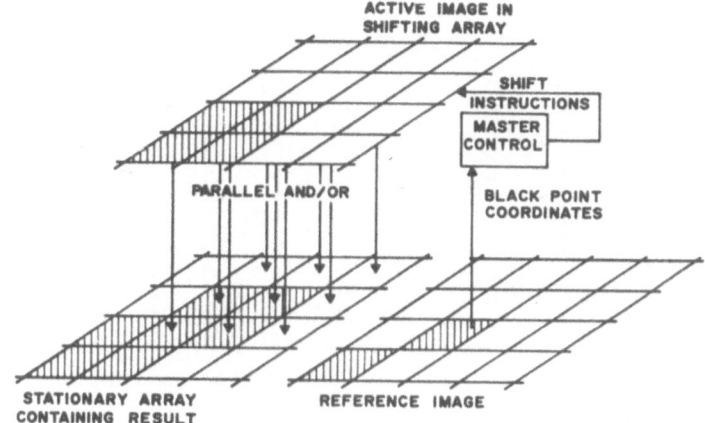

Fig. 5. Shifting array architecture for computing geometric operations on binary images.

in the digital reference image. Single time step translation of the active image in the shifting array by arbitrary displacements requires that every cell of the shifting array be interconnected to every other cell in the array. The expense of this universal interconnection is avoided by decomposing arbitrary active image shifts into a sequence of nearest neighbor shifts involving cells which are adjacently located in the shifting array. The required cellular interconnections are physically realizable in the plane, but computation time is adversely affected, particularly for reference images whose black points are widely disjoint.

A significant computational savings may be achieved in cases where the digital reference image may be expressed as the iterated dilation of two or more digital images. In such cases, the geometric operations can always be computed in at least as few steps as the total number of black points in the digital reference image, and frequently considerably fewer.

Dilating a first binary image iteratively by a second and then a third binary image is equivalent to dilating the first binary image by the dilation of the second with the third binary image. Eroding a first binary image iteratively by a second and then a third binary image is equivalent to eroding the first binary image by the dilation of the second with the third binary image (Figure 6).

The computational efficiency of effecting the dilation or erosion of an active binary image by the iteratively applied sub-images which compose the reference image through mutual dilation is enhanced as the decomposition allows for more subimages with

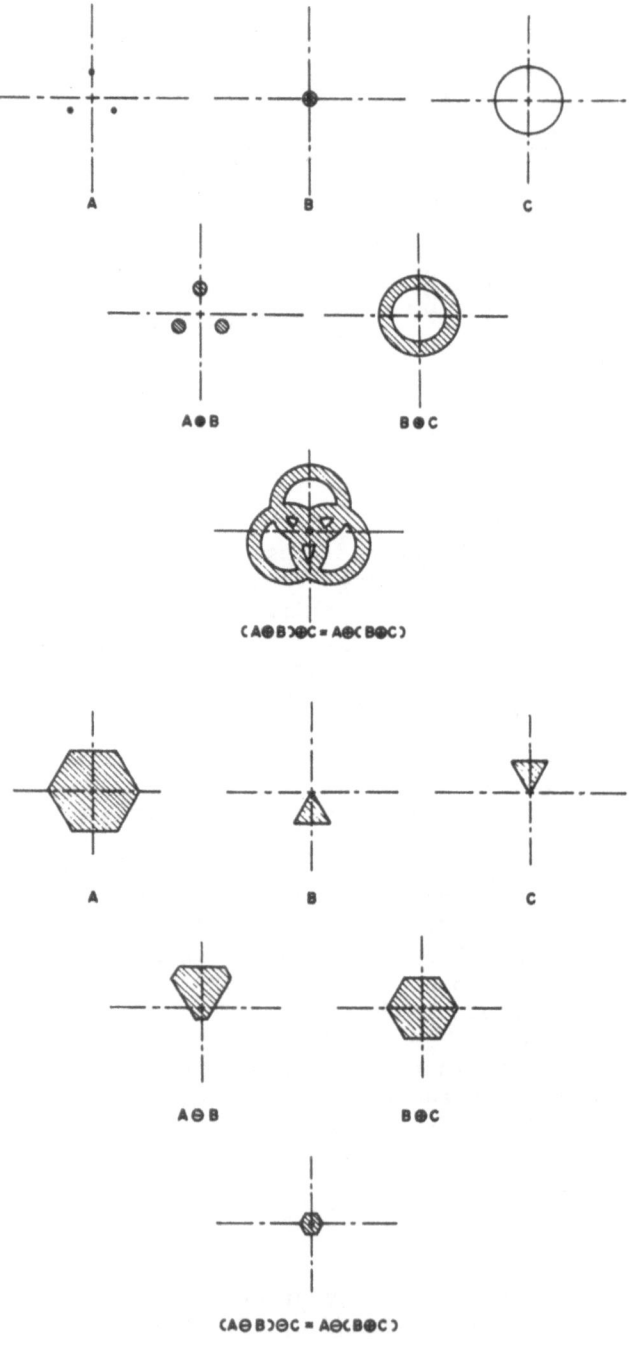

Fig. 6.   Iterated dilations and erosions of binary images。

corresponding smaller foreground regions. Of principal importance
is the question of whether a given reference image is decomposable
into subimages whose foregrounds are containable in binary images
of arbitrarily small foreground extent, called binary windows.

4.4 Window Configurations

A binary window is a binary image whose infinitely iterated
dilation with itself is the black unbounded plane. A digital binary
window is a binary image whose infinitely iterated dilation with itself is a binary lattice. An elemental digital binary window is a
digital binary window consisting of black points at the origin and
adjacent lattice points. The adjacency relationship may be formulated in several ways giving rise to the commonly implemented von
Neumann five point elemental window, the Golay seven point elemental
window and the Moore nine point window configurations (Figure 7).

Dilation and erosion by digital reference images which have
been decomposed into subimages on elemental window configurations
are readily implemented on digital electronic architectures which
are specifically adapted for that purpose. The logical array consists of identical binary cells arranged in the plane on binary
lattice centers with patterns of intercellular connections established about every cell as the pattern of points in an elemental
window. Each cell in the logical array simultaneously executes
the identical logical function of its selected inputs, the binary
value of the result becoming the new state of the cell. The pattern of selected inputs is the pattern of black points in the subimage on the elemental window, the OR function implementing dilation by the subimages of the decomposed reflected reference image
while the AND function implements erosion by the subimages of the
decomposed digital reference.

5.   ARRAY ARCHITECTURES

Array architectures for image processing require as many cells
in the array as there are pixels in the digitized image, frequently
resulting in prohibitive implementation cost. The logical array
architecture, however, may be modified to a pipeline configuration
which is efficient in its hardware requirements at the cost of increased computation time.

5.1 Pipelined Processor

The basic replicated element of the pipelined processor is the
stage. The function of the stage is to perform a dilation or

Fig. 7.   Elemental digital binary window configurations.

erosion of the entire active image by a subimage of the elemental
window.   Input to a stage is a sequence of active image binary
pixels arranged in raster scan format.   Shift register delays within the stage store contiguous scan lines while taps on the shift
register delays sequentially extract elemental window configurations
for input to the stage's logic module.   The logic module implements
the programmed logical function of its selected inputs as appropriate to dilation or erosion by subimages of elemental windows.

At each discrete time step, a new active image binary pixel is
clocked into the stage, while at the same instant, a single pixel
of the elementally dilated or eroded active image is passed out of
the stage from the logic module.   Simultaneously, the contents of
all shift register delay elements are shifted one element.   The
elemental window configuration extracted by the shift register taps
is thus sequentially scanned across the input array in raster scan
format.

Dilations and erosions of the active image by subimages on
elemental window configurations are computed within the basic clock
step permitting the output of a stage to occur at the same rate as
its input, thus permitting the serial concatenation of stages.
Real-time computation of the geometric operations by reference
images decomposed into subimages on elemental windows is accomplished by initially programming stage logic modules in subsequent
stages and then serially passing the scanned active image through
the programmed pipeline (Figure 8).

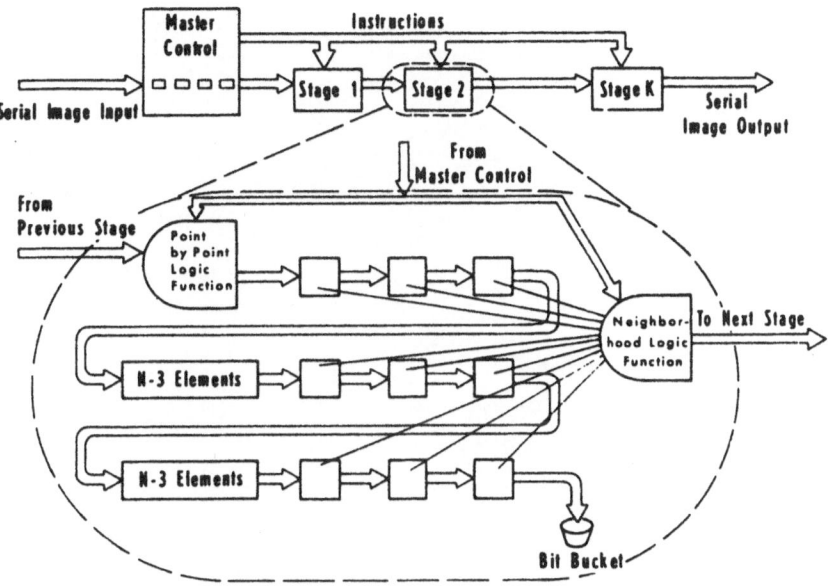

Fig. 8. Pipeline architecture for computing geometric operations on binary images.

6. CONCLUSION

In conclusion, architectures for image processing cannot be designed independent of the image processing operations which they are intended to implement. Highly parallel and conceptually simple arrangements of identical hardware modules represent the most cost efficient approach to image processing architectures, but the particular functions those modules implement and the methods of their interconnections are established through a basic understanding of the image manipulations which they are intended to perform.

Finally, computers specifically designed for image processing may be constructed on the basis of these architectures through the inclusion of appropriate input/output and control circuitry and the addition of image memory for storing partially computed results.

COLOR DISPLAY FOR IMAGE PROCESSING WITH MULTIPLE FUNCTIONS

M. Takagi and M. Onoe

Institute of Industrial Science

University of Tokyo, Tokyo 106, JAPAN

1. INTRODUCTION

Recently interest in digital image processing has been increasing rapidly. For the research, development, and practical applications of digital image processing, a device with half-tone or color display capability is essential, although not so easily available.

Taking this fact into account, we have developed various kinds of image input and output devices from the viewpoint that the introduction of technically well established television technology to digital image processing might make it possible to supply the means for real-time input and output, which is essential to image processing. We have worked intensively on input and measurement utilizing television techniques and paid attention to their application for display.

The recent intensive development of LSI technology has rapidly decreased the cost of semiconductor memory as the refresh memory for display systems. Also the price of a color display with a refresh memory is coming down within reach. In this chapter, assuming the utilization of semiconductor memory as a refresh memory for a color display, we discuss what kinds of capabilities should be incorporated in the display.

2.   DESIGN PRINCIPLES

When a color display in which an IC memory with considerable capacity is used as a refresh memory is installed in a digital image processing system, it is highly desirable to utilize the rather expensive memory not only for display but for other purposes. For example, the memory could be used as a large capacity peripheral memory of the computer to which the display is connected and also as a buffer memory for image input.

In a conventional system, image data are stored on a disc and a part of the image is transferred to the main memory of the computer. Consequently, the time for data transfer back and forth between the computer and the disc is very long. Whereas in our system the time for data transfer, including data input and output as well as processing time, can be shortened drastically.

Also, if the memory is used as a buffer memory for image input, the very high speed of the IC memory makes it possible to input a video signal within one frame of the television camera using an analog to digital converter at very high speed. Moreover, one image frame can be written into the memory from a file using a VTR or a video disc without a time-base corrector.

2.1 Multiple Image Storage in the IC Memory

If the number of bits per picture element and the number of picture elements for an image can be arbitrarily selected, a multiple number of images can be stored. The input and output of moving images are also possible by controlling the sequence of input and output.

2.2 Arithmetic Operations on Each Picture Element

In order to feed a standard television display, the image data stored in the IC memory are refreshed sequentially. If some simple arithmetic capabilities are added, each picture element may be processed at very high speed. This improves considerably on the problem of digital image processing, namely, that relatively long processing times are necessary. Also a table lookup memory may be provided to convert the monochrome data read out from the memory to corresponding red, green, and blue pseudo-color data. This auxiliary memory for table lookup can be also utilized for arithmetic operations. Addition, subtraction, and logical operations between images can be easily done. Such operations as nonlinear density curve correction, density histogram generation, and movement detection are possible using this auxiliary memory.

2.3 Access to the Displayed Data

Needless to say, the capability to select an arbitrary area
on a display using a joystick is necessary. Moreover, since it is
very hard to tell by visual observation the level of each picture
element, the capability to show only the data in an arbitrarily
selected range should be added to the display. Data on the value
of a selected picture element can also be displayed digitally on a
control panel.

2.4 Access from Multiple Computers

If this expensive display is connected to only one computer,
its idle time is lost. The display should be designed so as to be
accessed from multiple computers for efficient use of the display.
Each computer connected to the display should be able to request
use of the display if it is idle. Finally, data transfer between
computers is possible through the display memory.

3. SYSTEM CONFIGURATION

Figure 1 shows the block diagram of the system. As standard
television equipment is used, it is preferable to depend on the
standard television format. However, in digital image processing
the number of picture elements is usually selected as a power of
two. So, in our system the number of horizontal scanning lines
used is 563 lines to display 512 lines with interlace. The horizontal scanning time is selected as standard (63.5µs), so that it
takes 35.5ms to display one frame. The number of picture elements
for the vertical direction is 480 when input is from standard equipment such as a television camera, a VTR, or a video disc.

The image memory has the capacity of 262kW/18 bits to display
512x512 picture elements. The access time for one picture element
is 70ns in the display mode. But the cost of a large capacity IC
memory with this speed becomes too expensive. So, dynamic random
access memory chips with 800ns cycle time are used in this system.
In the display mode 16 words (144 bits) are read out in parallel,
stored in a buffer register, and fed to digital/analog converters
serially on a one pixel basis every 70ns. Also, when an image is
input, data is serial to parallel converted through a buffer register.

An auxiliary memory is provided that has the capacity of 256
word/18 bits and is a random access memory with 70ns cycle time.
Hence 8 bits of monochrome data can be converted to three 6 bits
(R, G, and B) of pseudocolor data. This auxiliary memory can also

be used as a lookup table for nonlinear density transformations, as
a memory for histogram generation, and as an address memory for extraction of data within a certain range.

This display system has three ports for minicomputers: two
for existing computers; one for a future control computer for a
special processor with higher arithmetic capabilities for image
processing which is under consideration. Video equipment such as
television cameras, monitors, etc. are connected to this system and
the connection of VTR's and a video disc is being planned. A photograph of the system is shown in Figure 2.

4.   UTILIZATION OF IMAGE MEMORY

An image memory with a capacity of 512x512 words (18 bits/
word) is used in this system. To utilize the memory efficiently
and to provide the capability of storing multiple images, one 18-
bit word can be divided into multiple partitions in depth and the
plane divided into multiple subareas. According to the number of
bits per picture element, one word can be divided in any one of
the following ways:

    1 bit x 18,

    4 bits x 4   + 2 bits,

    6 bits x 3,

or   8 bits x 2   + 2 bits,

The depth of 1 bit is used in the case of binary pictures such as
Chinese characters and so on. A depth of 4 bits is provided for
TV pictures of low quality and 6 bits used for R, G, and B color
display. The 2 extra bits in the case of a depth of 4 bits or 8
bits are used for the display of characters or pointers.

On the other hand, one 512x512 plane can be divided in the
following ways:

    32x32   picture elements 256 divisions,

    64x64   picture elements 64 divisions,

    128x128   picture elements 16 divisions,

    256x256   picture elements 4 divisions,

or   512x512   picture elements 1 division.

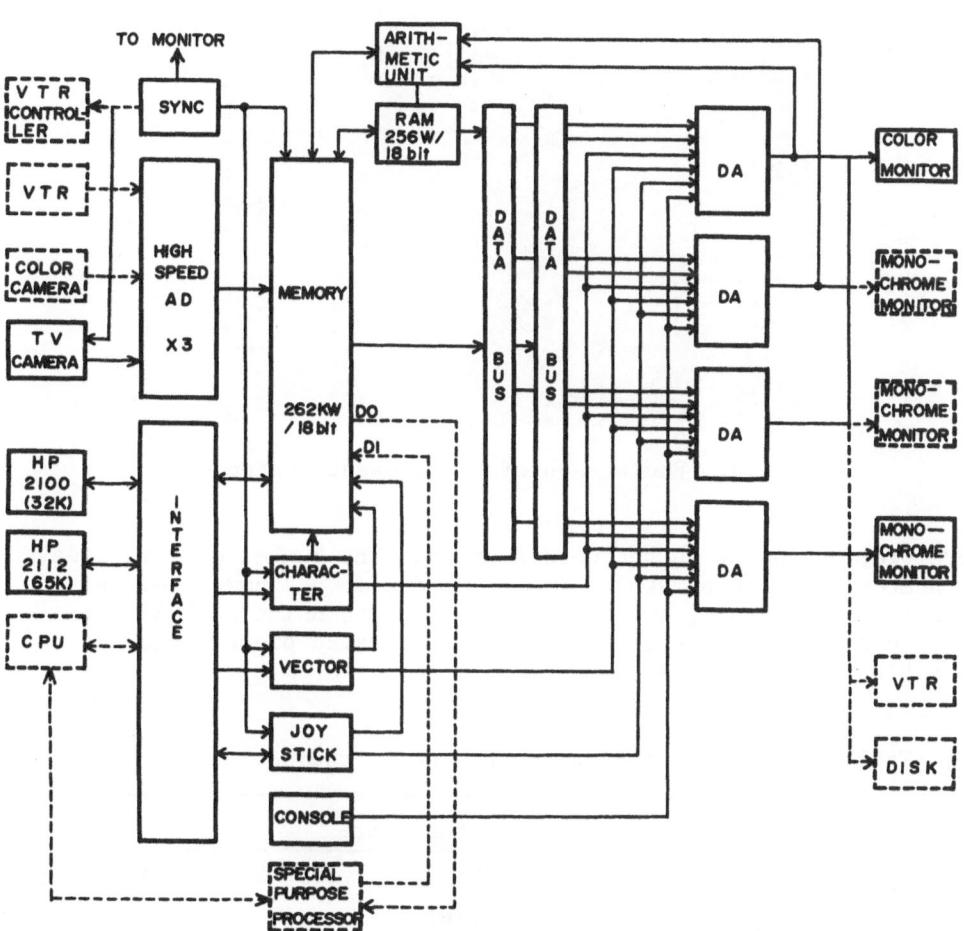

Fig. 1. Block diagram of the present system.

Fig. 2.   Photograph of the present system.

Therefore, depending on the word depth and the number of divisions,
the number of pictures stored in the memory is as summarized in
Table 1.   For example, 4608 Chinese characters using a 32x32 dot
matrix can be stored or 64 pictures with 128x128 picture elements
and 4 bits in gray scale, etc.

Table 1 – Number of Pictures Stored in the Whole Memory

Bits/Pixel	Picture Size				
	32x32 (256)	64x64 (64)	128x128 (16)	256x256 (4)	512x512 (1)
1	4608	1152	288	72	18
4	1024	256	64	16	4
6	768	192	48	12	3
8	512	128	32	8	2

5. CAPABILITIES

This section lists and briefly describes each of the many
capabilities of our system.

5.1 Data Input and Output

Data transfer to and from computers connected to this display
system is possible in either programmed transfer mode or DMA transfer mode. Special high-speed digital input and output channels
are provided for data transfer between the memory and the special
processor. Also, conventional manual input methods by a joystick,
a function keyboard, and the front panel are possible.

5.2 Switching of Monitors

One color monitor and one monochrome monitor are driven by 4
digital to analog converters. The data buses for the main memory
and the auxiliary memory, the selection of the memory and the division of the word, and the selection of monitor channels can be
arbitrarily done.

5.3 Pseudocolor Display

By referring to the auxiliary memory, 4-, 6-, or 8-bit monochrome picture data can be converted into 3 channel 6-bit data and
displayed on a color monitor.

5.4 Mosaic Display

By changing the mode of plane division, a picture with a small
number of picture elements can be displayed on a whole screen or
multiple pictures can be displayed in a mosaic. (See Color Plate 17*.)

5.5 Display of Moving Images

When the memory is used as a multiple image memory, images
can be displayed successively at an arbitrary frame interval using
as many as 512 frames in the sequence of word division, plane division, or their combination.

*The color plates will be found following page 46.

5.6 Roll-up

An image with specified size can be rolled up by one line on one word or on a one-word division basis.

5.7 Joystick

Pairs of coordinates for the upper left and bottom right corners of an image and that of an arbitrary picture element can be set from the console or the computer.

5.8 Vector Display

A line can be drawn between the two points specified by a joystick. The display level of the line corresponds to data at the specified starting point.

5.9 Character Display

If a command for character display is received, a character pattern generated by a ROM is stored on the 17th bit plane at the location specified by the joystick. The display of characters can be switched on and off by the command or the switch on the control panel.

5.10 Pointer Display

A flag "1" is written on the 16th bit of the memory at the address shown by the joystick. This flag turns the specified picture element into black or white by the inversion of the MSB of the data. The display is switched on and off by the command or the console panel and the display mode is specified by the console panel.

5.11 Variable Display Range

Each digital to analog converter has two comparators the upper and lower values of which are set by the command or the control panel. The data within the specified range are displayed as they are, but others are displayed in black to check quickly the data range of the objects under investigation.

5.12 Arithmetic Operations

By specifying two sources A and B an arithmetic operation is done and the result is stored at the destination D. The operation is executed only in the area of the properly divided plane, the divided part of the word, and the area specified by the joystick. The arithmetic operations shown in Table 2 are provided. Finally, it is also possible to write the data in the address corresponding to the result of the operation referring to the auxiliary memory into the destination plane. With these operations, addition and subtraction, logical operations, and nonlinear operations referring to the auxiliary memory can be executed on a picture-element-by-picture-element basis.

Table 2 – Arithmetic Operations

ADDITION	: WITH OVERFLOW WITHOUT OVERFLOW
SUBTRACTION	: 0 FOR NEGATIVE VALUE ADD 256 TO SHIFT RIGHT
DATA TRANSFER	: A $\overline{A}$
AND	: A B $\overline{A}$ B
OR	: A + B $\overline{A}$ + B
EXCLUSIVE OR INCLUSIVE OR	

Using the comparators of the third digital to analog converter
channel during the arithmetic operation, the address (18 bits) of
the main memory where the result of the operation falls in the
specified range can be written into the auxiliary memory. This
capability is very effective in detecting the change between two
images. Namely, after the subtraction of corresponding picture
elements between two images, the address of the picture element,
the value of which is in the specified range, is stored in the
auxiliary memory. Up to 256 such addresses can be stored in the
auxiliary memory. Reading out the auxiliary memory, the picture
elements with large changes can be extracted.

Moreover, the auxiliary memory can be used for histogram generation. During an arithmetic operation, assuming the data written
into the destination as the address of the auxiliary memory, a one
is added to the contents of the address. Then, while reading the
memory sequentially, the histogram of pixel values can be generated.

5.13 Image Input

The video signals are A/D converted by the externally supplied clock and stored into the specified area of the memory in
one frame. A VTR controller, which can select an arbitrary frame
from the recorded images and generate the sampling clock locked to
the VTR sync signal, is under development for the analysis of
moving images recorded on tape.

6.   CONCLUSION

A real-time color display is an essential tool for interactive
digital image processing. The capabilities to be given to such a
display have been discussed in this chapter. For the efficient
utilization of memory, the image memory can be utilized not only
as a refresh memory, but also as a peripheral memory for the associated computers and a buffer for high-speed image input. By dividing the memory in the plane as well as in the word depth, according to use, the memory can be used as a multiple image memory
and for the input and output of moving images.

Utilizing the fact that the memory is read out sequentially,
simple operations, such as addition, subtraction, logical operations, table lookup, histogram and change detection, can be processed quickly.

Development of a special processor for image processing and
the connection of a VTR to read and write on a one frame basis are
presently under consideration.

IMAGE PROCESSING ORIENTED MULTIPROCESSOR SYSTEM WITH A

MULTIPURPOSE VIDEO PROCESSOR

A. Tojo and S-i. Uchida

Electrotechnical Laboratory

Tsukuba Research City 305, JAPAN

1. INTRODUCTION

Although pictorial data processing has been considered to be one of the most promising areas of application of parallel and distributed computer systems for some time, the hardware realization of such systems for really serious applications has been quite difficult. The recent advent of LSI and microprocessor technologies, however, allows us to attempt implementing specialized image processing hardware with fairly large parallel processing capabilities with a relatively small effort. Some of the typical proposed or experimental systems have been reviewed by Fu (1978). There is always an increasing need for processing power both in space and time as well as flexible input/output capabilities. One of the main objectives of developing the system presented in this chapter is to ease this situation by taking advantage of current semiconductor device technologies.

Another main objective is to investigate flexible and efficient interconnection structures for a cost-effective multiprocessor system. Although distributed processing will undoubtedly be the most popular architecture in the future, using one or more of the various interconnections proposed, it is still not clear which one of these is most appropriate for a given set of problems. (See Anderson and Jensen, 1975.) In distributed or array processor systems their interconnection structures strongly affect their characteristics.

This chapter discusses the structure of the picture processing oriented poly-processor system (POPS) with a multipurpose video processor which is now in the final stage of the first phase of development. Some of the system features are: (1) a large shared memory to allow efficient access to raw picture data, (2) parallel processing between component processors utilizing a problem-oriented interconnection mechanism, (3) direct access to the local memory of each component processor, (4) color and half-tone graphic display with resolution and capabilities comparable to those of standard vector graphic displays, (5) high-resolution color or B&W image input in one TV frame time, and (6) firmware realization of low (signal) level picture processing operations.

2.   INTERCONNECTION STRUCTURE AND ORGANIZATION OF THE SYSTEM

To achieve high performance and flexibility in distributed computer systems, various interconnection structures have been proposed. The present POPS system employs a simple common bus structure as shown in Figure 1. Although this scheme has enough flexibility and modularity, it usually suffers from the limited bandwidth of the communication link because all the component processors share a single bus to communicate with each other.

In order to minimize the overheads caused by heavy traffic, component processors in POPS are connected to the common system bus via functional interface processors called Intelligent Couplers (ICPL). As is described in the next section, the number of component processors that can be included in the system while maintaining a high processor utilization ratio is quite large by virtue of this interconnection mechanism.

Functionally, the system is built around a large, high-speed shared memory with three independent ports which allow concurrent access from different sources. Two of these memory ports are standard random access ports for the system bus and the video processor. The third port is a high-speed sequential port for a scanned display or for video input from a TV camera or a VTR. Access control between these ports is conducted by the priority logic in the shared memory. The highest priority among them is assigned to the sequential-write request from port 3 and the lowest to the sequential-read request.

Figure 2 shows the configuration of the actual prototype system under development. The address space of the common system bus (CSB) is 4 million bytes and the shared memory and the local memory of each component computer are assigned unique addresses in this address space. The shared memory is used as a refresh memory and an image data buffer for the video system as well as a common data buffer between the component computers of POPS. It has a capacity of 1 million 9-bit bytes to accommodate a 1024x1024 image data

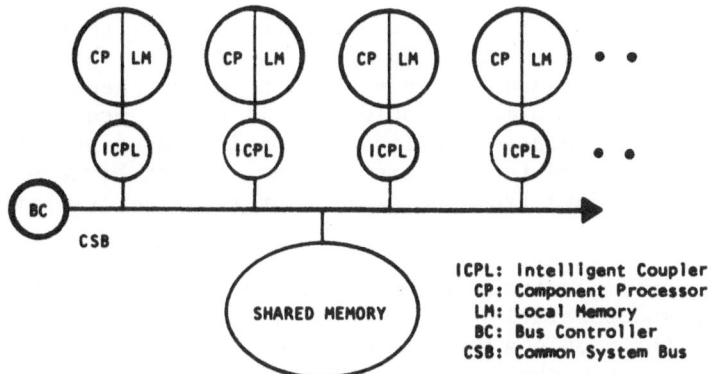

Fig. 1.   Simplified interconnection model of the POPS system.

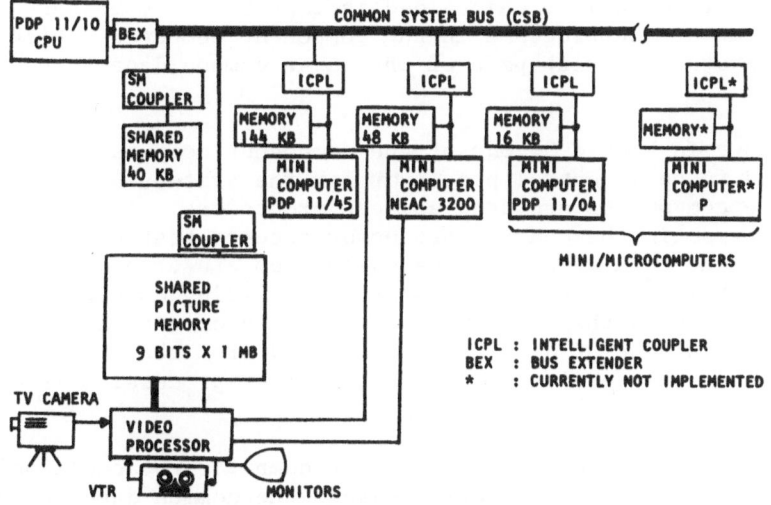

Fig. 2.   Block diagram of POPS.

field and has an average data transfer rate of 40MBytes/sec or
26ms/1024x1024-pixels to maintain an interlaced 60 Hz display refresh cycle at the high-speed sequential port.   (The peak transfer
rate is actually 45 MBytes/sec or 22ns/byte.)

    The video processor (VIP) in the video system is connected to
the host computers through separate local links.   It accepts various graphic and firmware start commands through this channel as a
normal I/O device, processes or generates picture data in the
shared memory, and sends back the results or sends input data from
the external devices.

3. INTELLIGENT COUPLER AND IMPLEMENTATION OF PARALLEL PROCESSING

In designing a multi-mini/microcomputer system, there are
several problems to be solved. In the case of POPS, which employs
a global bus structure and is intended to be mainly used in image
processing there are two main problems: (1) how to prevent the
degradation of the processor utilization ratio caused by bus saturation and (2) a method for efficient access to the large amounts
of two-dimensional data from the small address space of mini- or
microcomputers.

A microprogram controlled interface processor (ICPL) together
with the proposed parallel processing strategy described below can
solve most of the problems described by Uchida and Higuchi (1978).
In the present system the large two-dimensional data placed in the
shared memory are accessed only by an ICPL during the execution of
task programs and the task programs are not shared by the component
computers, but copies are maintained in each of the computers.
Task programs are subprograms in the component computers which
execute computations with part of the large common data to be processed.

As shown in Figure 3, one of the component computers in POPS
is assigned to be a control processor for the system while the
rest of the components are used as task processors. A special
monitor program in the control processor accepts task execution
requests in the form of subroutine calls from a user program and
dispatches them to the task processors recognizing concurrent processability between them. Using an ICPL, each of the task processors receives blocks of data to be processed from the shared memory
and sends back the results to the memory after processing is complete.

This strategy allows each component computer to access only
its local memory and not directly address the common data in the
shared memory. Since the access count to a data element in the
shared memory is greatly decreased by this method, the maximum
number of the component computers that can be included in the system to realize concurrent operations between the tasks without
saturating the system bus becomes quite large in certain types of
applications.

Moreover, because an ICPL can carry out fairly complex address
calculations or problem-oriented nonlinear address mappings during
the data transfer and these operations can be pipelined to the
process in the component computer attached to it, we can expect
another improvement in performance even with a single task-pro-
cessor-ICPL pair.

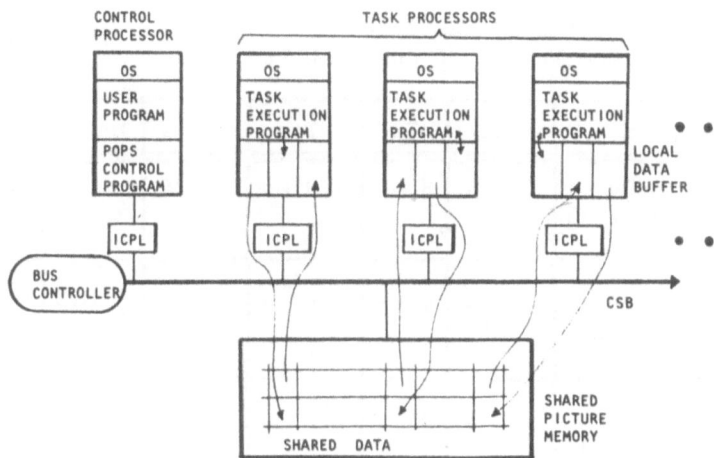

Fig. 3.   Control and data flow in POPS.

The maximum number of the component computers for efficient parallel processing can be estimated by the following conditions: (1) the control processor must be able to supply a sufficient number of tasks to the task processors to keep them busy and (2) the system bus must not be saturated by the data transfer carried out by an ICPL.   From these conditions, we can derive the following equations:

$$P < Tc/Ta \tag{1}$$
$$P < Tc/Tt$$

where

        P = number of the task processors,
        Tc = task execution time,
        Ta = task dispatching time, and
        Tt = bus occupancy time for data transfer.

Accordingly, the task dispatching time Ta is mostly determined by the time required for the analysis and recognition of parallel processability between the tasks.   This is true because a task program itself is maintained in each component computer and only a small number of control parameters are actually transferred from the control processor to each task processor to start a task.

Table 1 shows the estimation of the values of the quantities given in equation (1) for several types of problems.   As shown here, the maximum number of task processors is dependent on the degree of parallelism embedded in the given problem.

Table 1. Estimation of the
Maximum Number of the Task Processors

Problem	K	Tc	Ta	Tt	Tc/Ta	Tc/Tt	Comment
Complex FFT	1024 (1024 x 1)	2.6 s	8 ms	4.3 ms	325	600	Tc ∝ K*log K
Choleski's method	1024 (32 x 32)	655 ms	8 ms	6.5 ms	81	100	Tc ∝ K
8 x 8 local operation	1024 (32 x 32)	1.3 s	8 ms	2.2 ms	162	591	Tc ∝ K
Iterative method	4096 (64 x 64)	82 ms	8 ms	8.8 ms	10.3	9.3	Tc ∝ K

LEGEND

K : number of the data per task,
Tc: execution time of a task,
Ta: average recognition time plus dispatching time for a task,
Tt: CSB occupancy time for the data transfer required in a task.

Table 2 shows an example of performance improvement in performing the two-dimensional FFT obtained by a pipelined operation
of a single ICPL and a component computer. In this case, the ICPL
executes data transfers with index calculations and data shuffling
while the component computer carries out butterfly multiplications.
If the improvement factor is Q and a total of P task processors
are included in the system without bus saturation, we can obtain
an overall performance improvement factor of PxQ.

Table 2.   Improvement Factor Obtained by Pipe-Lining

(Example No.) device	Butterfly multiplication (microsecond)	FFT time for 1024 points (second)	FFT time for 1024 x 1024 points (minute)	Improvement ratio
(1) PDP 11/45 with DMA channel	505	3.5	118	1
(2) PDP 11/45 with ICPL	219	1.2	41	2.87

4.   FUNCTION AND ORGANIZATION OF THE VIDEO SYSTEM

The video system is composed of a microprogram controlled video processor (VIP) and a large, shared-picture memory. The system is designed to be used in a variety of applications and is equipped with such functions as high-resolution color, half-tone image display, and high-speed video data input. The system is also used as an external backup memory with two-dimensional windowing capability which is quite useful in a multiple image buffer environment and for two-dimensional paging of image data.

Figure 4 shows the physical organization and the data format for image data in the video system. A user address specified by a plane number and the X,Y coordinates of a pixel is automatically translated into an internal physical address compatible with sequential access from the interlaced video input/output control. As shown in Figure 4, the system has two resolution modes; high and standard. In the high-resolution mode, 1024x1024 pixels are displayed on the CRT monitors in contrast with 512x512 pixels in the standard mode. Interpretation of pixel data at the color channel is different for each mode because only 3 bits are assigned to each color in the high-resolution mode while a full byte is available in the standard mode. Except for this case, each byte assigned to a pixel is interpreted to have 1 control bit plus 8 brightness bits.

As a graphic display, the system accepts various graphic commands from a host computer to generate display data in the shared memory which is continuously refreshed on the high resolution CRT monitors. The same algorithm employed for drawing vectors is used to read out picture data from the shared memory for the host computers. Therefore, read-out operations can be performed in an

STANDARD RESOLUTION MODE

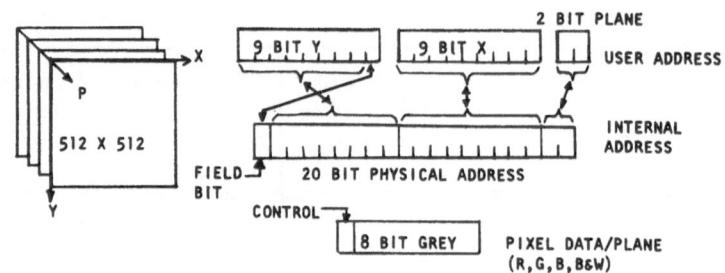

HIGH RESOLUTION MODE

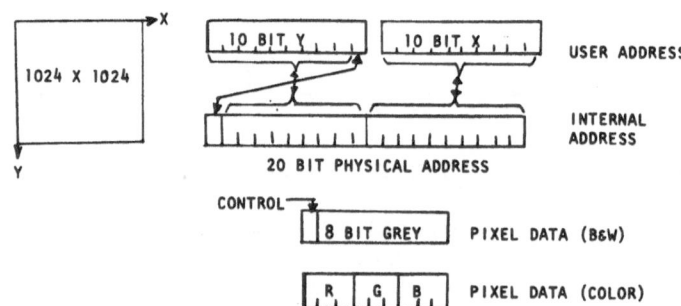

Fig. 4. Plane organization and data format in the video system.

arbitrary direction, which is quite convenient for obtaining cross-
sectional profiles of an image and for edge or line following.

Image data from video equipment having a 512x512 pixel format
can be written into the shared memory with 3/2 TV frame times or in
50 ms on the average. Users can define another window for this operation to mask out the input data outside the specified subregion.
The video input function is supported only in the standard mode.

The system accepts more than 30 basic commands to provide the
above mentioned functions. There are three data formats for picture data exchange between the video system and the host computers:
(1) normal 9 bit format, (2) packed 8-bit byte/character format (control bit discarded or cleared), and (3) packed 15-bit format (five
3-bit fields which is effective only in the high resolution mode).
These functions are performed at high speed by virtue of the employ-
ment of fast microprogrammable processing elements and ECL logic.

Figure 5 shows the organization of the video system. The system is equipped with a writable control store which can be loaded
with various application firmware. Using this facility, users can
define their own processing functions besides the basic functions
described above.

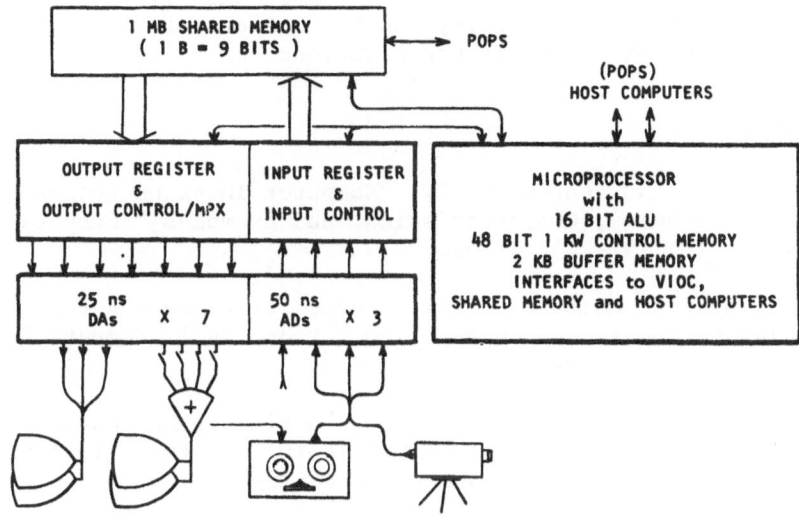

Fig. 5.   Organization of the video system.

5.   CONCLUSION

Since the start of the development of the POPS system in 1976, our main efforts have been devoted to the design and implementation of the problem-oriented interconnection mechanisms and of the multipurpose video system. (See Tojo et al., 1976.) At this moment, a three-processor mock-up of POPS is in its final stage of implementation and a set of prototype control programs is being designed. The multipurpose video system is partly operational and its basic firmware was completed at the end of 1978.

Various application programs for image processing experiments implemented on the host computers are now being modified to take advantage of the facilities supplied by the video system.

Although utilization of the parallel processing capabilities of the whole system in real applications may not be possible immediately, it is apparent that even partial use of the system functions will greatly improve the efficiency of image processing experiments. We expect that the appropriateness of the system organization will be proved through experience with these applications as well as by further study of the system characteristics.

6.   ACKNOWLEDGEMENTS

The authors are indebted to Dr. H. Nishino, the director of PIPS, and Dr. O. Ishii, the head of the Computer Science Division

for their support on this research. We would also like to thank
Mr. T. Higuchi of Keio University and the members of the Information Systems Section for their contributions and helpful discussions.

7.   REFERENCES

Anderson, G. A., and Jensen, E. D., "Computer Interconnection Struc-
    tures: Taxonomy, Characteristics, and Examples," Computing
    Surveys 7(4):197-213 (1975).

Fu, K. S., "Special Computer Architectures for Pattern Recognition
    and Image Processing--An Overview," Proc. Nat'l Comput. Conf.
    (1978), pp. 1003-1013.

Tojo, A., Uchida, S., and Fujimura, K., "A Picture Processing
    Experiment System Including A Multi-minicomputer," Info.
    Proc. Soc. Japan Bul. on Image Processing 6-1 (1976).

Uchida, S., and Higushi, T., "A Multi-minicomputer System for Pic-
    ture Processing Experiments and Its Interconnection Mechanism,"
    Proc. 1978 Conf. on Parallel Processing (1978).

APPENDIX

PARTICIPANTS

Hideo Aiso
Department of Electrical
    Engineering
Keio University
Keio, JAPAN

Haruo Asada
Toshiba Research and Development
    Center
1, Komukai Toshiba-cho,
    Saiwai-ku
Kawasaki City, Kanagawa 210
    JAPAN

N. S. Chang*
School of Electrical Engineering
Purdue University
West Lafayette, IN 47907 USA

Michael J. B. Duff
Department of Physics and
    Astronomy
University College London
London, WCIE GBT, ENGLAND

M. Ejiri*
Central Research Laboratory
Hitachi Ltd.
Kokubunji, Tokyo 185 JAPAN

Hajime Enomoto
Tokyo Institute of Technology
Ookayama, Tokyo, JAPAN

King-Sun Fu
School of Electrical Engineering
Purdue University
West Lafayette, IN 47907 USA

T. Fujisaki
IBM Japan Ltd.
Tokyo Scientific Center
Tokyo, JAPAN

Youji Fukada
Central Research Laboratory
Mitsubishi Electric Corp.
Amagasaki, JAPAN

*Non-participating co-author

Teruo Fukumura
Faculty of Engineering
Nagoya University
Furo-cho, Chikusa-ku
Nagoya 464 JAPAN

T. Hamada*
Meterological Satellite Center
235, Nakakiyoto 3-Chome,
  Kiyose-shi
Tokyo 180-04 JAPAN

S. Hanaki
Central Research Labs
Nippon Electric Co., Ltd.
4-1-1 Miyazaki, Takatsu-ku
Kawasaki City, JAPAN

Allen R. Hanson*
School of Language and Communi-
  cation
Hampshire College
Amherst, MA 01002 USA

Robert M. Haralick
Department of Computer Science
Virginia Polytechnic Institute
Blacksburg, VA 24061 USA

Mitsutoshi Hatori
Department of Electrical
  Engineering
University of Tokyo
Hongo, Bunkyo-ku
Tokyo, JAPAN

T. Hoshino*
Visual Communication Development
  Division
Yokosuka Electrical Communica-
  tion Laboratory
NTT, JAPAN

Tadao Ichikawa
Research and Development Labs
Kokusai Denshin and Denwa Co.,
  Ltd.
3-Nishishinjuku, 2-Chome
Shinjuku-ku, Tokyo 160 JAPAN

Yoshiki Ichioka
Faculty of Engineering
Osaka University
Suita, Osaka, JAPAN

T. Ida*
The Institute of Physical and
  Chemical Research
Wako-shi, Saitama 351 JAPAN

Masanori Idesawa*
The Institute of Physical and
  Chemical Research
Wako-shi, Saitama 351 JAPAN

J. Iisaka
IBM Japan Ltd.
Tokyo Scientific Center
Tokyo, JAPAN

Nobuyuki Inada
The Institute of Physical and
  Chemical Research
Wako-shi, Saitama 351 JAPAN

Mitsuru Ishizuka*
Institute of Industrial Science
University of Tokyo
Roppongi, Minato-ku
Tokyo 106 JAPAN

S. Ito
IBM Japan Ltd.
Tokyo Scientific Center
Tokyo, JAPAN

Takahiko Kamae
Visual Communication Development
  Division
Yokosuka Electrical Communica-
  tion Laboratory
NTT, JAPAN

T. Katayama
Tokyo Institute of Technology
Ookayama, Tokyo, JAPAN

K. Kato*
Meteorological Satellite Center
235, Nakakiyoto 3-Chome,
  Kiyose-shi
Tokyo 180-04 JAPAN

Satoshi Kawata
Faculty of Engineering
Osaka University
Suita, Osaka, JAPAN

Masatsugu Kidode
Toshiba Research and Development
  Center
1, Komukai Toshiba-cho,
  Saiwai-ku
Kawasaki City, Kanagawa 210
  JAPAN

Nobuhiko Kodaira
Meterological Satellite Center
235, Nakakiyoto 3-Chome,
  Kiyose-shi
Tokyo 180-04 JAPAN

Stefano Levialdi
Laboratorio di Cibernetica
80072 Arco Felice
Naples, ITALY

A. Maggiolo-Schettini*
Istituto Scienze dell'Infor-
  mazione
Corso Italia 40
Pisa, ITALY

H. Matsushima
Central Research Laboratory
Hitachi Ltd.
Kokubunji, Tokyo 185 JAPAN

I. Miyamura*
Tokyo Institute of Technology
Ookayama, Tokyo, JAPAN

Makoto Nagao
Electrical Engineering Depart-
  ment
Kyoto University
Kyoto, JAPAN

Paul A. Nagin*
Department of Ophthalmology
Tufts New England Medical
  Center
P.O. Box 450
171 Harrison Avenue
Boston, MA 02111 USA

M. Nagura*
Visual Communication Develop-
  ment Division
Yokosuka Electrical Communica-
  tion Laboratory
NTT, JAPAN

M. Napoli*
Laboratorio di Cibernetica
80072 Arco Felice
Naples, ITALY

M. Okada*
Visual Communication Develop-
  ment Division
Yokosuka Electrical Communica-
  tion Laboratory
NTT, JAPAN

Morio Onoe
Institute of Industrial Science
University of Tokyo
Roppongi, Minato-ku
Tokyo 106 JAPAN

Kendall Preston, Jr.
Department of Electrical
  Engineering
Carnegie-Mellon University
Pittsburgh, PA 15213 USA

Edward M. Riseman
Computer and Information Science
  Department
University of Massachusetts
Amherst, MA 01003 USA

Azriel Rosenfeld
Computer Vision Laboratory
Computer Science Center
University of Maryland
College Park, MD 20742 USA

Ken Sakamura*
Department of Electrical
  Engineering
Keio University
Keio, JAPAN

Hidenori Shinoda
Toshiba Research and Development
  Center
1, Komukai Toshiba-cho,
  Saiwai-ku
Kawasaki City, Kanagawa 210
  JAPAN

Takashi Soma
The Institute of Physical and
  Chemical Research
Wako-shi, Saitama 351 JAPAN

Stanley R. Sternberg
Environmental Research Insti-
  tute of Michigan
P.O. Box 8618
Ann Arbor, MI 48107 USA

Mikio Takagi
Institute of Industrial Science
University of Tokyo
Roppongi, Minato-ku
Tokyo 106 JAPAN

Y. Takao*
IBM Japan Ltd.
Tokyo Scientific Center
Tokyo, JAPAN

Y. Taki*
Department of Electrical
  Engineering
University of Tokyo
Hongo, Bunkyo-ku
Tokyo, JAPAN

Akio Tojo
Electrotechnical Laboratory
Nagata-cho, Chiyoda-ku
Tokyo 100 JAPAN

Jun-ichiro Toriwaki
Faculty of Engineering
Nagoya University
Furo-cho, Chikusa-ku
Nagoya 464 JAPAN

Kuniaki Tsuboi
Institute of Industrial Science
University of Tokyo
Roppongi, Minato-ku
Tokyo 106 JAPAN

G. Uccella*
Laboratorio di Cibernetica
80072 Arco Felice
Naples, ITALY

Shun-ichi Uchida
Electrotechnical Laboratory
Nagata-cho, Chiyoda-ku
Tokyo 100 JAPAN

T. Uno*
Central Research Laboratory
Hitachi Ltd.
Kokubunji, Tokyo 185 JAPAN

Sadakazu Watanabe
Toshiba Research and Development
  Center
1, Komukai Toshiba-cho, Saiwai-ku
Kawasaki City, Kanagawa 210
  JAPAN

Shigoki Yokoi
Faculty of Engineering
Mie University
Kamihama-cho, Tsu 514 JAPAN

N. Yonezaki
Tokyo Institute of Technology
Ookayam, Tokyo, JAPAN

AUTHOR INDEX

SUBJECT INDEX